Please detach and keep this tab for your records.
To obtain class tracking and score information within **15 business days** of your examination date, **record the class number on this answer sheet tab**. You will need this number to retrieve your score at **www.nraef.org/classes**.

Record your class number here.

National Restaurant Association
EDUCATIONAL FOUNDATION
175 West Jackson Boulevard, Suite 1500
Chicago, IL 60604-2814
www.nraef.org

ServSafe®

Coursebook

Third Edition

National Restaurant Association
EDUCATIONAL FOUNDATION

DISCLAIMER

The information presented in this book has been compiled from sources and documents believed to be reliable and represents the best professional judgment of the National Restaurant Association Educational Foundation. The accuracy of the information presented, however, is not guaranteed, nor is any responsibility assumed or implied, by the National Restaurant Association Educational Foundation for any damage or loss resulting from inaccuracies or omissions.

Laws may vary greatly by city, county, or state. This book is not intended to provide legal advice or establish standards of reasonable behavior. Operators who develop food safety-related policies and procedures as part of their commitment to employee and customer safety are urged to use the advice and guidance of legal counsel.

Copyright Permissions
National Restaurant Association Educational Foundation
175 West Jackson Boulevard, Suite 1500
Chicago, IL 60604-2814
Email: permissions@nraef.org

ISBN 1-58280-114-2 (Coursebook without Exam—CB3)
ISBN 1-58280-111-8 (Coursebook with Exam—CBX3)
ISBN 0-471-52208-2 (Wiley Coursebook without Exam—CB3-W)
ISBN 0-471-47802-4 (Wiley Coursebook with Exam—CBX3-W)

Printed in the U.S.A.

10 9 8 7 6

Table of Contents

INTRODUCTION

A MESSAGE FROM
The National Restaurant Association Educational Foundation

The National Restaurant Association Educational Foundation (NRAEF) is a not-for-profit organization dedicated to fulfilling the educational mission of the National Restaurant Association. Focusing on three key strategies of **risk management, recruitment,** and **retention,** the NRAEF is the restaurant and foodservice industry's premier provider of educational resources, materials, and programs. Sales from all NRAEF products and services benefit the industry by directly supporting the NRAEF's educational initiatives.

Risk management is critical to the success of every restaurant and foodservice operation, particularly the area of food safety. After all, **serving safe food is not an option**—it is an obligation of all restaurant and foodservice professionals. Proper training is one of the best ways to create a culture of food safety within your establishment.

By opening this book, you have made a significant commitment to promoting food safety. You are about to gain knowledge from the **industry standard** in food safety training. The ServSafe® program provides **accurate, up-to-date information** for all levels of employees on all aspects of handling food, from receiving and storing to preparing and serving—information your employees need to be part of the food safety team.

In the past year, the NRAEF has further enhanced the ServSafe program. Exciting new features include:

- **Third editions of *ServSafe Coursebook* and *ServSafe Essentials.*** With increased readability and a brand new look and format, your best source for food safety training is now even better. Both texts incorporate **the latest information from the 2003 supplement to the 2001 FDA Food Code** in a realistic, applicable manner.
- **American National Standards Institute (ANSI)–Conference for Food Protection (CFP) Accreditation**. The ServSafe Food Protection Manager Certification Program has been evaluated and recognized under ANSI–CFP requirements for the development, administration, and security of its examinations. The ServSafe Certification has been deemed valid, reliable, and legally defensible, as well as transferable between jurisdictions.
- **Online score access.** You are now able to obtain score information within fifteen days of your examination date. At the time of the examination, ask your instructor for your class identification number, which you will need to retrieve your score at www.nraef.org/classes.

We applaud you for making the commitment to food safety training. The NRAEF challenges you—and all managers—to share your food safety knowledge with your employees and help foster a food safety culture within your operation.

For more information on the NRAEF and its programs, please visit www.nraef.org.

The National Restaurant Association Educational Foundation's

International Food Safety Council®

The International Food Safety Council is a strategic initiative of the National Restaurant Association Educational Foundation created to heighten awareness of the importance of food safety education throughout the restaurant and foodservice industry.

Initiatives

The NRAEF's International Food Safety Council sponsors food safety events and activities to encourage a commitment to food safety education and ensure food safety is a priority.

- *Food Safety Illustrated* magazine
- Food Safety Summit—Washington, D.C.
- National Food Safety Education Month®—September, visit www.nraef.org/nfsem

For more information about the NRAEF's International Food Safety Council, sponsorship opportunities, and initiatives, please call 312.715.1010, ext. 744, or visit the NRAEF's Web site at www.nraef.org/ifsc.

Founding Sponsors

American Egg Board
Cattlemen's Beef Board and National Cattlemen's Beef Association
Ecolab Inc.
FoodHandler, Inc.
Heinz North America
San Jamar/Katch All
SYSCO Corporation
Tyson Foods, Inc.
UBF Foodsolutions North America

Campaign Sponsors

Bunzl Distribution, Inc.
Cargill, Inc.
Cintas Corporation
Colgate-Palmolive Company
Farquharson Enterprises, Ltd.
North American Association of Food Equipment Manufacturers
PepsiCo Foodservice Division
Produce Marketing Association

Acknowledgements

The development of the *ServSafe Coursebook* would not have been possible without the expertise of our many advisors and manuscript reviewers. The NRAEF is pleased to thank the following people for their time, effort, and dedication in making the refinements found in this third edition.

Bruno Abate
Follia Restaurant

Marie-Luise Baehr
Sodexho, Inc.

Harry D'Ercole
Enrico's Italian Dining

Jane Gibson
Cattlemen's Beef Board and National Cattlemen's Beef Association

Steven F. Grover, R.E.H.S.
National Restaurant Association

Bucky Gwartney, Ph.D.
National Cattlemen's Beef Association

Alice M. Heinze
American Egg Board

Todd McAloon
Cargill, Inc.

Anne Munoz-Furlong
Food Allergy and Anaphylaxis Network

Ann Rasor
North American Meat Processors Association

Mary Sandford
Burger King Corporation

SYSCO Corporation

Lacie Thrall
FoodHandler Inc.

Frank Yiannas
Walt Disney World Co.

FEATURES OF THE *SERVSAFE COURSEBOOK*

We have designed the *ServSafe Coursebook* to enhance your ability to learn and retain comprehensive food safety knowledge. Here are key features you will find in each chapter of this book.

Inside This Chapter

Chapter content is organized under the major headings identified in this section.

Learning Objectives

Learning objectives identify what you should be able to do after completing each chapter. These objectives are linked to the tasks required to keep your establishment safe.

Key Terms

These terms are important for a thorough understanding of the chapter content. They are **highlighted** throughout the chapter, where either they are explicitly defined or their meanings are made clear within the paragraphs in which they appear. Each key term is also defined in the Glossary.

Test Your Food Safety Knowledge

Each chapter begins with five True or False questions designed to test your prior knowledge of some of the concepts presented in the chapter. The answers to these questions appear in the Answer Key. If you wish to explore the concepts behind the questions further, see the page reference after each question.

Exhibits

Exhibits are placed throughout each chapter to visually reinforce the key concepts presented in the text. They are referenced by the chapter number followed by a letter, and they include charts, photographs, illustrations, and tables.

Key Point

Throughout each chapter, icons appear in the margins of the page. These icons emphasize concepts presented in the text that are important to your understanding of food safety.

Icons

Icons appear throughout each chapter in the margins of the page. They emphasize concepts presented in the text that are important to your understanding of food safety. While Key Point icons are the most common type, icons related to personal hygiene, cross-contamination, and time-temperature abuse are also included.

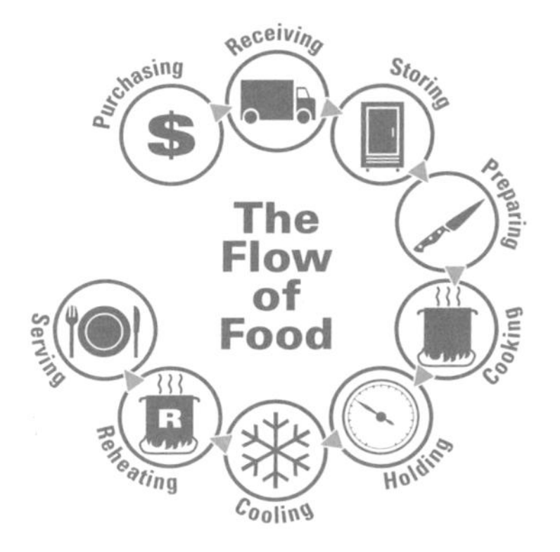

The Flow of Food Icon

An icon representing the various stages of the flow of food appears throughout Chapters 5–10 in the left margin of the page. As you read through these chapters, you will notice that the highlighted portion of the icon changes according to the area within the flow of food being discussed.

A Case In Point

These real-world scenarios give you the opportunity to apply some of the food safety concepts you have learned in each chapter. Answers to the study's questions are provided in Answer Key.

International Food Safety Icons

Icons have been developed by the International Association for Food Protection to provide the restaurant and foodservice industry with easily recognizable symbols that convey specific food safety messages to foodhandlers of all nationalities. One of these icons appears as part of each "A Case in Point" to help illustrate the primary principle addressed in the case study.

For more information on the International Food Safety Icons, visit www.foodprotection.org.

Discussion Questions

These questions are designed to make you think about some of the important food safety concepts presented in the chapter. Answers to the questions are provided in the Answer Key.

Multiple-Choice Study Questions

These questions are designed to test your knowledge of the food safety concepts presented in the chapter. If you have difficulty answering them, you should review the content again. Answers to these questions are provided in the Answer Key.

Additional Resources

In this section, you will find resources—books, articles, and Web sites—that will enable you to further explore the food safety concepts presented in each chapter.

Icons used with permission from the International Association for Food Protection

Unit 1

The Sanitation Challenge

1 Providing Safe Food

Inside this chapter:

- The Dangers of Foodborne Illness
- Preventing Foodborne Illness
- How Food Becomes Unsafe
- Key Practices for Ensuring Food Safety
- The Food Safety Responsibilities of a Manager

After completing this chapter, you should be able to:

- Analyze evidence to determine the presence of foodborne-illness outbreaks.
- Recognize risks associated with high-risk populations.
- Identify the characteristics of potentially hazardous food.
- Recognize a manager's responsibility to provide food safety training to employees.
- Identify the need to maintain food safety training records.
- Identify the appropriate training tools for teaching food safety.
- Ensure all foodservice employees are trained initially and on an ongoing basis.

Key Terms

Foodborne illness
Foodborne-illness outbreak
Warranty of sale
Reasonable care defense
Hazard Analysis Critical Control Point (HACCP)
FDA Food Code
Immune system
Potentially hazardous food
Heat-treated
Ready-to-eat food
Contamination
Biological, chemical, and physical hazards
Time-temperature abuse
Cross-contamination
Food-contact surface
Personal hygiene
Clean
Sanitary

Apply Your Knowledge

Check to see how much you know about the concepts in this chapter. Use the page references provided to explore the topic in each question.

Test Your Food Safety Knowledge

1. **True or False:** A foodborne-illness outbreak is confirmed when two or more people experience the same illness after eating the same food. *(See page 1-3.)*
2. **True or False:** Preschool-age children may be more likely than adults to become ill from contaminated food. *(See page 1-7.)*
3. **True or False:** It is the manager's responsibility to teach employees the food safety principles and practices learned in the ServSafe program. *(See page 1-6.)*
4. **True or False:** Employees only need to receive initial training in food safety. *(See page 1-6.)*
5. **True or False:** Potentially hazardous food is generally dry, contains protein, and is highly acidic. *(See page 1-8.)*

For answers, please turn to the Answer Key.

INTRODUCTION

When diners eat out, they expect safe food, clean surroundings, and well-groomed workers. Overall, the restaurant and foodservice industry does a good job of meeting these demands, but there is still room for improvement.

The risk of foodborne illness impacts the industry. Several factors account for this. They include:

- The emergence of new foodborne pathogens (disease-causing microorganisms)
- The importation of food from countries where food safety practices may not be well developed
- Changes in the composition of food, which may leave fewer natural barriers to the growth of microorganisms
- Increases in the purchase of take-out and home meal replacement (HMR) food

- Changing demographics, with an increased number of individuals at high risk for contracting foodborne illness
- Employee turnover rates that make it difficult to manage an effective food safety system

In the face of these challenges, all establishments must take the necessary steps to help ensure that the food they serve is safe. The first step is to develop a food safety system that includes effective and ongoing employee training.

THE DANGERS OF FOODBORNE ILLNESS

Health Alert

A foodborne-illness outbreak is an incident in which two or more people experience the same illness after eating the same food.

A **foodborne illness** is a disease carried or transmitted to people by food. The Centers for Disease Control and Prevention (CDC) define a **foodborne-illness outbreak** as an incident in which two or more people experience the same illness after eating the same food. A foodborne illness is confirmed when laboratory analysis shows that a specific food is the source of the illness.

Each year, millions of people are affected by foodborne illness, although the majority of cases are not reported and do not occur at restaurants and foodservice establishments. However, the cases that are reported and investigated help us understand some of the causes of illness, as well as what we, as restaurant and foodservice professionals, can do to control these causes in each of our establishments. The most commonly reported causes of foodborne illnesses are: failure to cool food properly, failure to cook and hold food at the proper temperature, and poor personal hygiene.

Fortunately, every restaurant and foodservice establishment, no matter how large or small, can take steps to ensure the safety of the food it prepares and serves to its customers.

The Costs of Foodborne Illness

National Restaurant Association figures show that a foodborne-illness outbreak can cost an establishment thousands of dollars. It can even be the reason an establishment is forced to close.

If your establishment is implicated in a foodborne-illness outbreak, your costs may include increased insurance premiums, as well as lawyer and court fees. You may have to pay for testing food supplies and employees, and may spend time and money retraining employees and cleaning and sanitizing the establishment. Food supplies that may or may not be contaminated will have to be discarded. Other costs include: lowered employee morale and absenteeism, embarrassment and bad publicity, loss of customers and sales, and loss of prestige and reputation. (See *Exhibit 1a.*)

Today, customers are very willing to sue to obtain compensation for injuries they feel they have suffered as a result of the food they were served. Under the federal Uniform Commercial Code, a plaintiff bringing about a lawsuit must prove all of the following:

- The food was unfit to be served.
- The food caused the plaintiff harm.
- In serving the food, the establishment violated the **warranty of sale,** that is, the rules stating how the food must be handled.

Exhibit 1a

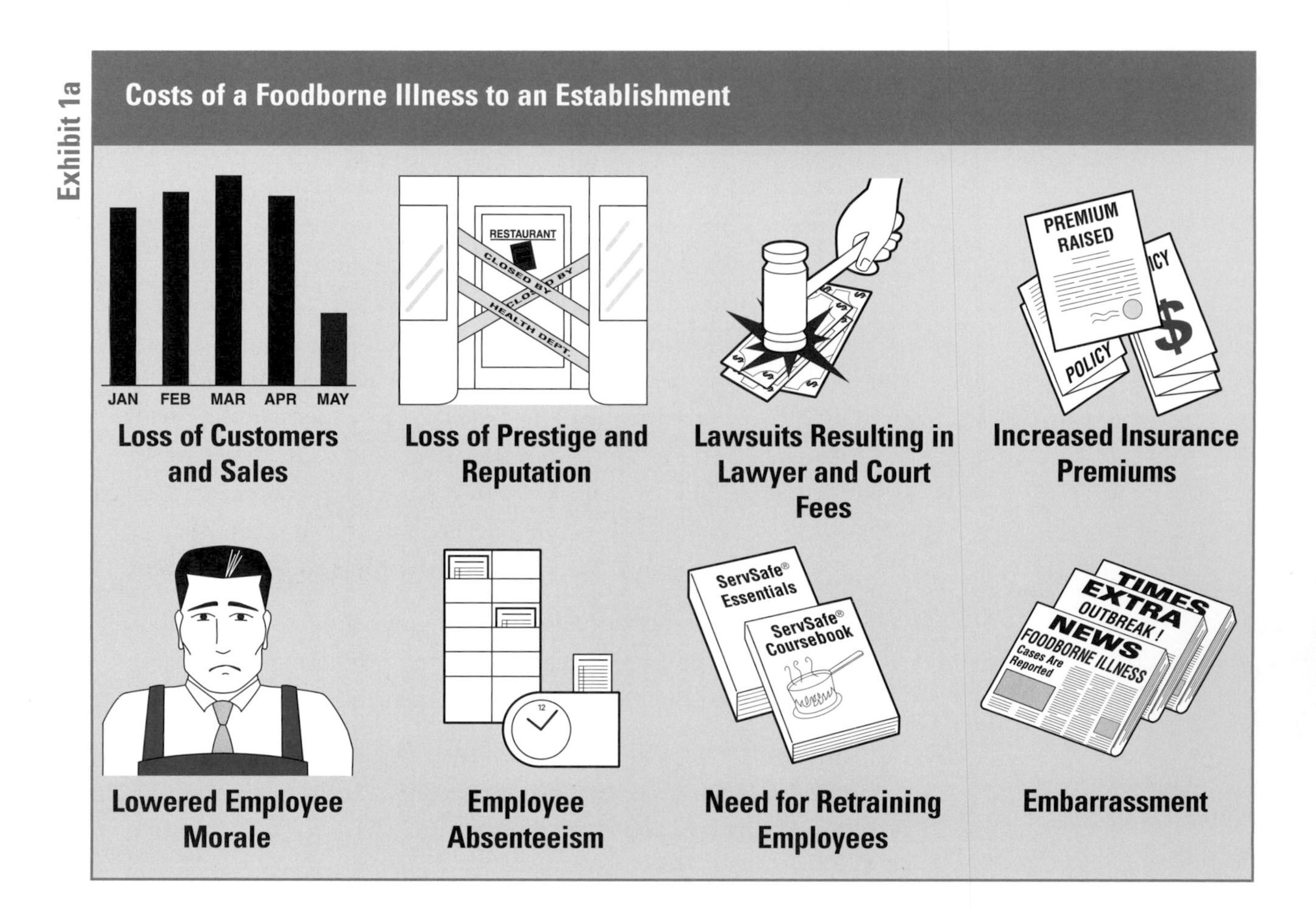

If the plaintiff wins the lawsuit, he or she can be awarded two types of damages. Compensatory damages can be awarded for lost work, lost wages, and medical bills. Punitive damages can be awarded in addition to normal compensation, punishing the defendant for wanton and willful neglect.

If you have a quality food safety system in place, however, you can use a **reasonable care defense** against a food-related lawsuit. A reasonable care defense requires proof that your establishment did everything that could be reasonably expected to ensure that the food served was safe. Evidence of written standards, training practices, procedures such as a HACCP plan with documentation, and positive inspection results are the keys to this defense. Since decisions in these suits follow the law in their respective states, check applicable state law as to the appropriate defense in any action.

The Benefits of a Food Safety System

Serving safe food is vital to your establishment's success. A well-designed, food safety system can help protect your establishment's employees, customers, and reputation. Repeat business from customers and increased job satisfaction among employees can lead to higher profits and better service. Your establishment may benefit directly from reduced or minimized insurance costs, and will most likely benefit by reducing health-code violations and becoming less open to lawsuits claiming injury and negligence.

An added benefit to a food safety system is that by handling food safely, you also preserve its quality. Safe foodhandling will help maintain the appearance, flavor, texture, consistency, and nutritional value of food. Food that is stored, prepared, and served properly is more likely to provide the quality your customers deserve and demand. Safe foodhandling can also lead to lower food costs due to less waste.

PREVENTING FOODBORNE ILLNESS

There are many challenges to preventing foodborne illness. These include high employee-turnover rates, service to an increasing number of high-risk customers, and the service of potentially hazardous food. Establishing a comprehensive food safety program, however, greatly reduces the likelihood of causing foodborne illness.

Training Employees in Food Safety

As a manager, it is your responsibility to ensure that the food safety principles you learn throughout the ServSafe program are practiced by everyone in your operation. All employees must be properly trained in food safety as it relates to their assigned job tasks. Food safety training should consist of the following:

- Programs for both new and current employees
- Assessment tools that identify ongoing food safety training needs for employees
- A selection of resources that includes books, videos, posters, and technology-based materials that meet your learners' needs
- Records documenting that employees have completed training

Food Safety Systems

Key Point

A food safety system is designed to prevent food safety hazards from causing foodborne illness.

A food safety management system will help you prevent foodborne illness by controlling hazards throughout the flow of food. A strong food safety system will incorporate the principles of active managerial control and HACCP.

Active managerial control focuses on establishing policies and procedures to control five common risk factors responsible for foodborne illness. The polices and procedures that an establishment puts in place will be the result of a careful analysis of potential breakdowns related to these five risk factors at each stage in the flow of food.

A **HACCP (Hazard Analysis Critical Control Point)** system focuses on identifying specific points within the flow of food through the operation that are essential to prevent, eliminate, or reduce a biological, chemical, or physical hazard to safe levels. To be effective, a HACCP system must be based on a plan specific to a facility's menu, customers, equipment, processes, and operation. The HACCP plan is developed following seven principles—sequential steps for building the food safety system. Active managerial control and HACCP will be covered in more detail in Chapter 10.

Both the National Restaurant Association and the Food and Drug Administration's (FDA) Food Code encourage an establishment to develop and use a food safety system to prevent foodborne illness. The **FDA Food Code** is a science-based reference for restaurants and retail establishments on how to prevent foodborne illness. Local, state, and federal regulators often use the FDA Food Code as a model when developing or updating their own food safety regulations, and to ensure consistency with national regulatory policy.

Populations at High Risk for Foodborne Illness

The demographics of our population show an increase in the percentage of people at high risk of contracting a foodborne illness, sometimes with serious consequences. (See *Exhibit 1b.*) They include:

- Infants and preschool-age children
- Pregnant women
- Elderly people
- People taking certain medications, such as antibiotics and immunosuppressants
- People who are ill (those who have recently had major surgery, are organ-transplant recipients, or who have pre-existing or chronic illnesses)

Young children are more at risk for contracting foodborne illnesses because they have not yet built up adequate **immune systems** (the body's defense system against illness) to combat some diseases.

Elderly people are more at risk because their immune systems and resistance may have weakened with age. In addition, as people age, their senses of smell and taste are diminished, so they may be

Exhibit 1b

Young Children

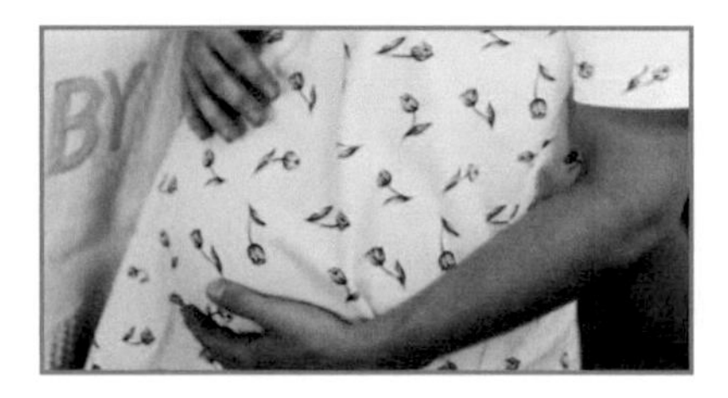

Pregnant Women

Elderly People

People Taking Medication

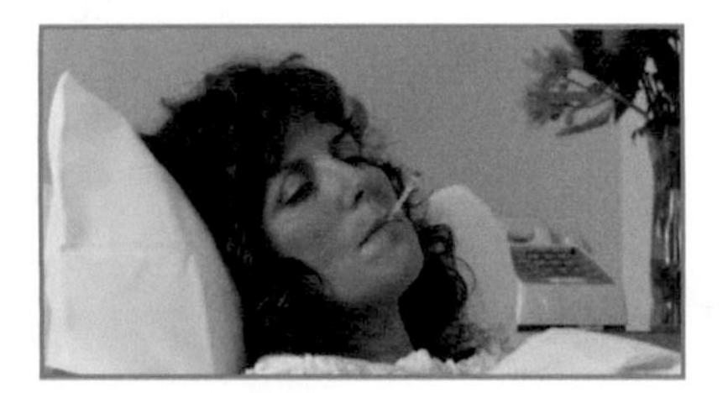

People Who Are Ill

less likely to detect "off" odors or tastes, which indicate that food may be spoiled.

Because these populations are susceptible to foodborne illness, it is of particular concern when they consume potentially hazardous food (or ingredients) that are raw or have not been cooked to the minimum internal temperatures identified in Chapter 8. In all cases, these high-risk consumers should be advised of any potentially hazardous foods (or ingredients) that are raw or not fully cooked. Tell them to consult a physician before regularly consuming this type of food. Additionally, check with your regulatory agency for specific requirements.

Food Most Likely to Become Unsafe

Although any food can become contaminated, most foodborne illnesses are transmitted through food in which microorganisms are able to grow rapidly. Such food is classified as **potentially hazardous food.** This food typically has a history of being involved in foodborne-illness outbreaks, has a natural potential for contamination due to production and processing methods, is often moist, contains protein, and has a neutral or slightly acidic pH.

Key Point

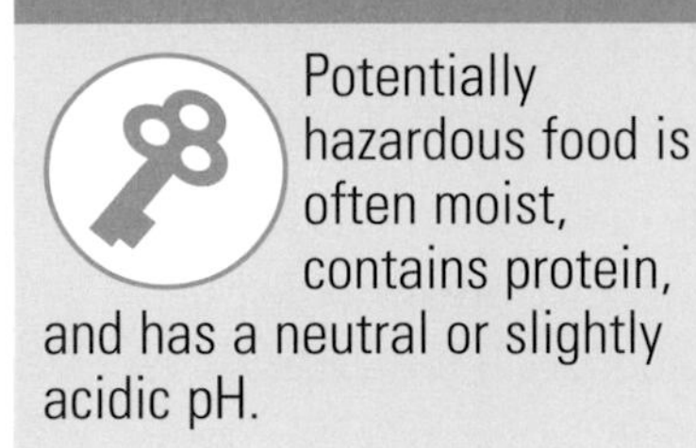

Potentially hazardous food is often moist, contains protein, and has a neutral or slightly acidic pH.

Exhibit 1c

Potentially Hazardous Food

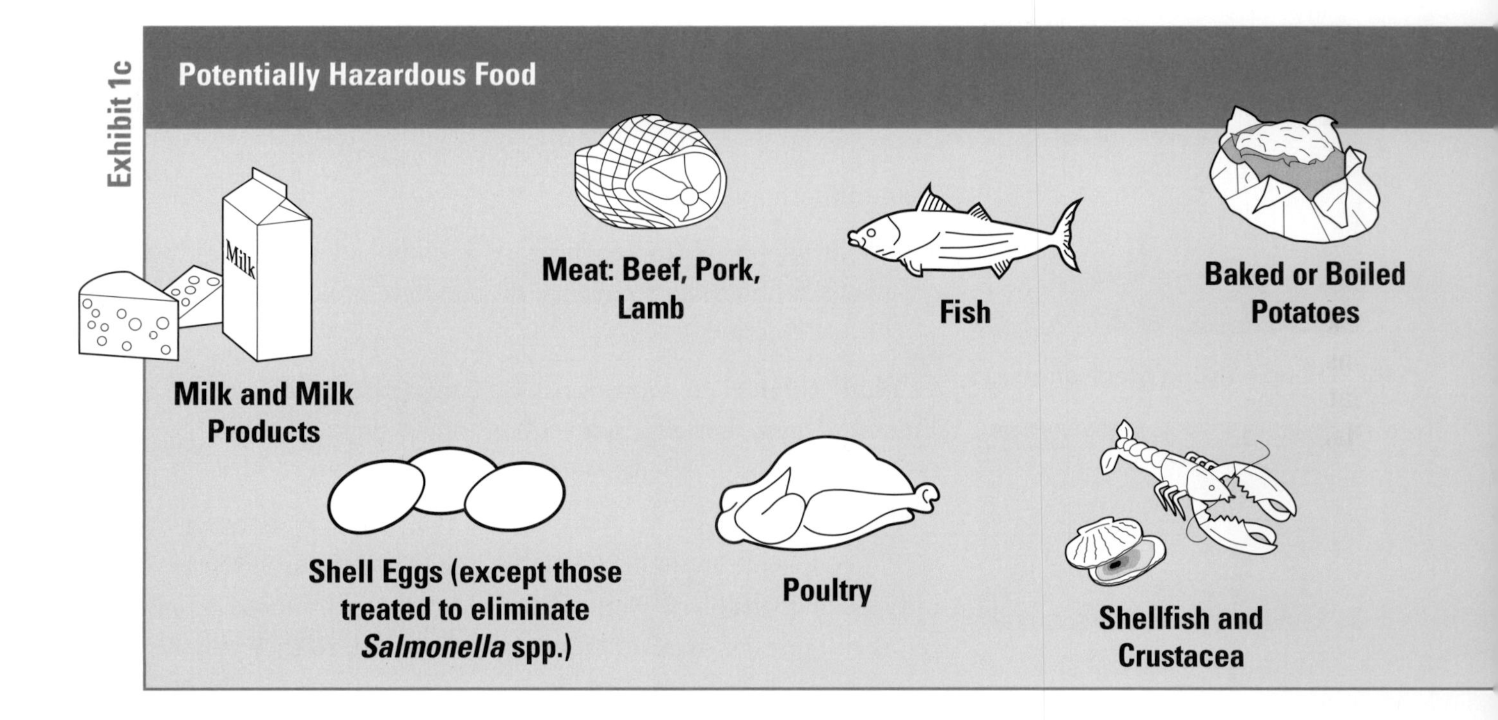

The FDA Food Code identifies potentially hazardous food (see *Exhibit 1c*) as any food that consists in whole, or in part, of the following:

- Milk and milk products
- Shell eggs (except those treated to eliminate *Salmonella* spp.)
- Meats, poultry, and fish
- Shellfish and edible crustacea (such as shrimp, lobster, and crab)
- Baked or boiled potatoes
- Tofu or other soy-protein food
- Garlic-and-oil mixtures
- Plant food—including fruit and vegetables—that has been **heat-treated** (cooked, partially cooked, or warmed)
- Sprouts and sprout seeds
- Sliced melons
- Synthetic ingredients (such as textured soy protein in meat alternatives)

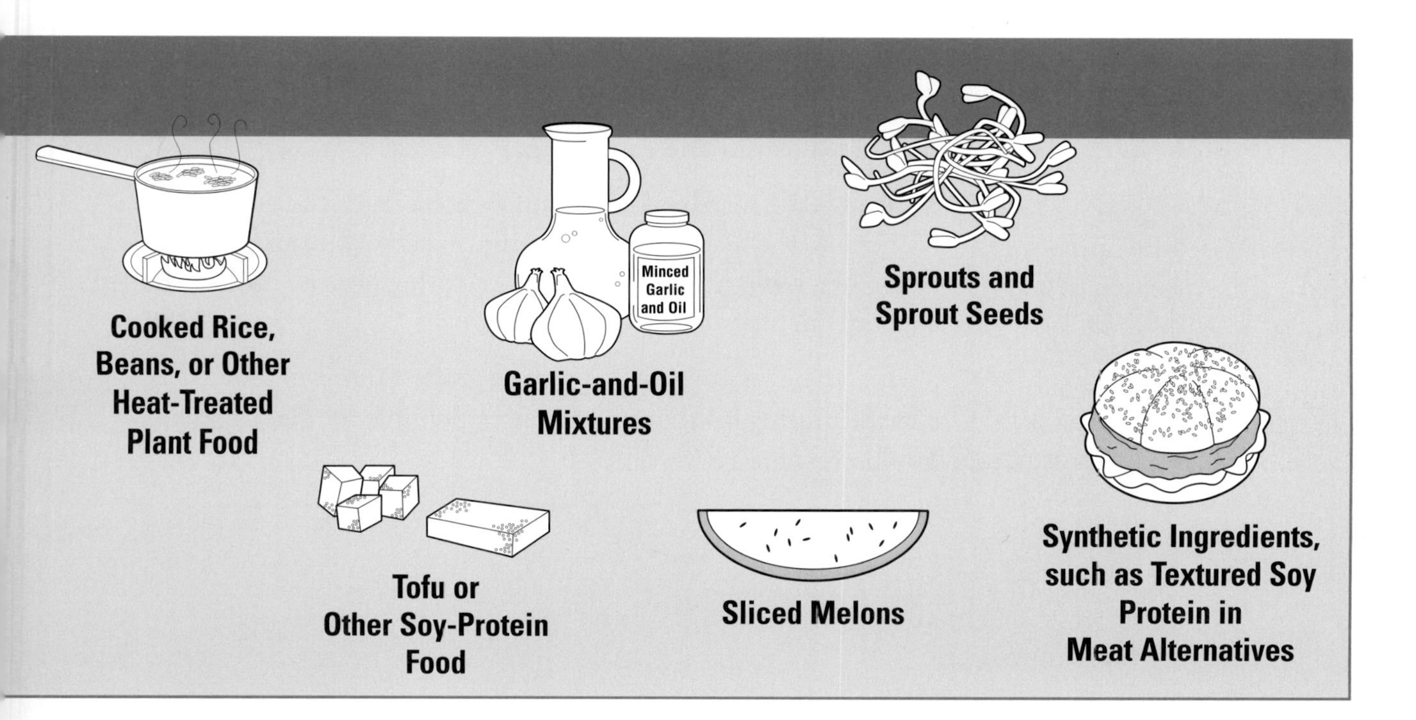

Care must be taken to prevent contamination when handling **ready-to-eat food,** which is any food that is edible without any further washing or cooking. Ready-to-eat food includes:

- Washed, whole or cut fruits
- Vegetables
- Deli meats
- Bakery items
- Spices, sugars, and seasonings
- Properly cooked food

Potential Hazards to Food Safety

Unsafe food usually results from **contamination,** which is the presence of harmful substances in the food. Some food safety hazards are introduced by humans or by the environment, and some occur naturally.

Food safety hazards are divided into three categories: **biological hazards, chemical hazards,** and **physical hazards.**

- **Biological hazards** include certain bacteria, viruses, parasites, and fungi, as well as certain plants, mushrooms, and fish that carry harmful toxins.
- **Chemical hazards** include pesticides, food additives, preservatives, cleaning supplies, and toxic metals that leach from cookware and equipment.
- **Physical hazards** consist of foreign objects that accidentally get into the food, such as hair, dirt, metal staples, and broken glass, as well as naturally occurring objects, such as bones in fillets.

By far, biological hazards pose the greatest threat to food safety. Disease-causing microorganisms are responsible for the majority of foodborne-illness outbreaks.

Key Point

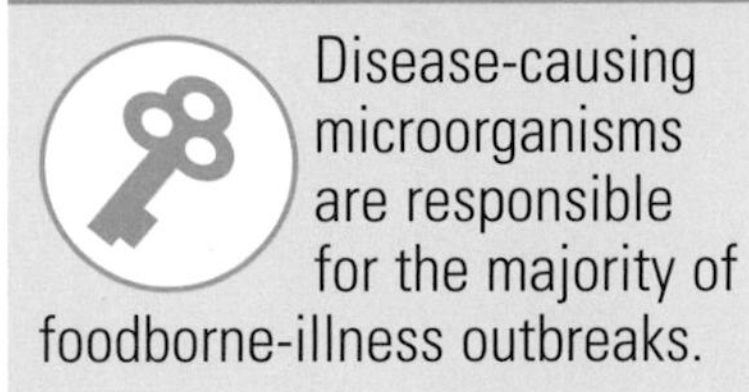

Disease-causing microorganisms are responsible for the majority of foodborne-illness outbreaks.

HOW FOOD BECOMES UNSAFE

The CDC have identified common factors that are responsible for foodborne illness. These include:

- Purchasing food from unsafe sources
- Failing to cook food adequately
- Holding food at improper temperatures
- Using contaminated equipment
- Poor personal hygiene

Key Point

A well-designed, food safety system will establish controls to prevent time-temperature abuse, cross-contamination, and poor personal hygiene.

Each of these common factors—with the exception of purchasing food from unsafe sources—is related to **time-temperature abuse, cross-contamination,** or poor **personal hygiene.** Reported cases of foodborne illness usually involve multiple causes. A well-designed, food safety system will control all of these factors.

Time-Temperature Abuse: Food has been time-temperature abused any time it has been allowed to remain too long at temperatures favorable to the growth of foodborne microorganisms. A foodborne illness can result if food is time-temperature abused in any of the following ways:

- Not held or stored at required temperatures
- Not cooked or reheated to temperatures that kill microorganisms
- Improperly cooled

Time-temperature abuse is also more likely to occur in food that has been prepared a day or more in advance of service since the food may go through several additional processes, including cooling, storage, and reheating.

Cross-Contamination

Cross-contamination occurs when microorganisms are transferred from one surface or food to another.

Cross-Contamination: Cross-contamination occurs when microorganisms are transferred from one surface or food to another. A foodborne illness can result if cross-contamination is allowed to occur in any of the following ways:

- Raw contaminated ingredients are added to food that receives no further cooking.
- Food-contact surfaces are not properly cleaned and sanitized before touching cooked or ready-to-eat food.
- Raw food is allowed to touch or drip fluids onto cooked or ready-to-eat food.

Exhibit 1d

Controlling Time and Temperature Throughout the Flow of Food

Receiving:
Receive and store food quickly.

▼

Storage:
Store food at its recommended temperatures.

▼

Preparation:
Minimize time spent in the temperature danger zone of 41°F (5°C) to 135°F (57°C).

▼

Cooking:
Cook food to its required minimum internal temperature for the appropriate amount of time.

▼

Holding:
Hold hot food at 135°F (57°C) or higher and cold food at 41°F (5°C) or lower.

▼

Cooling:
Cool cooked food from 135°F (57°C) to 70°F (21°C) within two hours and from 70°F (21°C) to 41°F (5°C) or lower within an additional four hours, for a total cooling time of six hours.

▼

Reheating:
Reheat food to an internal temperature of 165°F (74°C) for fifteen seconds within two hours.

- A foodhandler touches contaminated (usually raw) food and then touches cooked or ready-to-eat food.
- Contaminated cleaning cloths are not cleaned and sanitized before being used on other food-contact surfaces.

Poor Personal Hygiene: Individuals with poor personal hygiene can offend customers, contaminate food or food-contact surfaces, and cause illness. A foodborne illness can result if employees do any of the following:

- Fail to wash their hands properly after using the restroom or whenever they become contaminated
- Cough or sneeze on food
- Touch or scratch sores, cuts, or boils, and then touch food they are handling
- Come to work while sick

KEY PRACTICES FOR ENSURING FOOD SAFETY

The keys to food safety lie in controlling time and temperature throughout the flow of food, practicing good personal hygiene, and preventing cross-contamination. It is important to establish standard operating procedures that focus on these areas.

Controlling Time and Temperature

Microorganisms pose the largest threat to food safety. Like all living organisms, most cannot survive or reproduce outside certain temperature ranges. As outlined in *Exhibit 1d,* time and temperature must be controlled throughout the flow of food. Each of these steps will be discussed in further detail in later chapters.

Practicing Good Personal Hygiene

Training employees in good personal hygiene practices is the responsibility of every manager. Features of a good personal hygiene program include:

- **Proper handwashing.** Hands and fingernails should be washed and cleaned thoroughly before and after handling food, between each task, and before using food-preparation equipment.

- **Strictly enforced rules regarding eating, drinking, and smoking.** These activities should be prohibited while preparing or serving food, or while in areas used for washing and storing equipment and utensils.
- **Preventing employees who are ill from working with food.** All illnesses should be reported to the manager.
- **General cleanliness.** Insist on daily bathing, clean hair, and clean clothing.

Preventing Cross-Contamination

Employees must be carefully trained to recognize and prevent cross-contamination of microorganisms between food and food-contact surfaces. Some ways to prevent cross-contamination include the following:

- **Make sure employees wash their hands frequently when working with raw food.** They should never touch raw food and then touch ready-to-eat food without washing their hands. Hands must also be washed regularly to prevent cross-contamination.
- **Do not allow raw food to touch or drip fluids onto cooked or ready-to-eat food.**
- **Clean and sanitize cleaning cloths between each use.**
- **Clean and sanitize food-contact surfaces (such as equipment or utensils) that touch contaminated food before they come in contact with cooked or ready-to-eat food.**

Food-contact surfaces may be direct or indirect. A direct food-contact surface includes any equipment or utensil surface that normally touches food, such as tableware, cutting boards, knives, and other utensils used to prepare food, and counters where food is prepared. An indirect food-contact surface is a surface food might drain, drip, or splash onto during preparation, such as the backsplash of a counter. Food that splashes on an indirect food-contact surface may drain down onto a direct food-contact surface and contaminate it.

Food-contact surfaces, cleaning cloths, and sponges must be cleaned and sanitized to prevent cross-contamination. *Clean* simply means free of visible soil. *Sanitary*, on the other hand, means that the number of microorganisms on the surface has been reduced to safe levels.

Key Point

It is critically important for management to have a positive and supportive attitude toward food safety.

THE FOOD SAFETY RESPONSIBILITIES OF A MANAGER

The manager's basic food safety responsibilities are to serve customers safe food and to train employees in safe foodhandling practices. The manager's positive and supportive attitude toward food safety is critical. This attitude should be based on up-to-date knowledge of the regulations affecting the restaurant and foodservice industry.

Meeting Food Safety Regulations

To stay in operation, your establishment must comply with city, county, and state sanitation codes. The regulatory agency evaluating your establishment shares your commitment to serving safe food, and may have the authority to assess fines and close an establishment that serves unsafe food. Therefore, it is in your best interest to work with local authorities. During evaluations, the regulatory agency may identify deficiencies requiring your attention.

The FDA recommends that local and state health departments hold the person in charge of a restaurant or foodservice establishment responsible for knowing and demonstrating the following information:

- Diseases carried or transmitted by food and their symptoms
- Points in the flow of food where hazards can be prevented, eliminated, or reduced, and how procedures meet the requirements of the local code
- The relationship between personal hygiene and the spread of disease, especially concerning cross-contamination, hand contact with ready-to-eat food, and handwashing
- How to keep injured or ill employees from contaminating food or food-contact surfaces
- The need to control the length of time potentially hazardous food remains at temperatures at which disease-causing microorganisms can grow
- Hazards involved in the consumption of raw or undercooked meat, poultry, eggs, and fish
- Safe cooking temperatures and times for potentially hazardous food items—such as meat, poultry, eggs, and fish

- Safe temperatures and times for storing, holding, cooling, and reheating potentially hazardous food
- Correct procedures for cleaning and sanitizing utensils and food-contact surfaces of equipment
- Types of toxic materials used in the operation, and how to safely store, dispense, use, and dispose of them
- The need for equipment that is sufficient in number and capacity, and is properly designed, constructed, located, installed, operated, maintained, and cleaned
- Approved sources of potable water and the importance of keeping it clean and safe
- The principles of active managerial control or a HACCP-based food safety system
- Rights, responsibilities, and authorities the local code assigns to employees, managers, and the local health department

Marketing Food Safety

Make it clear to your employees and customers that your operation takes food safety very seriously. Show your employees through actions that top management is involved in and supports food safety policies, and that food safety training for managers and all employees is a high priority. Offer training courses, and update and evaluate them regularly. Discuss food safety expectations. Document foodhandling procedures, use them in regular inspections, and update them as necessary. Show employees that safe foodhandling is appreciated—consider awarding certificates for training and giving out small rewards for good food safety records. Set a good example by following all food safety rules yourself.

Show your customers that your employees know and follow safety rules. Make sure your employees' appearances reflect your concern for food safety. Consider having employees wear food safety pins or buttons, or use place mats and posters to reinforce your message. Be sure your employees can answer simple food safety questions from customers.

Marketing your food safety efforts will help assure your customers that you are committed to serving safe food.

SUMMARY

Foodborne illness is a major concern to the restaurant and foodservice industry. A foodborne illness is a disease carried or transmitted to people by food. A foodborne-illness outbreak involves two or more people experiencing the same illness after eating the same food. Some segments of the population are more susceptible to foodborne illnesses than others. These categories are called high-risk populations.

Some food has a history of involvement in foodborne-illness outbreaks. This is called potentially hazardous food. Typically having a natural potential for contamination due to production and processing methods, potentially hazardous food is often moist, contains protein, and has a neutral or slightly acidic pH.

An incident of foodborne illness can be very expensive for an establishment, including legal liability, damage to reputation, and other related factors. However, a well-designed, food safety system protects your customers, your employees, and your reputation. It will incorporate the principles of active managerial control and HACCP. Active managerial control focuses on establishing policies and procedures to control five common risk factors responsible for foodborne illness. A HACCP system focuses on identifying specific points within the flow of food through the operation that are essential to prevent, eliminate, or reduce a biological, chemical, or physical hazard to safe levels.

Key practices for ensuring food safety include controlling time and temperature, practicing strict personal hygiene, and preventing cross-contamination. It is also important to maintain food safety practices while receiving, storing, preparing, cooking, holding, serving, cooling, and reheating food using methods that maintain its safety.

People pose a major risk to safe food, especially foodhandlers who do not practice personal hygiene. You must carefully train, monitor, and reinforce food safety principles in your establishment. Establishing a well-designed, food safety system can help protect your customers by preventing foodborne-illness outbreaks, and can help the establishment avoid the potentially high costs associated with them.

Apply Your Knowledge

1. What did André do right?
2. What did he do wrong?
3. What should he do differently in the future?

For answers, please turn to the Answer Key.

A Case in Point

André, a restaurant manager, has been on the job for nine months at The Charter Café. In his first two weeks, ten new employees joined his team of twenty-five. To his credit, most of the thirty-five employees are still with the establishment.

André knew from his management training program that food safety training was important to the success of the restaurant. With that in mind, he made sure all of his employees went through a four-hour food safety training session that he presented himself. André tried his best to impart the knowledge he learned in his training to his employees. He even had them take a quiz that he created.

After five months on the job, André found himself responding to a foodborne-illness complaint and a failing health inspection score. He was confused about what could have gone wrong because all of his employees had been trained in food safety.

Apply Your Knowledge

Discussion Questions

Use these questions to review the concepts presented in this chapter.

1. What is the difference between a foodborne illness and a foodborne-illness outbreak?
2. What are the potential costs associated with foodborne-illness outbreaks?
3. Why are the elderly at higher risk for contracting foodborne illnesses?
4. What are the three major types of hazards to food safety?
5. A chef cuts up a salmon on a cutting board, then thoroughly rinses the cutting board and knife in warm water. She then uses the same cutting board and knife to slice fresh parsley. Is this an acceptable foodhandling practice? Why or why not? On what key food safety practice does this example focus?

For answers, please turn to the Answer Key.

Apply Your Knowledge

Use these questions to test your knowledge of the concepts presented in this chapter.

Multiple-Choice Study Questions

1. **Why do elderly people have a higher risk of contracting a foodborne illness?**
   A. They are more likely to spend time in a hospital.
   B. Their immune systems are likely to have weakened with age.
   C. Their allergic reactions to chemicals used in food production might be greater than those of younger people.
   D. They are likely to have diminished appetites.

2. **Which type of food would be the most likely to cause a foodborne illness?**
   A. Tomato juice
   B. Cooked rice
   C. Whole wheat flour
   D. Dry powdered milk

3. **Which of the following is *not* a common characteristic of potentially hazardous food?**
   A. They are moist.
   B. They are dry.
   C. They are neutral or slightly acidic.
   D. They contain protein.

4. **In order for a foodborne illness to be considered an "outbreak," how many people must experience the illness after eating the same food?**
   A. 1
   B. 2
   C. 10
   D. 20

5. **Food safety training should consist of all of the following *except***
   A. programs for new and current employees.
   B. assessment tools that identify training needs.
   C. records that document that training has occurred.
   D. methods for dealing with customers' complaints.

6. **Tools used for food safety training should**
   A. meet the needs of your learners.
   B. be inexpensive.
   C. be deliverable in five minutes.
   D. resemble a video game.

For answers, please turn to the Answer Key.

ADDITIONAL RESOURCES

Books and Periodicals

A fresh set of eyes: A professional food safety consultant might paint you a very different picture. 2001. *Food Safety Illustrated.* 1 (2):8–10.

Jay, J. M. 1996. *Modern Food Microbiology.* New York: Chapman & Hall.

McCoy, J. J. 1990. *How safe is our food supply?* New York: F. Watts.

National Research Council. 1998. *Ensuring safe food: From production to consumption.* Washington, D.C.: National Academy of Sciences.

Olson, D. G. 1998. Irradiation of food. *Food Technology,* 52 (1):56–65.

Web Sites

American Public Health Association (APHA)
www.apha.org
APHA is concerned with a broad set of issues affecting personal and environmental health, including programs and policies related to chronic and infectious diseases, federal and state funding for health programs, and professional education in public health.

Association for Food and Drug Officials (AFDO)
www.afdo.org
AFDO is a leader and trusted resource in developing strategies to resolve and promote public health and consumer protection issues related to the regulation of food, drugs, medical devices, and consumer products.

Centers for Disease Control and Prevention (CDC)
www.cdc.gov
The mission of the CDC is to promote health and quality of life by preventing and controlling disease, injury, and disability. To prevent and control foodborne illness, the CDC collect data on outbreaks. This Web site provides general information on foodborne illnesses and their prevention.

FDA Center for Food Safety and Applied Nutrition (CFSAN)

www.cfsan.fda.gov

As the center within the FDA responsible for food safety and nutrition, CFSAN promotes and protects public health by researching and implementing guidelines, policies, and standards to ensure that food is safe, nutritious, wholesome, and properly labeled. This Web site provides a wealth of information on food safety and sanitation, including corresponding guidelines, policies, and standards.

FDA Food Code

http://vm.cfsan.fda.gov/~dms/foodcode.html

As the basis for many local sanitation codes, as well as the basis for information in this textbook, the FDA Food Code, available at this Web address, is a useful resource for information relating to food safety for the restaurant and foodservice industry.

Institute of Food Technologists

www.ift.org

The Institute of Food Technologists is a scientific, educational society with an interest in providing a safe and wholesome food supply. Their goal is to provide guidance on relevant issues in the field of food. The Web site provides scientific articles on food and food safety, and lists daily happenings in the food industry.

International Association for Food Protection

www.foodprotection.org

Members of the International Association for Food Protection are a diverse group, representing all areas of food protection, industry, government, and academia. Members are kept current on rapidly changing technologies, innovations, and regulations in the area of food safety through two monthly scientific journals, as well as annual meetings.

International HACCP Alliance

www.haccpalliance.org

The International HACCP Alliance was formed to assist the meat and poultry industry in preparing for mandatory HACCP. The Alliance promotes public health and safety by facilitating uniform development and implementation of HACCP. This Web site houses documents relating to food safety, HACCP, and its implementation.

Morbidity and Mortality Weekly Report (MMWR)
www.cdc.gov/mmwr
The *MMWR* is prepared by the CDC and is available, free of charge, on this Web site. The reports contain data based on weekly accounts of reportable diseases from state health departments. Report databases can be searched for information on foodborne-illness outbreaks.

National Association of Convenience Stores (NACS)
www.cstorecentral.com **or** www.nacsonline.com
NACS is a proactive organization representing the convenience store segment. This informative Web site provides up-to-date industry happenings and contains reports outlining vital industry statistics.

National Center for Infectious Diseases (NCID)
www.cdc.gov/ncidod
NCID is one of the centers of the CDC. Its mission is to prevent illness, disability, and death caused by infectious diseases around the world. NCID accomplishes this mission by conducting surveillance, epidemic investigations, epidemiologic and laboratory research, and training. It also sponsors public education programs to develop, evaluate, and promote prevention and control strategies for infectious diseases. This Web site serves as another great resource for information on foodborne illness.

National Environmental Health Association (NEHA)
www.neha.org
The NEHA's goals include fostering cooperation and understanding among environmental health professionals while remaining solidly committed to improving the environment in cities, towns, and rural areas throughout the world. This Web site includes many useful resources for restaurant and foodservice inspectors, sanitarians, and the industry.

National Food Processors Association (NFPA)
www.nfpa-food.org
The NFPA is the voice of the food processing industry on scientific and public policy issues involving food safety, nutrition, technical and regulatory matters, and consumer affairs. Their Web site offers many resources on food safety and food processing for the entire food industry.

National Restaurant Association

www.restaurant.org

The National Restaurant Association is the leading business association for the restaurant industry. Together with the National Restaurant Association Educational Foundation, the Association's mission is to represent, educate, and promote the rapidly growing restaurant and foodservice industry. This Web site should be your starting place for all issues and concerns related to your restaurant. This Web site has it all, from tips for running your establishment to vital data on your customers' spending habits.

U.S. Department of Agriculture–Food Safety and Inspection Service (USDA–FSIS)

www.fsis.usda.gov

The Food Safety and Inspection Service (FSIS) is the public-health agency in the USDA responsible for ensuring that the nation's commercial supply of meat, poultry, and egg products is safe, wholesome, and correctly labeled and packaged. This Web site contains a wealth of information on food safety relating to meat, poultry, and eggs.

10/0.25
160/0.17

The Microworld

Inside this chapter:

- Microbial Contaminants
- Bacteria
- Viruses
- Parasites
- Fungi
- Foodborne Infection vs. Foodborne Intoxication
- Emerging Pathogens and Issues
- Hazards Associated with Fresher and Healthier Food
- Technological Advancements in Food Safety

After completing this chapter, you should be able to:

- Identify factors that affect the growth of foodborne pathogens (FAT TOM).
- Differentiate between foodborne intoxication, infections, and toxin-mediated infections.
- Identify major foodborne illnesses and their symptoms.
- Identify characteristics of major foodborne pathogens including sources, foods involved in outbreaks, and methods of prevention.

Key Terms

Microorganisms
Pathogens
Toxins
Bacteria
Virus
Parasite
Fungi
Spoilage microorganism
Spore
pH
Lag phase
Log phase
Vegetative microorganism
Stationary phase
Death phase
FAT TOM
Temperature danger zone
Water activity (a_w)
Host
Mold
Yeast
Foodborne infection
Foodborne intoxication
Foodborne toxin-mediated infection
Food irradiation

Apply Your Knowledge

Check to see how much you know about the concepts in this chapter. Use the page references provided to explore the topic in each question.

Test Your Food Safety Knowledge

1. **True or False:** *Anisakis simplex* is often found in raw seafood. *(See page 2-20.)*
2. **True or False:** A foodborne intoxication occurs when a person eats food containing pathogens, which then grow in the intestines and cause illness. *(See page 2-22.)*
3. **True or False:** A person with listeriosis may experience bloody diarrhea. *(See page 2-12.)*
4. **True or False:** Cooling rice properly can help prevent an outbreak of *Bacillus cereus* Gastroenteritis. *(See page 2-13.)*
5. **True or False:** Highly acidic food typically does not support the growth of foodborne microorganisms. *(See page 2-6.)*

For answers, please turn to the Answer Key.

INTRODUCTION

In the previous chapter, you learned that foodborne microorganisms pose the greatest threat to food safety, and that disease-causing microorganisms are responsible for the majority of foodborne-illness outbreaks. In this chapter, you will learn about the microorganisms that cause foodborne illness, as well as the conditions they require in order to grow. When you understand these conditions, you will begin to see how the growth of foodborne microorganisms can be controlled, a topic that will be covered in greater detail in later chapters.

Microorganisms are small, living beings that can be seen only with a microscope. While not all microorganisms cause disease, some do. These are called **pathogens.** Eating food contaminated with foodborne pathogens or their **toxins** (poisons) is the leading cause of foodborne illness.

MICROBIAL CONTAMINANTS

There are four types of microorganisms that can contaminate food and cause foodborne illness: **bacteria, viruses, parasites,** and **fungi.**

These microorganisms can be arranged into two groups: **spoilage microorganisms** and pathogens. Mold is an example of a spoilage microorganism. While moldy food has an unpleasant appearance, smell, and taste, it seldom causes illness. However, pathogens such as *Salmonella* spp. and the hepatitis A virus cannot be seen, smelled, or tasted, but food contaminated by these pathogens often causes some form of illness when ingested.

BACTERIA

Of all microorganisms, bacteria are of greatest concern to the manager. Knowing what bacteria are and understanding the environment in which they grow is the first step in controlling them.

Basic Characteristics of Bacteria that Cause Foodborne Illness

Bacteria that cause foodborne illness have some basic characteristics.

- They are living, single-celled organisms.
- They may be carried by a variety of means: food, water, soil, humans, or insects.
- Under favorable conditions, they can reproduce very rapidly.
- Some can survive freezing.
- Some turn into **spores,** a change that protects the bacteria from unfavorable conditions.
- Some cause illness by producing toxins as they multiply, die, and break down. These toxins are not typically destroyed by cooking.

Bacterial Growth

To grow and reproduce, bacteria need the following:

- Food
- Appropriate level of acidity **(pH)**
- Proper temperature
- Adequate time
- The necessary level of oxygen
- Ample moisture

Their growth can be broken down into four progressive stages (phases): lag, log, stationary, and death. (See *Exhibit 2a.*)

When bacteria are first introduced to food, they go through an adjustment period, called the **lag phase.** In this phase, their number is stable as they prepare for growth. To control their number and prevent food from becoming unsafe, it is important to prolong the lag phase as long as possible. You can accomplish this by controlling the bacteria's requirements for growth in food: time, temperature, moisture, oxygen, and pH. For example, by refrigerating food, you can keep bacteria in the lag phase. If these conditions are not controlled, bacteria can enter the next phase, the **log phase,** where they will grow remarkably fast.

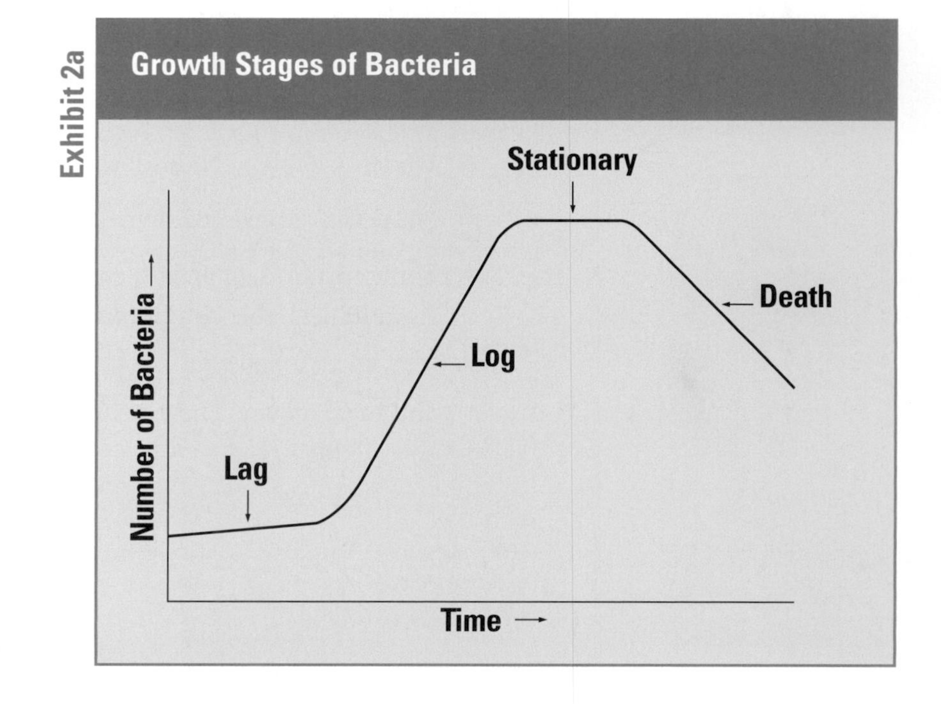

Exhibit 2b

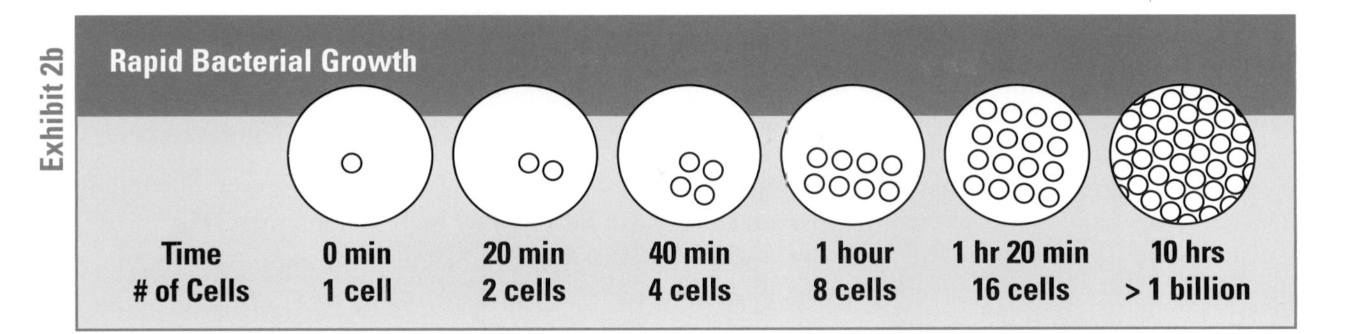

Bacteria reproduce by splitting in two. Those in the process of reproduction are called **vegetative microorganisms.** As long as conditions are favorable, bacteria can grow and multiply very rapidly, doubling their number as often as every twenty minutes. (See *Exhibit 2b.*) This is called exponential growth, and it occurs in the log phase. Food will rapidly become unsafe during the log phase.

Bacteria can continue to grow until nutrients and moisture become scarce, or conditions become unfavorable. Eventually the population reaches a **stationary phase,** in which just as many bacteria are growing as are dying. When the number of bacteria dying exceeds the number growing, the population declines. This is called the **death phase.**

Exhibit 2c

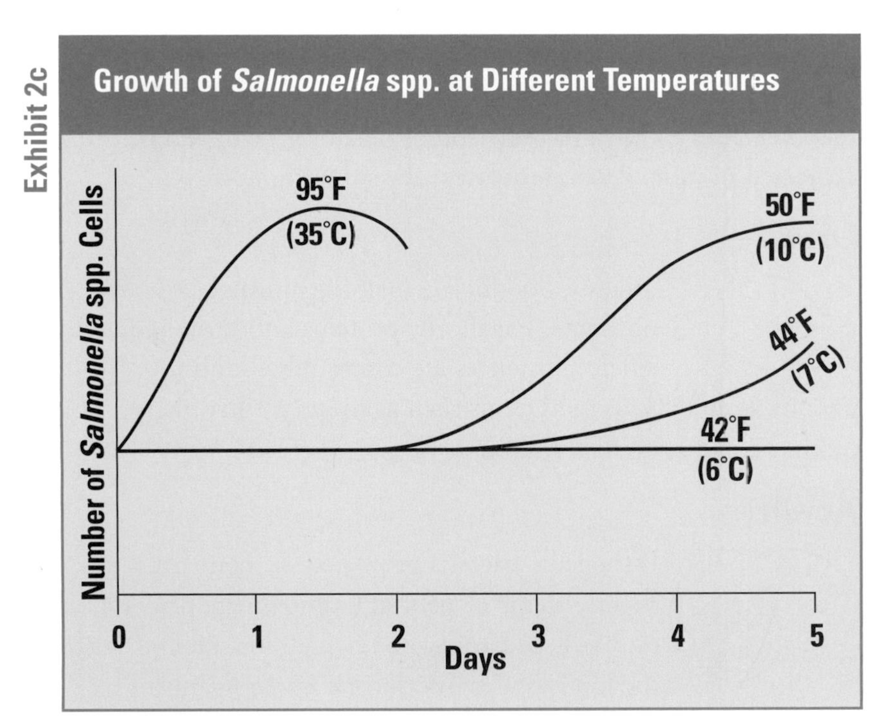

The time required for bacteria to adapt to a new environment (lag phase) and to begin a rapid rate of growth (log phase) depends on several factors, such as temperature. *Exhibit 2c* shows how different temperatures affect the growth rate of *Salmonella* spp. As the graph shows, at warmer temperatures (95°F [35°C]), *Salmonella* spp. grows more quickly than at colder temperatures (44°F and 50°F [7°C and 10°C]). At even colder temperatures (42°F [6°C]), *Salmonella* spp. does not grow at all—but notice that it does not die either (prolonging the lag phase). This is why refrigerating food properly helps keep it safe.

Vegetative Stages and Spore Formation

Although vegetative bacteria may survive low—even freezing—temperatures, they can be killed by high temperatures. For example, pathogenic bacteria can be killed during proper cooking. *(This will be discussed in Chapter 8.)* Some types of bacteria, however, have the ability to change into a different form, called a spore. The spore's thick wall protects the bacteria against unfavorable conditions, such as high or low temperature, low moisture, and high acidity.

While a spore cannot reproduce, it is capable of turning back into a vegetative organism when conditions again become favorable. For example, bacteria in food may form a spore when exposed to freezer temperatures, allowing the bacteria to survive. As the food thaws and conditions improve, the spore can turn back into a vegetative cell and begin to grow in the food.

Since spores are so difficult to destroy, it is important to cook, cool, and reheat food properly.

FAT TOM: What Microorganisms Need to Grow

The conditions that favor the growth of most foodborne microorganisms can be remembered by the acronym **FAT TOM**. (See *Exhibit 2e.*) Each of these conditions for growth will be explained in more detail in the next several paragraphs.

Food

To grow, foodborne microorganisms need nutrients, specifically proteins and carbohydrates. These substances are commonly found in potentially hazardous food items such as meat, poultry, dairy products, and eggs. (See *Exhibit 1c* on page 1-8.)

Acidity

Pathogenic bacteria grow best in food that is slightly acidic or neutral (approximate pH of 4.6 to 7.5—see *Exhibit 2d*), which includes most of the food we eat. Foodborne microorganisms

Exhibit 2d

pH Scale and Bacterial Growth

Acidic
0
Neutral 7
4.6–7.5 ideal for growth
Alkaline
14

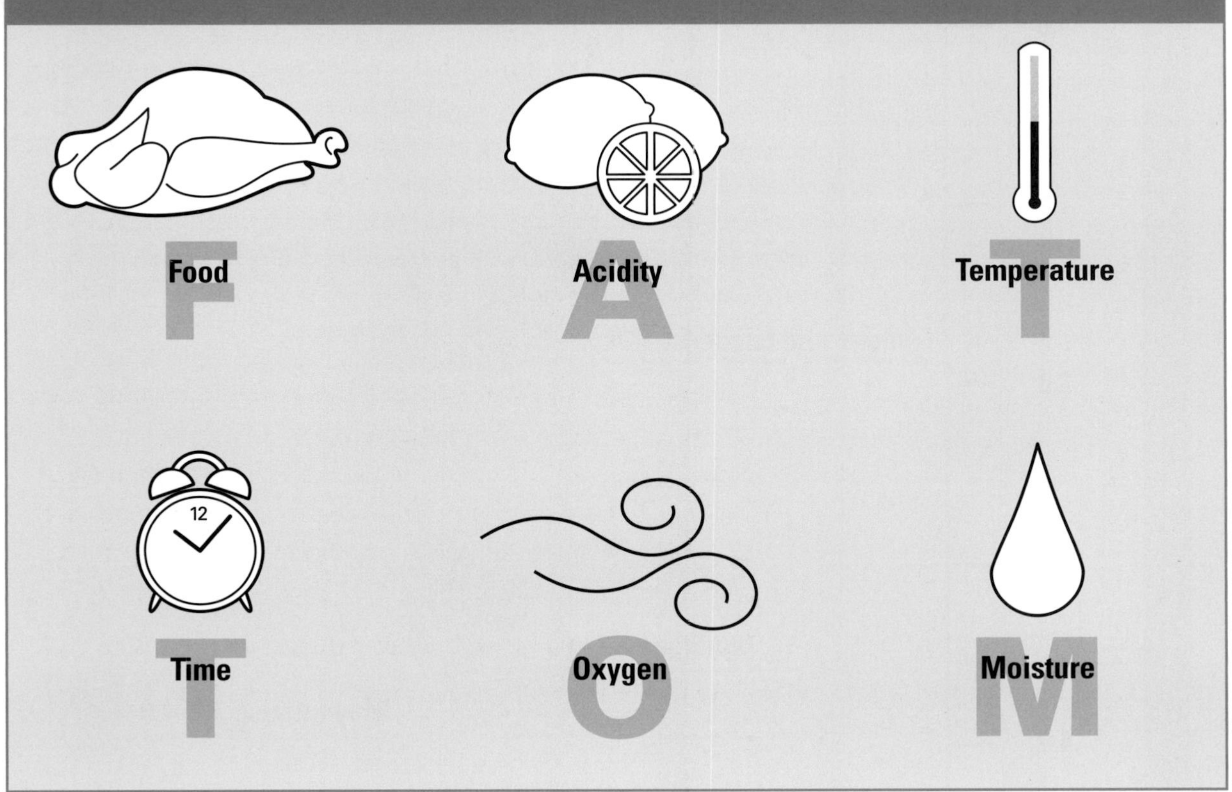

typically do not grow in alkaline or highly acidic foods, such as crackers or lemons. (See *Exhibit 2f* on the next page.)

Temperature

Most foodborne microorganisms grow well between the temperatures of 41°F and 135°F (5°C and 57°C). (See *Exhibit 2g* on page 2-9.)

This range is known as the **temperature danger zone**. However, exposing microorganisms to temperatures outside the danger zone does not necessarily kill them. Refrigeration temperatures, for example, may only slow their growth. Some bacteria—such as *Listeria monocytogenes* and *Yersinia enterocolitica*—are able to grow at refrigeration temperatures. Bacterial spores can often survive extreme heat and cold. Food must be handled very carefully when it is thawed, cooked, cooled,

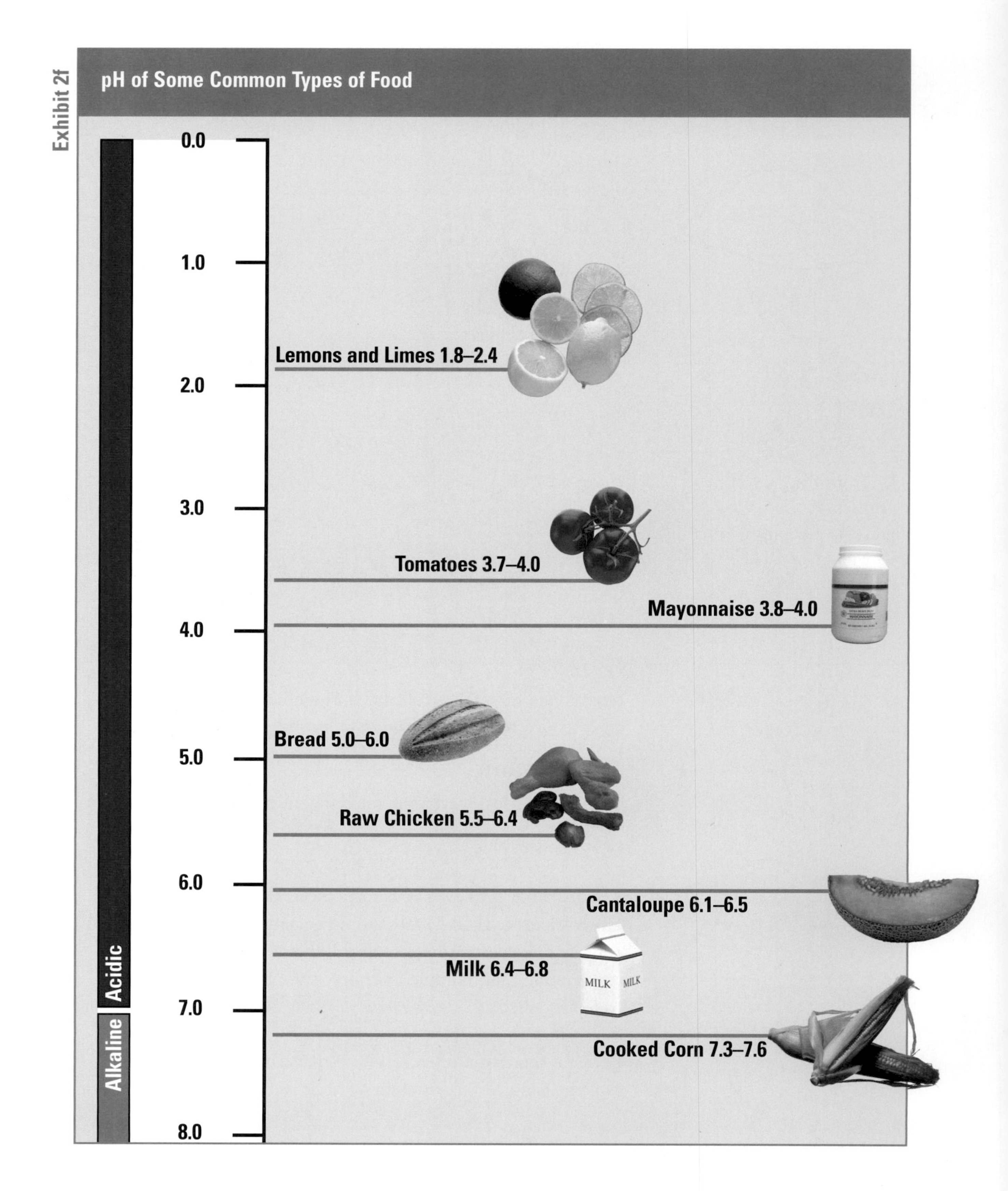
Exhibit 2f
pH of Some Common Types of Food
0.0
1.0
2.0
3.0
4.0
5.0
6.0
7.0
8.0
Acidic
Alkaline
Lemons and Limes 1.8–2.4
Tomatoes 3.7–4.0
Mayonnaise 3.8–4.0
Bread 5.0–6.0
Raw Chicken 5.5–6.4
Cantaloupe 6.1–6.5
Milk 6.4–6.8
MILK
MILK
Cooked Corn 7.3–7.6

and reheated since it can be exposed to the temperature danger zone during these times.

Time

Foodborne microorganisms need sufficient time to grow. This means that even under favorable conditions, microorganisms need enough time to move from the lag phase (slow growth) to the log phase (rapid growth). Keep in mind that some bacteria can double their population every twenty minutes.

If potentially hazardous food remains in the temperature danger zone for four hours or more, pathogenic microorganisms can grow to levels high enough to make someone ill. Therefore, it is important to control the amount of time potentially hazardous food remains in the temperature danger zone.

Oxygen

Some pathogens require oxygen to grow while others grow when oxygen is absent. Pathogens that grow without oxygen can occur in cooked rice, untreated garlic-and-oil mixtures, and foil-wrapped baked potatoes that have been temperature abused.

Moisture

Because most foodborne microorganisms require water to grow, they grow well in moist food. The amount of moisture available in a food for microorganisms to grow is called its **water activity (a_w)**. It is measured on a scale of 0 through 1.0, with water having a water activity of 1.0. Most microorganisms that cause foodborne illness grow best in food with water activities between .85 and .97, although some can grow in food with lower water-activity levels. Potentially hazardous food items typically have a water activity of .85 or higher. *Exhibit 2h* on the next page presents the water activity levels of some common food items.

Exhibit 2g

Temperature and Bacterial Growth

Most foodborne microorganisms grow well at temperatures between 41°F and 135°F (5°C and 57°C).

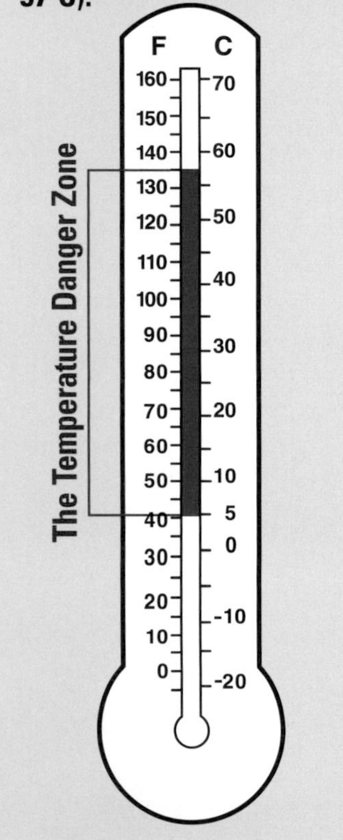

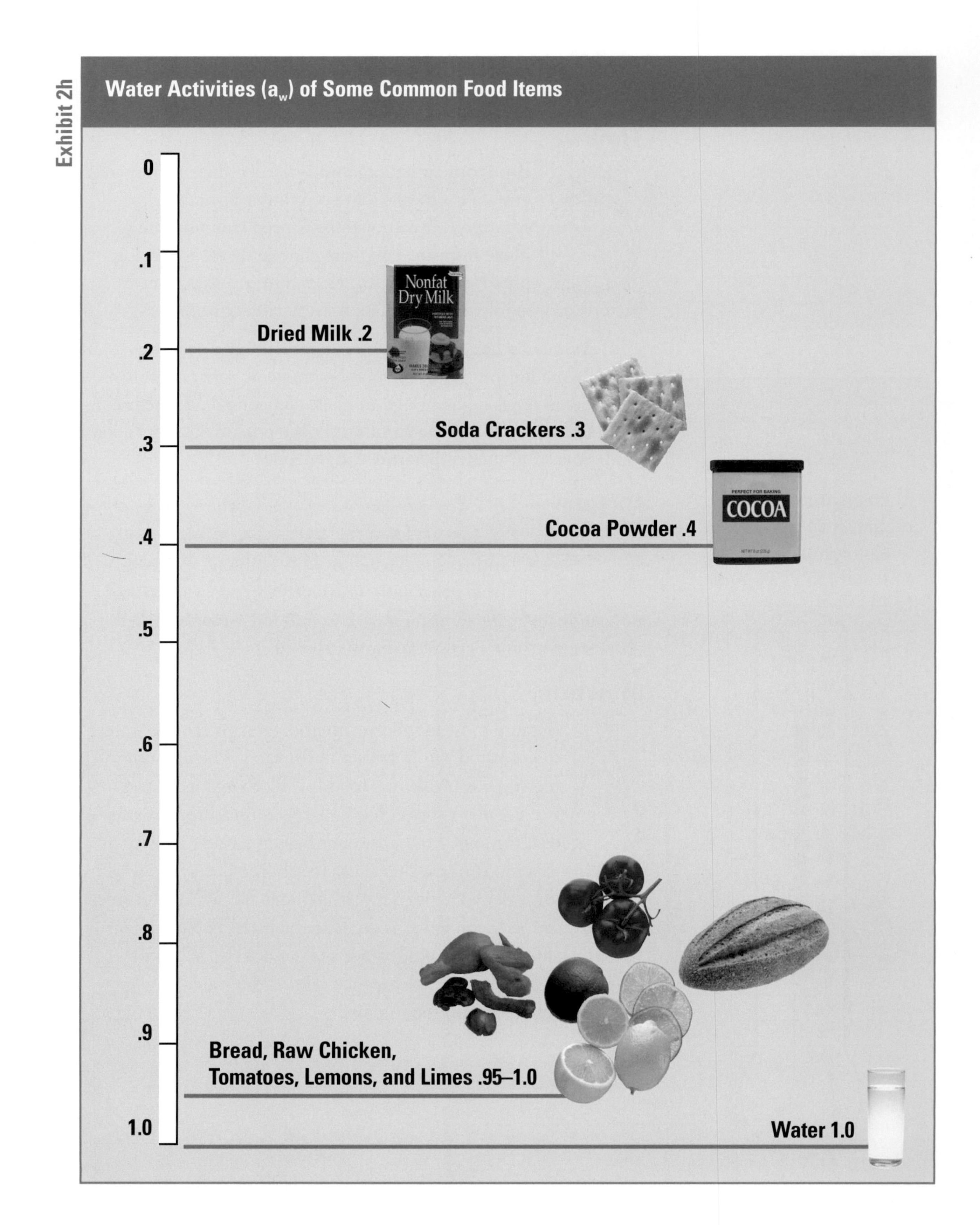
Exhibit 2h
Water Activities (a_w) of Some Common Food Items
0
.1
.2
.3
.4
.5
.6
.7
.8
.9
1.0
Dried Milk .2
Nonfat Dry Milk
Soda Crackers .3
Cocoa Powder .4
PERFECT FOR BAKING
COCOA
Bread, Raw Chicken,
Tomatoes, Lemons, and Limes .95–1.0
Water 1.0

Controlling the Growth of Microorganisms

FAT TOM is the key to controlling the growth of microorganisms in food since denying any one of these requirements can prevent growth. Food processors use several methods to keep microorganisms from growing, including:

- Adding lactic or citric acid to food to make it more acidic
- Adding sugar, salt, alcohol, or acid to a food to lower its water activity
- Using vacuum packaging to deny oxygen

While these prevention methods may not be practical for your establishment, there are two important requirements for growth that you can control—time and temperature. Remember that microorganisms grow well at temperatures between 41°F (5°C) and 135°F (57°C). Move food out of this temperature range by cooking it to the proper temperature, freezing it, or by refrigerating it at 41°F (5°C) or lower. In addition, foodborne microorganisms need sufficient time to grow. When preparing food, limit the amount of time it spends in the temperature danger zone and prepare it in small batches, as close to service as possible.

Information about major foodborne illnesses caused by bacteria, as well as ways to prevent them, is presented in *Exhibit 2i* on the next page. *(Biological toxins are discussed in Chapter 3.)*

Exhibit 2i

Major Foodborne Illnesses Caused by Bacteria

Foodborne Illness	Salmonellosis (nontyphoid)	Shigellosis (bacillary dysentery)	Listeriosis
Bacteria	*Salmonella* spp.	*Shigella* spp.	*Listeria monocytogenes*
Symptoms	Nausea, vomiting, abdominal cramps, headache, fever, and diarrhea; may cause severe dehydration in infants and elderly	Diarrhea (may be bloody), abdominal pain, fever, nausea, cramps, vomiting, chills, fatigue, dehydration	Fever and diarrhea are common in individuals who are not immuno-compromised; septicemia, meningitis, encephalitis may result in those who are immuno-compromised (elderly, pregnant women, newborns); may result in stillbirth or abortion of fetuses
Source	Water, soil, insects, domestic and wild animals, and the human intestinal tract; widespread in poultry and swine	Human intestinal tract; flies; frequently found in water polluted by feces	Soil, water, plants; cold damp environments; humans; domestic and wild animals
Food Involved In Outbreaks	Raw poultry and poultry salads; raw meat and meat products; fish; shrimp; milk and dairy products; shell eggs and egg products, such as improperly cooked custards, sauces, and pastry creams; tofu and other protein foods; sliced melons, sliced tomatoes, raw sprouts, and other fresh produce	Salads (potato, tuna, shrimp, chicken, and macaroni); lettuce; raw vegetables; milk and dairy products; poultry	Unpasteurized milk and soft cheeses; raw vegetables; poultry and meat; seafood and seafood products; prepared and chilled ready-to-eat food (e.g., soft cheese, deli foods, pâté, hot dogs)
Preventive Measures	Thoroughly cook poultry to at least 165°F (74°C) for at least 15 seconds and cook other food to required minimum internal temperatures; avoid cross-contamination; properly refrigerate food; properly cool cooked meat and meat products; properly handle and cook eggs; ensure that employees practice good personal hygiene	Ensure that employees practice good personal hygiene when handling ready-to-eat food; avoid cross-contamination; use sanitary food and water sources; control flies; properly cool food	Use only pasteurized milk and dairy products; cook food to required minimum internal temperatures; avoid cross-contamination; clean and sanitize surfaces; thoroughly wash raw vegetables

Major Foodborne Illnesses Caused by Bacteria *continued*			
Foodborne Illness	***Staphylococcal* Gastroenteritis**	***Clostridium perfringens* Gastroenteritis**	***Bacillus cereus* Gastroenteritis**
Bacteria	*Staphylococcus aureus*	*Clostridium perfringens*	*Bacillus cereus*
Symptoms	Vomiting/retching, nausea, diarrhea, abdominal cramps; in severe cases—headache, muscle cramping, changes in blood pressure and pulse rate	Abdominal pain and cramping, diarrhea, nausea (fever, headache, and vomiting usually absent)	Vomiting and nausea, sometimes abdominal cramps or diarrhea (emetic); watery diarrhea, abdominal cramps, pain, nausea (diarrheal)
Source	Humans: nose, skin, hair, throat, and infected sores; animals	Humans and domestic animals (intestinal tracts), soil	Soil and dust; cereal crops
Food Involved In Outbreaks	Reheated or improperly hot-held ready-to-eat food; meat and meat products; poultry, egg products, and other protein food; sandwiches, milk and dairy products; cream-filled pastries; salads (egg, tuna, chicken, potato, and macaroni)	Meat, meat dishes such as stew and gravy, poultry, beans that have been temperature abused	Rice products; starchy food (pasta, potatoes, and cheese products); food mixtures, such as sauces, puddings, soups, casseroles, pastries, salads, and dairy products (emetic); meats; milk and dairy products, vegetables; and fish (diarrheal)
Preventive Measures	Avoid contamination of food from unwashed bare hands; practice good personal hygiene; exclude employees with skin infections from foodhandling and preparation tasks; properly refrigerate food; rapidly cool prepared food	Use careful time and temperature control when holding, cooling, and reheating cooked food	Practice careful time and temperature control when holding, cooling, and reheating cooked food; cook food to required minimum internal temperatures

Major Foodborne Illnesses Caused by Bacteria *continued*

Foodborne Illness	Botulism	Campylobacteriosis	Hemorrhagic Colitis
Bacteria	*Clostridium botulinum*	*Campylobacter jejuni*	Shiga toxin-producing *Escherichia coli,* including O157:H7 and O157:NM
Symptoms	Fatigue, weakness, vertigo followed by blurred or double vision, difficulty speaking and swallowing, dry mouth; eventually leading to paralysis and death	Diarrhea (watery or bloody); fever and nausea; abdominal pain, headache, and muscle pain	Diarrhea (watery, may become bloody); severe abdominal cramps and pain, vomiting, mild or no fever; may cause kidney failure in the very young; symptoms more severe in the immuno-compromised
Source	Present on almost all food of either animal or vegetable origin; soil; water	Poultry and other animals; unpasteurized milk; unchlorinated water	Animals; particularly found in the intestinal tracts of cattle and humans; raw unpasteurized milk
Food Involved In Outbreaks	Food that was under-processed or temperature abused in storage, improperly canned foods, untreated garlic-and-oil mixtures, temperature-abused sautéed onions in butter, leftover baked potatoes, stews, meat/poultry loaves; risk for MAP and *sous vide* products	Unpasteurized milk and dairy products, raw poultry, nonchlorinated or fecal-contaminated water	Raw and undercooked ground beef, unpasteurized milk and apple cider/juice, beef, improperly cured dry salami, lettuce, nonchlorinated water, alfalfa sprouts
Preventive Measures	Do not serve home-canned products; use careful time and temperature control for all bulky thick foods, purchase only acidified garlic-and-oil mixtures; sauté onions to order or hold them properly; properly cool leftovers	Thoroughly cook food, especially poultry, to required minimum internal temperatures; use pasteurized milk and treated water; avoid cross-contamination	Thoroughly cook ground beef to at least 155°F (68°C) for 15 seconds; avoid cross-contamination, practice good personal hygiene, use only pasteurized milk, dairy products, and juices

Major Foodborne Illnesses Caused by Bacteria *continued*

Foodborne Illness	*Vibrio Parahaemolyticus* Gastroenteritis	*Vibrio vulnificus* Primary Septicemia	Yersiniosis
Bacteria	*Vibrio parahaemolyticus*	*Vibrio vulnificus*	*Yersinia enterocolitica*
Symptoms	Diarrhea, abdominal cramps, nausea, vomiting, headache	Fever, chills, nausea, hypotension, skin lesions may develop	Vary by age group, but diarrhea is common; symptoms may mimic appendicitis
Source	Crabs, clams, oysters, shrimp, lobster, scallops	Raw oysters, particularly those harvested during warmer months; clams, crabs	Domestic animals, soil, water
Food Involved In Outbreaks	Raw or partially cooked oysters, raw or partially cooked shellfish (clams and mussels); cross-contaminated crabs, lobster, shrimp	Raw or partially cooked oysters	Contaminated pasteurized milk, raw unpasteurized milk; tofu; nonchlorinated water; meat (pork, beef, lamb); oysters; fish
Preventive Measures	Tell high-risk populations to consult a physician before regularly consuming raw or partially cooked oysters; purchase seafood from approved suppliers; avoid cross-contamination; maintain time and temperature control	Tell high-risk populations to consult a physician before regularly consuming raw or partially cooked oysters; purchase seafood from approved suppliers; avoid cross-contamination; maintain time and temperature control	Use only pasteurized milk; minimize cross-contamination; thoroughly cook food to required minimum internal temperatures; ensure that utensils and equipment are properly sanitized; use only sanitary, chlorinated water supplies

VIRUSES

Viruses are the smallest of the microbial contaminants. They consist of genetic material wrapped with an outer layer of protein. While a virus cannot produce outside a living cell, once inside a human cell, it will produce more viruses. Viruses are responsible for several foodborne illnesses, such as hepatitis A and infections caused by Norovirus and Rotavirus.

Basic Characteristics of Viruses

Key Point

Viruses can be transmitted from person to person, from people to food, and from people to food-contact surfaces.

Health Alert

Practicing good personal hygiene and minimizing bare-hand contact with ready-to-eat food are important ways to prevent the contamination of food by foodborne viruses.

Viruses share some basic characteristics.

- Unlike bacteria, they rely on a living cell to reproduce.
- They are not complete cells.
- Unlike bacteria, they do not reproduce in food.
- Some may survive freezing and cooking.
- They can be transmitted from person to person, from people to food, and from people to food-contact surfaces.
- They usually contaminate food through a foodhandler's improper personal hygiene.
- They can contaminate both food and water supplies.

Practicing good personal hygiene is an important way to prevent the contamination of food by foodborne viruses. It is especially important to minimize bare-hand contact with ready-to-eat food. Information about major foodborne illnesses caused by viruses, as well as ways to prevent them, is presented in *Exhibit 2j.*

PARASITES

Parasites are organisms that need to live in a host organism to grow. A **host** is a person, animal, or plant on which another organism lives and takes nourishment. Parasites can live inside many animals that humans use for food, such as cattle, poultry, pigs, and fish. Foodborne parasites include protozoa and roundworms. Parasites require many different hosts to carry out their life cycles; however, they are typically passed to humans through the meat of an animal host. To prevent foodborne illness

Exhibit 2j

Major Foodborne Illnesses Caused by Viruses

Foodborne Illness	Hepatitis A	Norovirus Gastroenteritis	Rotavirus Gastroenteritis
Virus	*Hepatovirus* or hepatitis A virus	Norovirus (formerly called Norwalk virus)	Rotavirus
Symptoms	Sudden onset of fever; fatigue, nausea, loss of appetite, vomiting, abdominal pain, and jaundice after several days; children often exhibit no symptoms	Nausea, vomiting (more common in children), watery diarrhea with abdominal cramps, mild fever	Vomiting, watery diarrhea, abdominal pain, and mild fever (illness more common in children than adults)
Source	Human intestinal tract; feces-contaminated water	Human intestinal tract and feces-contaminated water	Human intestinal tract; feces-contaminated water
Food Involved In Outbreaks	Shellfish; salads; cross-contaminated deli meats and sandwiches; fruit and fruit juices; milk and milk products; vegetables; any food that will not receive a further heat treatment; water and ice	Ready-to-eat food including salads, sandwiches, and bakery products; liquid items such as salad dressing or cake icing; oysters from contaminated waters; contaminated raspberries; contaminated well water	Water and ice, raw and ready-to-eat food (salads, fruit), contaminated water
Preventive Measures	Obtain shellfish from approved sources; ensure foodhandlers practice good personal hygiene; prevent cross-contamination from hands; clean and sanitize food-contact surfaces; use sanitary water sources	Ensure foodhandlers practice good personal hygiene; obtain shellfish from approved sources; use sanitary, chlorinated water	Ensure foodhandlers practice good personal hygiene; prevent cross-contamination from hands; thoroughly cook food to required minimum internal temperatures; use sanitary, chlorinated water

caused by parasites, make sure food comes from an approved source and has been properly frozen, use proper cooking techniques, avoid cross-contamination, use sanitary water supplies, and follow proper handwashing procedures, especially after using the restroom.

Basic Characteristics of Foodborne Parasites

Key Point

Parasites are typically passed to humans through an animal host.

Parasites share some basic characteristics.

- They are living organisms that need a host to survive.
- They grow naturally in many animals—such as pigs, cats, rodents, and fish—and can be transmitted to humans.
- Most are very small, often microscopic, but larger than bacteria.
- They pose hazards to both food and water.

Information about major foodborne illnesses caused by parasites, as well as ways to prevent them, is presented in *Exhibit 2l* on page 2-20.

FUNGI

Fungi range in size from microscopic, single-celled organisms to very large, multicellular organisms. They are found naturally in air, soil, plants, water, and some food. **Molds, yeasts,** and mushrooms are examples of fungi. The fungi of concern to restaurants and foodservice establishments are molds and yeasts. *(Mushroom toxins are discussed in Chapter 3.)*

Molds

Individual mold cells can usually be seen only with a microscope. However, fuzzy or slimy mold colonies, consisting of a large number of cells, are often visible to the naked eye. Bread mold is an example. The spores produced by molds are not the same as the spores produced by bacteria. Molds use spores for reproduction.

Molds are responsible for the spoilage of food. This spoilage results in discoloration and the formation of odors and off-flavors. Molds are able to grow on almost any food at almost any storage temperature. They can also grow in environments that are moist or dry, have a high or low pH, and are salty or sweet. They typically prefer to grow in and on acidic food with low water activity. Molds often spoil fruit, vegetables, meat, cheese, and bread because of their water activity and pH.

Some molds produce toxins that can cause allergic reactions, nervous system disorders, and kidney and liver damage. For example, aflatoxin, produced by the molds *Aspergillus flavus* and *Aspergillus parasticus,* can cause liver disease.

Food such as corn and corn products, peanuts and peanut products, cottonseed, milk, and tree nuts (such as Brazil nuts, pecans, pistachio nuts, and walnuts) have been associated with aflatoxins.

Basic Characteristics of Foodborne Molds

Molds share some basic characteristics.

- They spoil food and sometimes cause illness.
- They grow under almost any condition, but grow well in acidic foods with low water activity.
- Freezing temperatures prevent or reduce the growth of molds, but do not destroy them.
- Some molds produce toxins such as aflatoxins.

Although the FDA recommends cutting away any moldy areas in hard cheese—at least one inch (2.5 centimeters) around them—to avoid illnesses caused by mold toxins, throw out all moldy food, unless the mold is a natural part of the food (e.g., cheeses such as Gorgonzola, Bleu, Brie, and Camembert).

While mold cells and spores can be killed by heating them, toxins that may be present are not destroyed by normal cooking methods. Food with molds that are not a natural part of the product should always be discarded. (See *Exhibit 2k.*)

Exhibit 2k

Mold on Cheese

Food with mold that is not a natural part of the product should always be discarded.

Exhibit 2I

Major Foodborne Illnesses Caused by Parasites

Foodborne Illness	Trichinosis	Anisakiasis	Giardiasis
Parasite	*Trichinella spiralis*	*Anisakis simplex*	*Giardia duodenalis* (also called *G. lamblia)*
Symptoms	Nausea, vomiting, diarrhea, fever, and fatigue followed by facial swelling and muscle pain	Tingling or tickling sensation in throat, vomiting or coughing up worms; severe abdominal pain, vomiting, nausea, diarrhea	Intestinal gas, diarrhea, abdominal cramps, nausea, weight loss, fatigue
Source	Domestic pigs; wild game, such as bears and walruses	Marine fish (saltwater species only)	Intestinal tract of humans; contaminated water; inadequately treated water
Food Involved In Outbreaks	Raw and undercooked pork or pork products (particularly sausage), raw and undercooked wild game	Raw, undercooked, or improperly frozen seafood, especially cod, haddock, fluke, Pacific salmon, herring, flounder, halibut monkfish, mackerel and fish used for sashimi and ceviche	Contaminated water and ice, salads and (possibly) other raw vegetables washed with contaminated water
Preventive Measures	Cook pork and game meat to required minimum internal temperatures; wash, rinse, and sanitize equipment, such as sausage grinders and utensils used in the preparation of raw pork and other meats; purchase meat and meat products from approved suppliers; ensure employees practice good personal hygiene	Obtain seafood from approved sources; when serving raw or undercooked seafood, only use sashimi-grade fish that has been properly treated to eliminate parasites; fish intended to be eaten raw should be frozen at –4°F (–20°C) or lower for 7 days in a freezer, or at –31°F (–35°C) or lower for 15 hours in a blast chiller	Use sanitary water supplies; ensure that foodhandlers practice good personal hygiene; wash raw produce carefully

Major Foodborne Illnesses Caused by Parasites *continued*

Foodborne Illness	Toxoplasmosis	Intestinal Cryptosporidiosis	Cyclosporiasis
Parasite	*Toxoplasma gondii*	*Cryptosporidium parvum*	*Cyclospora cayetanensis*
Symptoms	Often, there are no symptoms; when symptoms occur, they include enlarged lymph nodes in head and neck, severe headaches, severe muscle pain, and rash; most commonly affects fetuses	Mild to severe nausea, abdominal cramping, watery diarrhea	Onset of symptoms is sudden; mild to severe nausea, abdominal cramping, mild fever, watery diarrhea
Source	Animal feces (especially felines), mammals	Intestinal tract of humans, cattle, and other domestic animals; drinking water contaminated with run-off from farms or slaughterhouses	Intestinal tract of humans; contaminated water supplies
Food Involved In Outbreaks	Contaminated water; raw or undercooked meat—especially pork, lamb, wild game, and poultry	Water; salads and raw vegetables; milk; unpasteurized apple cider; ready-to-eat food	Water, raw produce, marine fish, raw milk
Preventive Measures	Properly wash hands if they come in contact with soil, raw meat, cat feces, or raw vegetables; avoid raw or undercooked meat (especially lamb, wild game, or poultry); cook meat to the required minimum internal temperature	Ensure that foodhandlers practice good personal hygiene; thoroughly wash produce; use sanitary water sources	Ensure that foodhandlers practice good personal hygiene; thoroughly wash produce; use sanitary water sources

Exhibit 2m

Yeast on Jelly

Food that has been spoiled by yeast should be discarded.

Yeasts

Some yeasts are known for their ability to spoil food rapidly. Carbon dioxide and alcohol are produced as yeast slowly consumes food. Yeast spoilage may, therefore, produce a smell or taste of alcohol. Yeast may appear as a pink discoloration or slime and may bubble.

Yeasts are similar to molds in that they grow well in acidic food with low water activity, such as jellies, jams, syrup, honey, and fruit juice. (See *Exhibit 2m.*) Food that has been spoiled by yeast should be discarded.

FOODBORNE INFECTION VS. FOODBORNE INTOXICATION

Foodborne diseases are classified as infections, intoxications, or toxin-mediated infections. Each occurs in a different way.

- **Foodborne infections** result when a person eats food containing pathogens, which then grow in the intestines and cause illness. Typically, symptoms of a foodborne infection do not appear immediately in reasonably healthy people.
- **Foodborne intoxications** result when a person eats food containing toxins that cause illness. The toxin may have been produced by pathogens found on the food or may be the result of a chemical contamination. The toxin might also be a natural part of the plant or animal consumed. Typically, symptoms of a foodborne intoxication appear quickly, within a few hours.
- **Foodborne toxin-mediated infections** result when a person eats food containing pathogens, which then produce illness-causing toxins in the intestines.

EMERGING PATHOGENS AND ISSUES

Progress has been made in preventing foodborne diseases. For example, typhoid fever (caused by *Salmonella typhii*) was very common in the early twentieth century. Typhoid is now almost forgotten in the United States due to improved methods of treating drinking water and sewage, milk sanitation and pasteurization, and shellfish sanitation. Similarly, cholera and trichinosis, once very common, are now also rare in the U.S.

In the past, foods involved in foodborne-illness outbreaks were often raw or inadequately cooked meat, poultry, seafood, or unpasteurized milk. Today, some foods previously considered safe have become vehicles for foodborne pathogens. For example, it was assumed for centuries that the internal contents of an egg were sterile and safe to eat raw. Now, however, there are clear indications that, in rare cases, eggs can be contaminated internally with *Salmonella* Enteritidis. A number of outbreaks of salmonellosis have been traced to undercooked, contaminated eggs included in traditional recipes, such as eggnog, hollandaise sauce, Caesar salad, lightly cooked omelets, French toast, and even meringue.

Produce and highly acidic foods, once considered safe, are now considered potential vehicles for foodborne pathogens. As detection techniques and foodborne-illness investigations improve, new pathogens causing foodborne illness are being recognized.

These trends have emerged for a variety of reasons. As the population continues to grow, more food is needed. Food is produced in greater quantities and at centralized sources. More food is shipped from greater distances, across state and national borders. It remains important for establishments to purchase food from reputable suppliers and to keep food safe by using proper storage, preparation, and serving practices.

HAZARDS ASSOCIATED WITH FRESHER AND HEALTHIER FOOD

Many changes are occurring that can greatly affect food safety. Natural or organic food without additives or preservatives is becoming more available, along with low-fat, low-calorie versions of everyday food. Also, food from all over the world is becoming more accessible. In response to the public's health consciousness, many products commonly seen on grocery store shelves are now being altered.

Altering the makeup of food is not without risk to its safety. If the stability of a product was traditionally due to the water activity or pH (i.e., the product was too dry or acidic to allow microorganisms to grow), changing its makeup may eliminate that barrier, causing it to become unsafe.

Organic products might also have food safety risks. Organic food is grown and processed without the use of pesticides, additives, or preservatives. Lacking these traditional barriers to microbial growth, organic products might have reduced shelf lives and allow spoilage microorganisms to grow.

TECHNOLOGICAL ADVANCEMENTS IN FOOD SAFETY

Several new technologies are emerging to reduce pathogenic and spoilage microorganisms in food. These include food irradiation, electron pasteurization, high-pressure processing, bacteriocins, and various light treatments, including laser light, ultraviolet light, and pulsed light.

Food irradiation, often called cold pasteurization, involves exposing food to an electron beam or gamma rays, similar to the way microwaves are used to heat food. The FDA has approved the use of irradiation on certain foods in the U.S. To date, irradiation is allowed for raw meat and meat products (to eliminate shiga toxin-producing *E.coli* and to reduce levels of *Listeria* spp., *Salmonella* spp., and *Campylobacter* spp.), pork (to control the parasite *Trichinella spiralis*), poultry (to control bacterial contaminants), fruit and vegetables (to prevent premature maturation and to control insects), and on herbs, spices, teas, and other dried vegetable substances (to control various microbial contaminants).

Benefits of food irradiation include:

- Reduction or elimination of pathogens and spoilage microorganisms
- Replacement of chemical treatment of food
- Extended shelf life of food

Irradiated foods have only recently become widely available in the U.S., despite regulatory approval and repeated endorsements by professional health and nutrition organizations. No evidence exists of harmful effects resulting from the amount or types of radiation used, and nutritional value, appearance, and taste are virtually not affected. However, some consumers remain apprehensive about food exposed to radiation, due to their limited knowledge of the irradiation process. Studies show that once the process and benefits are explained to them, consumers are more receptive to the idea of irradiation and would purchase irradiated products.

It is important to note that food irradiation does not replace proper food storage and handling practices, since food could still become unsafe if cross-contaminated. Therefore, when handling irradiated food, you should follow the same food safety practices you would follow with other food.

SUMMARY

Microbial contaminants are responsible for the majority of foodborne illnesses. Because bacteria, viruses, parasites, and fungi can be introduced at any point in the flow of food, they must be monitored and controlled throughout it. Understanding how these microorganisms grow, reproduce, contaminate food, and infect humans is critical to understanding how to prevent the foodborne illnesses they cause.

Bacteria are living, single-celled organisms that can be carried by a variety of means. Some cause food spoilage; others cause illness. Some bacteria cause illness by producing toxins as they multiply, die, and break down. These toxins are not typically destroyed by cooking. Under favorable conditions, bacteria can reproduce very rapidly. Although vegetative bacteria may be resistant to low—even freezing—temperatures, they can be killed by high temperatures, such as those reached during cooking. Some types of bacteria, however, have the ability to form spores, which protect the bacteria from unfavorable conditions. Since spores are so difficult to destroy, it is important to thaw, cook, cool, and reheat food properly to keep bacteria from growing to harmful levels.

The acronym FAT TOM—which stands for Food, Acidity, Time, Temperature, Oxygen, and Moisture—might help you remember the conditions that promote the growth of foodborne microorganisms. Needing nutrients to grow, specifically proteins and carbohydrates, microorganisms grow best in food with a slightly acidic-to-neutral pH. Most foodborne microorganisms grow well between the temperatures of 41°F and 135°F (5°C and 57°C). They also need sufficient time within these temperatures to grow. If potentially hazardous food remains in the temperature danger zone for four hours or more, pathogenic microorganisms can grow to levels high enough to make someone ill. Some pathogens require oxygen to grow while others grow when oxygen is absent. Most require the moisture in food to grow. FAT TOM is the key to controlling the growth of microorganisms in food since denying any one of these requirements can prevent growth. There are two important requirements that you can control—time and temperature. Cook food to the proper temperature, freeze it, or refrigerate it at 41°F (5°C) or lower. When preparing food, limit the amount of time it spends in the temperature danger zone, and prepare it in small batches as close to service a possible.

Viruses are the smallest of the microbial contaminants. While a virus cannot reproduce in food, once ingested it will cause illness. Viruses can be transmitted from person to person, from people to food, and from people to food-contact surfaces. They can contaminate both food and water supplies. Some may survive freezing and cooking. Practicing good personal hygiene and minimizing bare-hand contact with ready-to-eat food is an important defense against foodborne illness from viruses.

Parasites are organisms that need to live in a host organism to survive. They can live inside many animals humans eat, such as cattle, poultry, pigs, and fish. They can be killed by proper cooking and freezing.

Fungi are mostly responsible for the spoilage of many kinds of food, while some molds can produce harmful toxins. Molds are able to grow in a variety of environments, but they typically prefer to grow in and on acidic food with low water activity. Yeasts are known for their ability to spoil food rapidly. They are similar to molds in that they grow well in acidic food with low water activity.

Foodborne diseases are classified as infections, intoxications, or toxin-mediated infections. Each occurs in a different way. Foodborne infections result when a person eats food containing pathogens, which then grow in the intestines and cause illness. Typically, symptoms do not appear immediately. Foodborne intoxications result when a person eats food containing illness-causing toxins produced by pathogens found on the food or as a result of a chemical contamination. The toxin might also come from a plant or animal that was eaten. Typically, symptoms of foodborne intoxication appear quickly, within a few hours. Foodborne toxin-mediated infections result when a person eats food that contains pathogens, which then produce illness-causing toxins in the intestines.

Our ever-changing world is opening new doors for microorganisms. Many new pathogens are emerging. New product formulations are creating niches for microorganisms that may not have been able previously to grow in the product. However, many new technologies are available and others are being developed to prevent the growth of pathogenic microorganisms.

Apply Your Knowledge

1. Based on the information given, was the illness caused by bacteria, a virus, a parasite, or fungi?
2. What is the name of the microorganism most likely to have caused the outbreak?
3. Is this illness an infection or an intoxication?

For answers, please turn to the Answer Key.

A Case in Point

A day-care center is serving stir-fried rice for lunch. The rice was cooked to the proper temperature for the proper amount of time at 1:00 P.M. The covered rice was then placed on the countertop and allowed to cool to room temperature. At 6:00 P.M., the cook placed it in the refrigerator. At 9:00 A.M. the following day, the rice was combined with the other ingredients for stir-fried rice and cooked to 165˚F (74˚C) for at least fifteen seconds. The cook covered the stir fried rice and left it on the range until she gently reheated it at noon. Within an hour of eating the stir-fried rice, however, several of the children began vomiting, and a few had diarrhea. Samples from some of the children revealed the rice as the probable cause of the outbreak.

Apply Your Knowledge

Use these questions to review the concepts presented in this chapter.

Discussion Questions

1. Why shouldn't potentially hazardous food be left in the temperature danger zone?
2. What is the difference between a foodborne infection, a foodborne intoxication, and a toxin-mediated infection?
3. How can an outbreak of salmonellosis be prevented?
4. What are some basic characteristics of foodborne viruses?
5. What two FAT TOM requirements for the growth of microorganisms are easiest for an establishment to control? Give examples of how each requirement can be controlled.

For answers, please turn to the Answer Key.

Apply Your Knowledge

Use these questions to test your knowledge of the concepts presented in this chapter.

Multiple-Choice Study Questions

1. **Foodborne microorganisms grow well at temperatures between**
   - A. 41°F and 135°F (5°C and 57°C).
   - B. 32°F and 70°F (0°C and 21°C).
   - C. 38°F and 155°F (3°C and 68°C).
   - D. 70°F and 165°F (21°C and 74°C).

2. **All of the following conditions typically support the growth of microorganisms *except***
   - A. moisture.
   - B. a protein or carbohydrate food source.
   - C. high acidity.
   - D. time.

3. **A person who has a foodborne infection most likely has eaten a food containing**
   - A. a ciguatoxin.
   - B. histamine.
   - C. a plant toxin.
   - D. a live pathogen.

4. **Which food is most likely to transmit parasites to humans?**
   - A. Improperly cooked eggs
   - B. Improperly frozen sashimi
   - C. Improperly refrigerated milk
   - D. Unpasteurized apple juice

5. **Which of the following is not a basic characteristic of foodborne mold?**
   - A. It grows well in acidic food with low water activity.
   - B. Freezing temperatures prevent or slow its growth, but do not destroy it.
   - C. Its cells and spores may be killed by heating, but the toxins it produces may not be destroyed.
   - D. It needs a host to survive.

Continued on next page...

Apply Your Knowledge — Multiple-Choice Study Questions *continued*

6. **Which of the following statements regarding foodborne-intoxication is true?**
   A. Symptoms of intoxication often appear days after exposure.
   B. Medical treatment for intoxication can be painful.
   C. Foodborne intoxication is more common than foodborne infection.
   D. Symptoms of intoxication appear quickly, within a few hours.

7. **Which of the following microorganisms is associated with unpasteurized milk and soft cheeses?**
   A. *Vibrio parahaemolyticus*
   B. *Listeria monocytogenes*
   C. *Trichinella spiralis*
   D. *Clostridium botulinum*

8. **Which of the following microorganisms is found in cereal crops such as rice?**
   A. *Staphylococcus aureus*
   B. Hepatitis A virus
   C. *Campylobacter jejuni*
   D. *Bacillus cereus*

9. **A person who has a severe case of *Staphylococcal* Gastroenteritis may experience**
   A. changes in blood pressure and pulse rate.
   B. hallucinations.
   C. a tingling or tickling sensation in the throat.
   D. facial swelling.

10. **Which of the following practices can help prevent hepatitis A?**
    A. Obtaining shellfish from approved sources
    B. Cooking pork to the proper internal temperature
    C. Practicing careful time and temperature control for all thick foods
    D. Avoiding the use of home-canned food

For answers, please turn to the Answer Key.

ADDITIONAL RESOURCES

Books and Periodicals

American Public Health Association. 2000. *Control of communicable diseases in man,* 17th ed. Washington, D.C.: American Public Health Association.

Buchanan, R. L. and M. P. Doyle. 1997. Foodborne disease significance of *Escherichia coli* O157:H7 and other enterohemorrhagic *E. coli. Food Technology.* 51 (10):69–76.

Bush, R. K. 1995. Seafood allergy and allergens: A review. *Food Technology.* 49 (10):103–116.

Cliver, D. O. 1997. Virus transmission via food. *Food Technology.* 51 (4):71–78.

Cliver, D. O., ed. 1990. *Foodborne diseases.* San Diego, CA: Academic Press.

Featsent, A. W. 1998. Food fright! Consumers' perceptions of food safety versus reality. *Restaurants USA.* 18 (6):30.

Grossbauer, S. 1998. Food safety on the Web. *FoodService Director.* 11 (2):140.

Hernandez, J. 1997. Eliminating the *E. coli* threat. *Food Management.* 32 (12):67.

International Commission on Microbiological Specifications for Food (ICMSF). 1988. *Microorganisms in foods 1: Their significance and methods of enumeration,* rev. ed. Toronto: University of Toronto Press.

ICMSF. 1997. *Microorganisms in foods 3: Vol. 1. Factors affecting life and death of microorganisms: Vol. 2. Food commodities: Vol. 3. Microbial ecology of foods.* New York: Academic Press.

ICMSF. 1996. *Microorganisms in foods 5: Characteristics of microbial pathogens.* London: Blackie Academic & Professional.

Jay, J. M. 1996. *Modern food microbiology.* New York: Chapman & Hall.

Longree, K. and G. Armbruster. 1996. *Quantity food sanitation,* 5th ed. New York: Wiley.

Marriott, N. G. 1994. *Principles of food sanitation.* New York: Chapman & Hall.

National Restaurant Association. 1996. *Sanitation survival kit.* Washington, D.C.: National Restaurant Association.

Rehe, S. 1990. *Preventing foodborne illness: A guide to safe foodhandling.* Washington, D.C.: U.S. Department of Agriculture Food Safety & Inspection Service.

Web Sites

American Council on Science and Health (ACSH)
www.acsh.org/food
ACSH is a consumer education consortium concerned with issues related to food, nutrition, chemicals, pharmaceuticals, lifestyle, the environment, and health. ACSH is an independent, nonprofit, tax-exempt organization. This Web site provides practical, matter-of-fact information on foodborne-illness stories you might hear about in the news.

American Society of Microbiology (ASM)
www.asm.org
The mission of the ASM is to promote the microbiological sciences and their applications for the common good. To this end, ASM publishes journals and books, and conducts and supports education, training, and public information programs to facilitate the dissemination and application of new microbiological knowledge affecting public interest. This Web site contains excellent information on foodborne illness and teaching microbiology.

Centers for Disease Control and Prevention (CDC)
www.cdc.gov
The mission of the CDC is to promote health and quality of life by preventing and controlling disease, injury, and disability. To prevent and control foodborne illness, the CDC collect data on outbreaks. This Web site provides general information on foodborne illnesses and their prevention.

Compendium of Fish and Fishery Product Processes, Hazards, and Controls
http://seafood.ucdavis.edu/haccp/compendium/compend.htm
Developed by the FDA, the compendium available at this Web site provides useful information on biological, chemical, and physical hazards associated with fish and fish products.

Council for Biotechnology Information
www.whybiotech.com
The Council for Biotechnology Information was founded by leading biotechnology companies to create a comprehensive communication campaign about biotechnology. The Council is committed to providing objective, balanced information to help the public better understand and appreciate the benefits of biotechnology.

FDA Bad Bug Book
http://vm.cfsan.fda.gov/~mow/intro.html
Produced by the FDA's Center for Food Safety and Nutrition (FDA CFSAN), this online handbook provides basic facts regarding pathogenic foodborne microorganisms and natural toxins. It brings together information from the FDA, the CDC, the USDA Food Safety Inspection Service, and the National Institutes of Health.

FDA Center for Food Safety and Applied Nutrition (CFSAN)
www.cfsan.fda.gov
As the center within the FDA responsible for food safety and nutrition, CFSAN promotes and protects the public health by researching and implementing guidelines, policies, and standards to ensure that food is safe, nutritious, wholesome, and properly labeled. This Web site provides a wealth of information on food safety and sanitation, including corresponding guidelines, policies, and standards.

FDA Food Code
http://vm.cfsan.fda.gov/~dms/foodcode.html
As the basis for many local sanitation codes, as well as the basis for information in this textbook, the FDA Food Code, available at this Web address, is a useful resource for information relating to food safety information for the restaurant and foodservice industry.

Hepatitis Control Report

www.hepatitiscontrolreport.com

This site is an online quarterly newsletter devoted to news on the public-health control of viral hepatitis. Its mission is to provide accurate and balanced reporting of developments in hepatitis epidemiology, control programs, and public policy.

International Atomic Energy Agency (IAEA)

www.iaea.org

As a specialized agency within the United Nations system, IAEA serves as the world's central intergovernmental forum for scientific and technical cooperation in the nuclear field. This Web site houses up-to-date information on all issues concerning atomic energy, including irradiation.

International Food Information Council (IFIC) Foundation

www.ific.org

IFIC collects and disseminates scientific information on food safety, nutrition, and health. They work with an extensive roster of scientific experts to help translate research into understandable and useful information for opinion leaders and, ultimately, consumers. This Web site provides easy-to-understand information on foodborne illness and health-related stories circulating in the news.

Journal of Emerging Infectious Diseases

www.cdc.gov/ncidod/eid

This journal provides information on emerging infections and diseases. It is published by the National Center for Infectious Diseases and is available, free of charge, on this site in several languages. The journal databases can be searched for foodborne pathogen-related information.

Monsanto Company

www.monsanto.com

The Monsanto Company is a leading provider of agricultural solutions to growers worldwide. They provide top-quality, cost-effective, and integrated approaches to help farmers improve their productivity and produce better quality food. This Web site contains information on agricultural biotechnology.

Morbidity and Mortality Weekly Report (MMWR)
www.cdc.gov/mmwr

The *MMWR* is prepared by the CDC and is available, free of charge, on this Web site. The reports contain data based on weekly accounts of reportable diseases from state health departments. Report databases can be searched for information on foodborne-illness outbreaks.

National Center for Infectious Diseases (NCID)
www.cdc.gov/ncidod

NCID is one of the centers of the CDC. Its mission is to prevent illness, disability, and death caused by infectious diseases around the world. NCID accomplishes this mission by conducting surveillance, epidemic investigations, epidemiologic and laboratory research, and training. It also sponsors public education programs to develop, evaluate, and promote prevention and control strategies for infectious diseases. This Web site serves as another great resource for information on foodborne illness.

National Restaurant Association
www.restaurant.org

The National Restaurant Association is the leading business association for the restaurant industry. Together with the National Restaurant Association Educational Foundation, the Association's mission is to represent, educate, and promote the rapidly growing restaurant and foodservice industry. This Web site should be your starting place for all issues and concerns related to your restaurant. This Web site has it all, from tips for running your establishment to vital data on your customers' spending habits.

Apply Your Knowledge

Notes

10
HOTEL SIZE
SOAP
PADS
Manual Pot and Pan Detergent
Détergent pour le nettoyage à la main des casseroles
Detergente para el Lavado a Mano de Ollas y Sartenes
3.78 L / 1 GALLON (US)

3 Contamination, Food Allergens, and Foodborne Illness

Inside this chapter:

- Types of Foodborne Contamination
- The Deliberate Contamination of Food
- Food Allergens

After completing this chapter, you should be able to:

- Identify biological, chemical, and physical contaminants.
- Identify methods to prevent biological, chemical, and physical contamination.
- Identify the eight most common allergens, associated symptoms, and methods of prevention.

Key Terms

Biological toxins
Chemical toxins
Ciguatera poisoning
Scombroid poisoning
Histamine
Toxic-metal poisoning
Food security
Food allergy

Apply Your Knowledge

Check to see how much you know about the concepts in this chapter. Use the page references provided to explore the topic in each question.

Test Your Food Safety Knowledge

1. **True or False:** Fish that has been properly cooked will be safe to eat. *(See page 3-5.)*
2. **True or False:** Cooking can destroy the toxins in toxic wild mushrooms. *(See page 3-6.)*
3. **True or False:** Copper utensils and equipment can cause an illness when used to prepare acidic food. *(See page 3-8.)*
4. **True or False:** Cleaning products may be stored with packages of food. *(See page 3-8.)*
5. **True or False:** A person who is allergic to food may experience tightening in the throat. *(See page 3-13.)*

For answers, please turn to the Answer Key.

INTRODUCTION

Food is considered contaminated when it contains hazardous substances. These substances may be biological, chemical, or physical. The most common food contaminants are biological contaminants that belong to the microworld—bacteria, parasites, viruses, and fungi. Most foodborne illnesses result from these contaminants, but **biological toxins** and **chemical toxins** are also responsible for many foodborne illnesses. While biological and chemical contamination pose a significant threat to food, the danger from physical hazards should also be recognized.

TYPES OF FOODBORNE CONTAMINATION

Ensuring the safety of food is the manager's most important job. A thorough understanding of the causes and prevention of various types of contamination can help you keep food safe.

Biological Contamination

As you learned in Chapter 2, a foodborne intoxication occurs when a person eats food containing toxins. The toxin may have been produced by pathogens found on the food or may be the result of a chemical contamination. The toxin could also come from a plant or animal that was eaten. Toxins in seafood, plants, and mushrooms are responsible for many cases of foodborne illness in the U.S. each year. Most of these biological toxins occur naturally and are not caused by the presence of microorganisms. Some occur in animals as a result of their diet.

Seafood Toxins

The ciguatera toxin occurs in certain predatory tropical reef fish, such as amberjack, barracuda, grouper, and snapper. (See *Exhibit 3a.*) Ciguatera accumulates in the tissue of these large, predatory fish after they eat smaller fish that have fed upon certain species of toxic algae. When a person eats fish containing this toxin, an illness may result, requiring weeks or months of recovery. Symptoms of **ciguatera poisoning** include vomiting, severe itching, nausea, dizziness, hot and cold flashes, temporary blindness, and, sometimes, hallucinations. Because the ciguatera toxin cannot be smelled or tasted and is not destroyed by cooking,

Exhibit 3a

Ciguatera Toxin

Because the ciguatera toxin cannot be smelled or tasted and is not destroyed by cooking, predatory tropical reef fish, such as snapper, should be purchased only from approved suppliers.

it is very important to purchase predatory tropical reef fish only from approved suppliers.

Shellfish may contain toxins that occur because of the algae upon which they feed. Illnesses caused by shellfish poisoning vary and are specific to the type of toxin consumed. Paralytic shellfish poisoning (PSP), the most serious type, may lead to respiratory failure and even death if respiratory support is not provided. PSP is generally associated with mussels, clams, cockles, and scallops. Since cooking may not destroy shellfish toxins, it is important to purchase shellfish from approved suppliers who can certify that the shellfish have been harvested from safe waters. (See *Exhibit 3b.*)

Shellfish Toxins

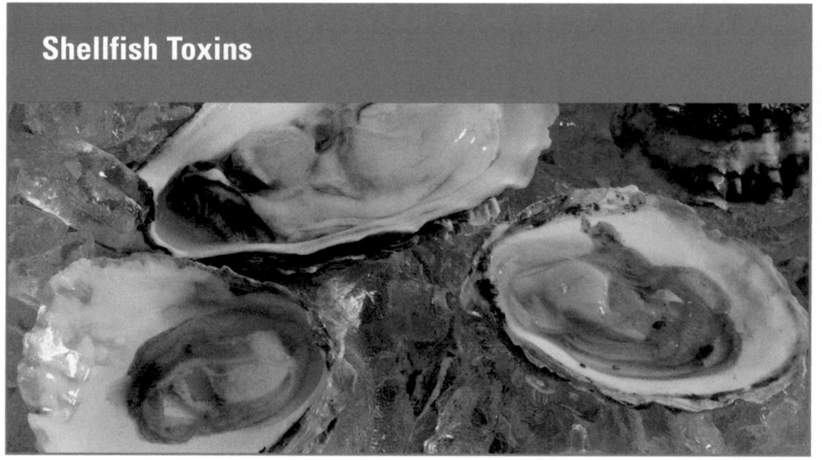

Purchase shellfish from approved suppliers who can certify the shellfish have been harvested from safe waters.

Key Point

Purchase scombroid fish, such as tuna, from reputable suppliers who practice strict time-temperature controls.

Scombroid poisoning is one of the more common forms of illness caused by fish toxins in the U.S. It occurs when scombroid species of fish—such as tuna, mackerel, bluefish, skipjack, and bonito—are time-temperature abused. Under these conditions, bacteria associated with the fish produce the toxin **histamine.** This odorless, tasteless chemical causes scombroid intoxication when consumed. Symptoms of the illness include flushing and sweating, a burning, peppery taste in the mouth, dizziness, nausea, and headache. Sometimes a facial rash, hives, edema, diarrhea, and abdominal cramps will follow. Scombroid poisoning, also known as histamine poisoning, has been associated with other fish species as well, such as mahi-mahi, marlin, and sardines. Histamine is not destroyed by cooking or freezing. Since time-temperature abuse during the harvesting process may cause scombroid fish to become unsafe, it is important to purchase these fish from reputable suppliers who practice strict time-temperature controls.

Some fish toxins are systemic—that is, they occur as a natural part of the fish. An example of a potentially toxic fish is pufferfish, which contains tetrodotoxin in its liver, skin, and other organs.

Consuming the toxin in this fish can produce rapid and violent death. Few cases of this type of poisoning have been reported in the U.S. Cooking may not destroy systemic fish toxins. Pufferfish should be eaten only if they are handled and prepared by a trained and licensed chef.

The following general procedures should be practiced to guard against a seafood-specific foodborne illness.

- Purchase fish from reputable suppliers who maintain strict time-temperature controls.
- Refuse fish that has been thawed and refrozen.
- Check temperature. The temperature of fresh fish must be 41°F (5°C) or lower upon arrival. However, this temperature standard will not prevent a scombroid poisoning if the fish has been temperature-abused during the harvesting process or prior to being received. This is why it is critical to purchase fish from reputable suppliers.
- Thaw frozen fish at refrigerator temperatures of 41°F (5°C) or lower.

Remember that toxins are not living organisms, so cooking or freezing may not destroy them. Therefore, most of the preventive measures taken against biological toxins in seafood must be implemented when purchasing and receiving. Chapter 6 will discuss proper receiving practices in greater detail.

Plant Toxins

Plant toxins are another form of biological contamination. Most poisonings caused by plants result when toxic plants have been used in medicinal home remedies. Foodborne illnesses have occurred after people have consumed rhubarb leaves, jimsonweed, and the root of water hemlock. Some illnesses have occurred after animals have eaten toxic plants and people have consumed the by-products of those animals. For example, people have become ill after consuming milk from cows that had eaten snakeroot. People have also become ill from consuming honey produced by bees that had gathered nectar from mountain laurel or rhododendrons.

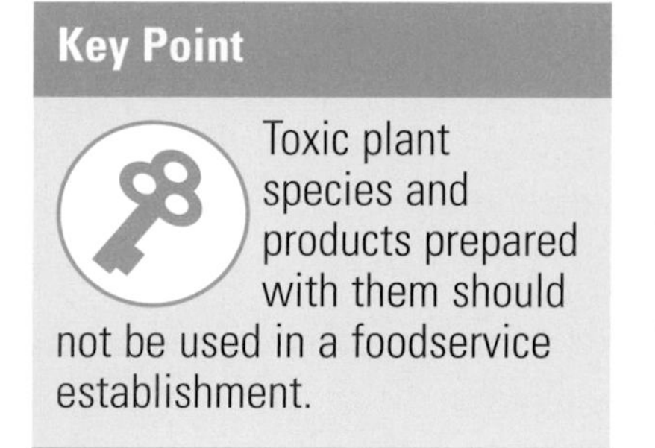

Key Point

Toxic plant species and products prepared with them should not be used in a foodservice establishment.

Some plants may be toxic in their raw state but safe when properly cooked. For example, fava beans and red kidney beans may be unsafe if eaten when raw or undercooked. In general, toxic plant species and products prepared with them should be avoided. Only commercially processed honey and properly cooked beans should be used.

Mushroom Toxins

Foodborne-illness outbreaks associated with mushrooms are almost always caused by the consumption of wild mushrooms collected by amateur mushroom hunters. Most cases occur when toxic mushroom species are confused with edible species. The symptoms of intoxication vary depending upon the species consumed. Some mushroom toxins will destroy internal organs; others cause convulsions, hallucinations, and coma; still others produce nausea, vomiting, abdominal cramping, and diarrhea.

Exhibit 3c

Mushroom Toxins

Mushrooms must be purchased from approved suppliers.

Cooking or freezing will not destroy the toxins produced by toxic wild mushrooms. Establishments should not use mushrooms picked in the wild or products made with them unless the mushrooms have been purchased from approved suppliers. (See *Exhibit 3c.*) Establishments that serve mushrooms picked in the wild should have written buyer specifications that:

- identify the mushroom's common name, and the Latin binomial and its author.
- ensure that the mushroom was identified in its fresh state.
- indicate the name of the person who identified the mushroom and include a statement regarding his/her qualifications.

Exhibit 3d summarizes information about common biological toxins, the sources of these toxins, the food with which they are often associated, and preventive measures that can be taken to keep these toxins from causing foodborne illness.

Exhibit 3d

Biological Toxins (Biological Contaminants)

Biological Toxin		Source of Contamination	Associated Food	Preventive Measures
Seafood Toxins	Ciguatera Toxin	Fish that have eaten algae containing the toxin	Predatory tropical reef fish, such as amberjack, barracuda, grouper, and snapper	Cooking does not destroy these toxins; purchase predatory tropical reef fish only from approved suppliers
	Scombroid Toxin (histamine)	Histamine produced by bacteria in some fish when they are time-temperature abused	Primarily occurs in tuna, bluefish, mackerel, skipjack, roundfish, and bonito; other fish, such as mahi-mahi, marlin, and sardines, have also been implicated in histamine poisoning	Cooking does not destroy histamine; because time-temperature abuse during the harvesting process may cause the fish to become unsafe, it is important to purchase from reputable suppliers
	Shellfish Toxins	Shellfish that have eaten a type of algae containing the toxin	Shellfish, especially mollusks, such as mussels, clams, cockles, and scallops	Cooking may not destroy these toxins; purchase these shellfish from approved suppliers who can certify they are harvested from safe waters
	Systemic Fish Toxins	Toxins that are a natural part of some fish	Pufferfish, moray eels, and freshwater minnows	Cooking may not destroy systemic fish toxins; pufferfish should be handled and prepared by properly trained chefs
Plant Toxins		Toxins that are a natural part of some plants	Fava beans, rhubarb leaves, jimsonweed, water hemlock, and apricot kernels; honey from bees that have gathered nectar from mountain laurel; milk from cows that have eaten snakeroot, jimsonweed, and other toxic plants	Cooking may not destroy these toxins; avoid these plant species and products prepared with them
Fungal Toxins		Toxins that are a natural part of some varieties of fungi	Poisonous varieties of mushrooms and other fungi	Cooking does not destroy these toxins; do not use mushrooms picked in the wild or products made with them unless the mushrooms have been purchased from approved suppliers

Chemical Contamination

Chemical contaminants are responsible for many cases of foodborne illness. Contamination can come from a variety of substances normally found in restaurant and foodservice establishments. These include toxic metals, pesticides, and chemicals.

Toxic Metals

Key Point

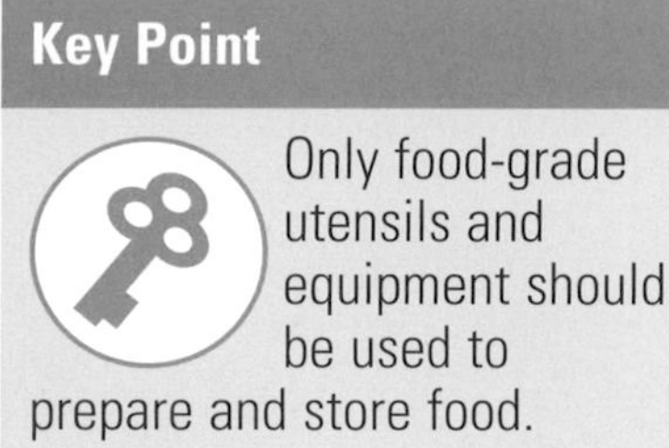

Only food-grade utensils and equipment should be used to prepare and store food.

Utensils and equipment that contain toxic metals—such as lead, copper, brass, zinc, antimony, and cadmium—can cause a **toxic-metal poisoning.** If acidic food is stored in or prepared with this type of equipment, it can leach these metals from the item and become contaminated. For example, storing tomato sauce in a copper pot or lemonade in a pewter pitcher could lead to a foodborne illness. Only food-grade utensils and equipment should be used to prepare and store food.

Improperly installed carbonated-beverage dispensers can also create a hazard. If carbonated water is allowed to flow back into the copper supply lines, it could leach copper from the line and contaminate the beverage. Beverage-dispensing systems should be installed and maintained only by professionals who will ensure that a proper backflow-prevention device is installed.

Key Point

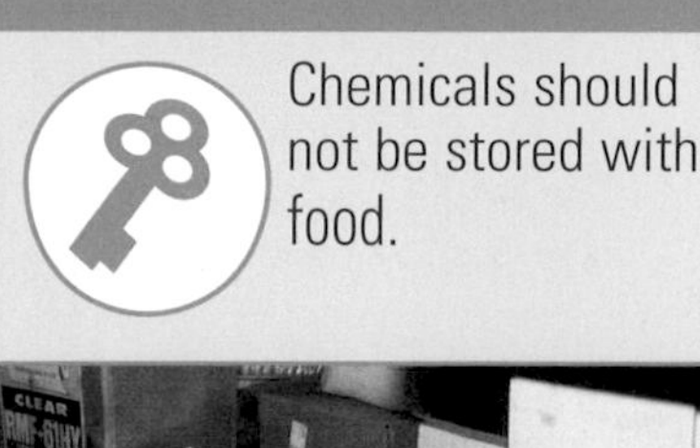

Chemicals should not be stored with food.

Chemicals and Pesticides

Chemicals such as cleaning products, polishes, lubricants, and sanitizers can contaminate food if they are improperly used or stored. Follow the directions supplied by the manufacturer when using these chemicals. Exercise caution when using chemicals during operating hours to prevent contamination of food and food-preparation areas. Store chemicals away from food, utensils, and equipment used for food. Keep them in a locked storage area in their original containers. If chemicals must be transferred to smaller containers or spray bottles, label each container appropriately.

Key Point

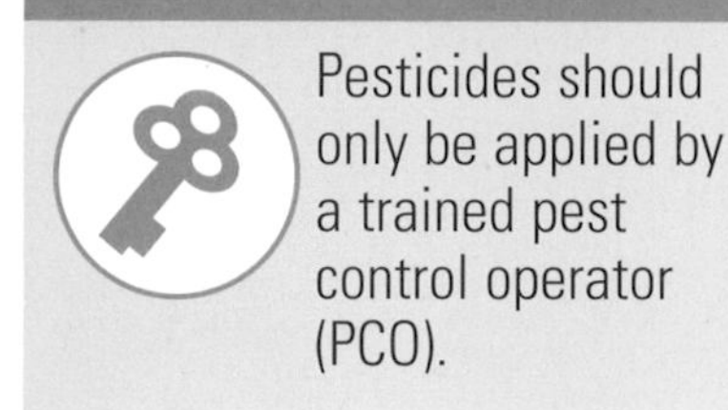

Pesticides should only be applied by a trained pest control operator (PCO).

If used, pesticides should be applied only by a licensed pest control operator (PCO). All food should be wrapped or stored prior to application of the pesticide. Pesticides should be stored with the same care as other chemicals used in the establishment.

Exhibit 3e summarizes information about some common chemical contaminants, their sources, the food with which they

Exhibit 3e

Chemical Contaminants

Chemical Toxin	Source of Contamination	Associated Food	Preventive Measures
Toxic Metals	Utensils and equipment containing potentially toxic metals, such as lead, copper, brass, zinc, antimony, and cadmium	Any food, but especially high-acid food, such as sauerkraut, tomatoes, and citrus products; the acidity of this food can cause metal ions to leach into its liquids Carbonated beverages; the carbonated water used to make the beverage might leach copper ions from copper water-supply lines	Use only food-grade storage containers Use metal and plastic containers only for their intended use Use only food-grade brushes on food; do not use paintbrushes or wire brushes Do not use enamelware, which may chip and expose the underlying metal Do not use equipment or utensils made of materials that contain lead (such as pewter) for food preparation Do not use zinc-coated (galvanized) equipment or utensils for food preparation Use a backflow-prevention device to prevent carbonated water in soft drink-dispensing systems from flowing back into the copper water-supply lines
Chemicals	Cleaning products, polishes, lubricants, and sanitizers		Follow manufacturers' directions for storage and use; use only recommended amounts Store away from food, utensils, and equipment used for food Store in a dry cabinet in original, labeled containers, apart from other chemicals that might react with them Tools used for dispensing chemicals should never be used on food If chemicals must be transferred to smaller containers or spray bottles, label each container appropriately Use only food-grade lubricants or oils on kitchen equipment or utensils
Pesticides	Used in kitchens and food-preparation and storage areas to control pests, such as rodents and insects		Pesticides should only be applied by a licensed professional; wrap or store all food before pesticides are applied

are often associated, and measures that can be taken to prevent them from causing foodborne illness.

Physical Contamination

Physical contamination can occur when foreign objects are accidentally introduced into food, or when naturally occurring objects, such as bones in fillets, pose a physical hazard. Common physical contaminants include metal shavings from cans (see *Exhibit 3f*), staples from cartons, glass from broken light bulbs, blades from plastic or rubber scrapers, fingernails, hair, bandages, dirt, and bones. Closely inspect the food you receive and take steps to ensure that it will not be physically contaminated during the flow of food in your operation.

Exhibit 3f

Physical Contamination

Metal shavings in an opened can might contaminate the food inside.

THE DELIBERATE CONTAMINATION OF FOOD

While the principles of food safety help an establishment address the accidental contamination of food, managers must also be aware of how to prevent or eliminate the deliberate contamination of food, known as **food security.** In addition to biological, chemical, and physical contaminants, nuclear and radioactive contaminants are also a concern.

Threats to food security in the restaurant and foodservice industry might occur at any level in the food-supply chain and are the result of criminal activity. These attacks are usually focused on a specific food item, process, company, or business. Those who would knowingly contaminate a food product include, but are not limited to, organized terrorist or activist groups, individuals posing as customers, current or former employees, vendors, and competitors.

The key to protecting food is to make it as difficult as possible for even a single tampering to occur. An effective food security program will consider all of the points where food is vulnerable to intentional contamination. Potential threats can come from the following three areas:

- Human elements
- Interior elements
- Exterior elements

Managers must ensure that all employees are aware of their roles in keeping food secure in the operation by developing procedures and training that address each potential threat. *Exhibit 3g* on the next page lists elements to consider when determining how to handle food security in your establishment.

Exhibit 3g

Addressing Food Security Threats in Your Operation*

Human Elements	▶ Learn about your applicants—ask for references, verify references, and check identification. ▶ Train employees in food security and establish food security awareness with all employees. ▶ Train your employees to report any suspicious activity. ▶ Establish a system to identify employees on duty. Only on-duty employees should be allowed in work areas. ▶ Establish rules for the opening of the back door—determine those employees that are authorized to open these doors and under what circumstances. ▶ Control customer and non-employee access to food-production areas. ▶ Allow employees to bring only essential items to work. These items might include uniforms, name badges, and anything else necessary for job functions. ▶ Consider a two-employee rule during food preparation—no single employee should be allowed in a production area by him/herself. Having more than one employee around is a built-in check system. ▶ Supervise and survey production areas on a regular basis. This can be accomplished via video cameras, windows, other employees, or management.
Interior Elements	▶ Limit access to doors, windows, roofs, and food-storage areas. Good lighting is important too. ▶ Keep food display, storage, and kitchen entrances and exits controlled or under surveillance. ▶ Eliminate hiding places in all areas of the operation. Make sure there are no places for an intruder to hide until after working hours. ▶ Identify and inspect all incoming food items, and never accept suspect food. Have a specific food inspection program in place, and make sure that employees are familiar with it. ▶ Restrict traffic in food-prep and food-storage areas. ▶ Monitor all customer self-service and displayed food items, such as salad bars, condiments, and exposed tableware.
Exterior Elements	▶ Ensure that the building's exterior is well lit. There should be no areas where an intruder could remain unseen. ▶ Control access to ventilation system to prevent tampering. ▶ Identify all company food suppliers and consider using tamper-evident packages. Check the identification of the delivery person, the scheduled times of delivery, and document those deliveries. ▶ Tell your suppliers that food security is a priority, and ask what steps they are taking to ensure their products are secure. ▶ Verify and pre-approve all service personnel and providers. ▶ Prevent unmonitored access to facilities by non-employees after normal business hours.

*For more information on food security, visit www.nraef.org/foodsecurity.

FOOD ALLERGENS

Six to seven million Americans have food allergies. A **food allergy** is the body's negative reaction to a particular food protein. Depending on the person, allergic reactions may occur immediately after the food is eaten or several hours later. The reaction could include some or all of the following symptoms:

- Itching in and around the mouth, face, or scalp
- Tightening in the throat
- Wheezing or shortness of breath
- Hives
- Swelling of the face, eyes, hands, or feet
- Gastrointestinal symptoms, including abdominal cramps, vomiting, or diarrhea
- Loss of consciousness
- Death

Employees should be aware of the most common food allergens, including milk and dairy products, eggs and egg products, fish, shellfish, wheat, soy and soy products, peanuts, and tree nuts.

Your employees should be able to inform customers of menu items that contain these potential allergens. Designate one person per shift to answer customers' questions regarding menu items. To help customers with allergies enjoy a safe meal at your establishment, keep the following points in mind.

- Be able to fully describe each of your menu items when asked. Tell customers how the item is prepared and identify any "secret" ingredients used.
- If you do not know if an item is free of an allergen, tell the customer. Urge the customer to order something else.
- When preparing food for a customer with allergies, ensure that the food makes no contact with the ingredient to which the customer is allergic. Make sure all cookware, utensils, and tableware are allergen-free to prevent food contamination.
- Serve menu items as simply as possible to customers with allergies. Sauces and garnishes are often the source of allergic reactions. Serve these items on the side.

SUMMARY

Biological and chemical toxins are responsible for many foodborne-illness outbreaks. Most occur naturally and are not caused by the presence of microorganisms. Some occur in the animal as a result of its diet. Since toxins are not living organisms, cooking or freezing typically will not destroy them. Most measures taken to prevent foodborne intoxication center on proper purchasing and receiving.

There are several things a manager can do to help prevent seafood-specific toxins from causing foodborne illness. Purchase seafood from reputable suppliers who maintain strict time-temperature controls and can certify the seafood has been harvested from safe waters. Make sure fish is received at 41°F (5°C) or lower, and refuse fish that has been thawed and refrozen. Thaw fish at refrigeration temperatures.

Toxins are a natural part of certain plants. Some plants may be toxic in their raw state but safe when properly cooked. In general, avoid toxic plant species and products prepared with them. Foodborne-illness outbreaks associated with mushrooms are almost always caused by the consumption of misidentified wild mushrooms. Establishments should not use mushrooms picked in the wild or products made with them unless the mushrooms have been purchased from approved suppliers.

Chemical contaminants are responsible for many cases of foodborne illness. Contamination can come from a variety of substances normally found in restaurant and foodservice establishments, such as toxic metals, pesticides, and cleaning chemicals. Use only food-grade utensils and equipment to prepare and store food. Cleaning products, polishes, lubricants, and sanitizers should be used as directed. Be cautious when using these chemicals during operating hours and store them properly. If used, pesticides should be applied only by a licensed pest control operator (PCO).

Physical contamination can occur when physical objects are accidentally introduced into food or when naturally occurring objects, such as the bones in fish, pose a physical hazard. Closely inspect the food you receive and take steps to ensure food will not become physically contaminated during its flow through your operation.

Food security addresses the prevention or elimination of the deliberate contamination of food. Contamination can occur in biological, chemical, physical, nuclear, or radioactive form. The key to protecting food is to make it as difficult as possible for even a single tampering to occur. Managers must develop and maintain a food security program that focuses on the potential threats posed by the interior, exterior, and human elements of their establishment.

Many people have food allergies. Employees should be aware of the most common food allergens, which include milk and dairy products, eggs and egg products, fish, shellfish, wheat, soy and soy products, peanuts, and tree nuts. You should be able to inform customers of these and other potential food allergens that may be included in food served at your establishment.

Apply Your Knowledge — A Case in Point 1

Roberto receives a shipment of frozen mahi-mahi steaks. The steaks are frozen solid at the time of delivery, and the packages are sealed and contain a lot of large ice crystals, indicating they have been time-temperature abused. Roberto accepts the mahi-mahi steaks and thaws them in a refrigerator at a temperature of 38°F (3°C). The thawed fish steaks are then held at this temperature during the evening shift and are cooked to order. The cooks follow all appropriate guidelines for preparing, cooking, and serving the fish, monitoring time and temperature throughout the process. Unfortunately, these fish steaks are implicated in an outbreak of scombroid intoxication.

1. Explain why this may have happened.

For answers, please turn to the Answer Key.

Apply Your Knowledge — A Case in Point 2

A busboy, Mark, has been asked to spray for ants in the dry-storage room. He sprays along the ants' trail and then places the can of spray on the top shelf in the dry-storage room, along with the other pesticides and foodservice chemicals. Mark then returns to busing tables.

1. What has Mark done correctly?
2. What unsafe practices are evident?

For answers, please turn to the Answer Key.

Apply Your Knowledge

Use these questions to review the concepts presented in this chapter.

Discussion Questions

1. How does the scombroid toxin differ from the other seafood toxins listed in *Exhibit 3d?*

2. What preventive measures should be taken to guard against a seafood-specific foodborne illness?

3. A restaurant is serving broiled red snapper along with side dishes of asparagus and rice with wild mushrooms. Which food could contain biological toxins, and why?

4. Why is serving orange juice from an enamelware pitcher not recommended?

5. Mike is serving drinks and cannot find the ice scoop. He uses a clean, sanitary glass tumbler as a temporary ice scoop. Why is this not a safe practice?

For answers, please turn to the Answer Key.

Apply Your Knowledge — Multiple-Choice Study Questions

Use these questions to test your knowledge of the concepts presented in this chapter.

1. **You have ordered frozen tuna steaks for your restaurant. When the delivery arrives, you notice there is excessive frost and ice in the package, which indicates they have been time-temperature abused. You refuse the delivery. Why?**
   A. You suspect the steaks may have been contaminated with a cleaning compound.
   B. You suspect the steaks may contain ciguatera toxins.
   C. You suspect the steaks may cause scombroid poisoning if you serve them.
   D. You believe the supply company may have treated the steaks with an unauthorized preservative.

2. **A customer becomes ill after eating grouper. It is discovered that the shipment of grouper contained the ciguatera toxin. This is an example of**
   A. chemical contamination.
   B. physical contamination.
   C. biological contamination.

3. **Which of the following is not a common food allergen?**
   A. Eggs
   B. Dairy products
   C. Peanuts
   D. Beef

4. **You find a piece of glass at the bottom of your ice storage bin. This is an example of**
   A. chemical contamination.
   B. physical contamination.
   C. biological contamination.

5. **Which of the following fish is associated with ciguatera?**
   A. Mahi-mahi
   B. Pufferfish
   C. Snapper
   D. Fresh water minnow

Continued on next page...

Multiple-Choice Study Questions *continued*

6. **An establishment should do all of the following to guard against a seafood-specific foodborne illness *except***
   A. thaw fish at temperatures higher than 41°F (5°C).
   B. purchase seafood from suppliers who practice time-temperature control.
   C. refuse fish that has been thawed and refrozen.
   D. accept fish that has been received at 41°F (5°C) or lower.

7. **An establishment should do all of the following to prevent contamination *except***
   A. store food away from chemicals.
   B. keep high-acid food separate from other types of food.
   C. purchase food products from approved suppliers.
   D. use food-grade storage containers.

8. **Which of the following is an example of a physical contaminant?**
   A. A virus present in shellfish
   B. Dirt on a head of lettuce
   C. Sanitizer residue left on a cutting board
   D. Parasites present in a raw fish fillet

9. **All of the following can lead to the chemical contamination of food *except***
   A. cooking tomato sauce in a copper pot.
   B. storing orange juice in a pewter pitcher.
   C. using a backflow-prevention device on a carbonated beverage dispenser.
   D. serving fruit punch in a galvanized tub.

10. **Which of the following statements is true about fish containing ciguatera or scombroid (histamine) toxins?**
    A. Freezing will destroy these toxins.
    B. Cooking will not destroy these toxins.
    C. You can see and smell these toxins.
    D. Cooking will destroy these toxins.

For answers, please turn to the Answer Key.

ADDITIONAL RESOURCES

Books and Periodicals

Chin, J., ed. 2000. *Control of communicable diseases in man,* 17th ed. Washington, D.C.: American Public Health Association.

Food Allergy & Anaphylaxis Network. 2001. *Food Allergy Training Guide.* Washington, D.C.: National Restaurant Association.

Help prevent food allergies. 2003. *Food Safety Illustrated.* 3 (2):14.

International Commission on Microbiological Specifications for Food. 1996. *Microorganisms in foods 5: Characteristics of microbial pathogens.* London: Blackie Academic & Professional.

National Restaurant Association. 1998. *Sanitation survival kit.* Washington, D.C.: National Restaurant Association.

Web Sites

Centers for Disease Control and Prevention (CDC)
www.cdc.gov
The mission of the CDC is to promote health and quality of life by preventing and controlling disease, injury, and disability. To prevent and control foodborne illness, the CDC collect data on foodborne-illness outbreaks. This Web site provides general information on foodborne illnesses and their prevention.

Compendium of Fish and Fishery Product Processes, Hazards, and Controls
http://seafood.ucdavis.edu/haccp/compendium/compend.htm
Developed by the FDA, the compendium available at this Web site provides useful information on biological, chemical, and physical hazards associated with fish and fish products.

Ecolab, Inc.
www.ecolab.com
An excellent source of information on housekeeping and sanitation supplies for the restaurant and foodservice industry from the world's leading sanitation product supplier.

The Food Allergy and Anaphylaxis Network (FAAN)

www.foodallergy.org

FAAN was established in 1991 and works to build public awareness of food allergies through media, education, publishing, advocacy, and research efforts. Its membership includes families, dietitians, nurses, physicians, school staff, representatives from government agencies, and the food and pharmaceutical industries. This Web site contains a wealth of information related to food allergies, including scientific studies, product recall reports, and recipes.

FDA Bad Bug Book

http://vm.cfsan.fda.gov/~mow/intro.html

Produced by the FDA's Center for Food Safety and Nutrition (FDA CFSAN), this online handbook provides basic facts regarding pathogenic foodborne microorganisms and natural toxins. It brings together information from the FDA, the CDC, the USDA Food Safety Inspection Service, and the National Institutes of Health.

FDA Center for Food Safety and Applied Nutrition (CFSAN)

http://vm.cfsan.fda.gov

As the center within the FDA responsible for food safety and nutrition, CFSAN promotes and protects the public health by researching and implementing guidelines, policies, and standards to ensure that food is safe, nutritious, wholesome, and properly labeled. This Web site provides a wealth of information on food safety and sanitation, including corresponding guidelines, policies, and standards.

FDA Food Code

http://vm.cfsan.fda.gov/~dms/foodcode.html

As the basis for many local sanitation codes, as well as the basis for information in this textbook, the FDA Food Code, available at this Web address, is a useful resource for information relating to food safety for the restaurant and foodservice industry.

FDA Seafood Information and Resources

http://vm.cfsan.fda.gov/seafood1.html

The FDA operates an oversight compliance program for fishery products under which responsibility for product safety, wholesomeness, identity, and economic integrity rests with the processor or importer, who must comply with regulations under the Federal Food, Drug, and Cosmetic (FD&C) Act. This Web site houses information on the seafood program, foodborne pathogens and contaminants associated with seafood, and HACCP compliance.

International Food Information Council (IFIC) Foundation
http://ific.org
IFIC collects and disseminates scientific information on food safety, nutrition, and health. They work with an extensive roster of scientific experts to help translate research into understandable and useful information for opinion leaders and, ultimately, consumers. This Web site provides easy-to-understand information on foodborne illness and health-related stories circulating in the news.

Morbidity and Mortality Weekly Report (MMWR)
www.cdc.gov/mmwr
The *MMWR* is prepared by the CDC and is available, free of charge, on this Web site. The reports contain data based on weekly accounts of reportable diseases from state health departments. Report databases can be searched for information on foodborne-illness outbreaks.

The Mushroom Council
www.mushroomcouncil.com
This interesting and useful Web site from The Mushroom Council contains recipes and information about how to handle and store mushrooms to ensure safety and quality.

National Center for Infectious Diseases (NCID)
www.cdc.gov/ncidod
NCID is one of the centers of the CDC. Its mission is to prevent illness, disability, and death caused by infectious diseases around the world. NCID accomplishes this mission by conducting surveillance, epidemic investigations, epidemiologic and laboratory research, and training. It also sponsors public education programs to develop, evaluate, and promote prevention and control strategies for infectious diseases. This Web site serves as another great resource for information on foodborne illness.

National Shellfish Sanitation Program Manual of Operations
http://vm.cfsan.fda.gov/~ear/nsspman.html
The National Shellfish Sanitation Program (NSSP) is a voluntary cooperative program between the federal and state governments and industry. The program relies on regulatory controls by the State Shellfish Authority (SSA) to ensure safe mollusks and

shellfish. The NSSP Manual of Operations, published by the FDA and available at this Web site, is a guide for establishing state shellfish laws and regulations.

Orkin Commercial
www.orkin.com
From the leader in pest control, Orkin's Web site contains useful information on how to prevent and control pests in your foodservice establishment. It also contains a bug guide, providing all the information you will ever want to know about ants, rodents, flies, and birds.

4

The Safe Foodhandler

Inside this chapter:

- How Foodhandlers Can Contaminate Food
- Diseases Not Transmitted Through Food
- Components of a Good Personal Hygiene Program
- Management's Role in a Personal Hygiene Program

After completing this chapter, you should be able to:

- Identify personal behaviors that can contaminate food.
- Identify proper handwashing procedures.
- Identify when hands should be washed.
- Identify appropriate hand sanitizers and when to use them.
- Identify hand maintenance requirements.
- Identify the proper procedure for covering cuts, wounds, and sores.
- Identify procedures that must be followed when using gloves.
- Identify jewelry that poses a hazard to food safety.
- Identify requirements for employee work attire.
- Identify the regulatory exceptions for allowing bare-hand contact with ready-to-eat and cooked food.
- Identify criteria for excluding employees from the establishment or restricting them from working with or around food.
- Identify criteria for excluding or restricting employees from working within establishments that serve high-risk populations.
- Identify illnesses that are required to be reported to the health agency.
- Identify policies that should be implemented at the establishment regarding eating, drinking, and smoking while working with food.

Key Terms

Gastrointestinal illness
Infected lesion
Carriers
Single-use paper towel
Hand sanitizers
Finger cot
Hair restraint
Hepatitis A
Jaundice

Apply Your Knowledge

Check to see how much you know about the concepts in this chapter. Use the page references provided to explore the topic in each question.

Test Your Food Safety Knowledge

1. **True or False:** During handwashing, foodhandlers must vigorously scrub their hands and arms for two minutes. *(See page 4-5.)*
2. **True or False:** Gloves should be changed at least every four hours during continual use. *(See page 4-9.)*
3. **True or False:** Foodhandlers must wash their hands after smoking. *(See page 4-7.)*
4. **True or False:** A foodhandler diagnosed with salmonellosis cannot continue to work at an establishment for the duration of the illness. *(See page 4-13.)*
5. **True or False:** Establishments should only use hand sanitizers that have been approved by the FDA. *(See page 4-7.)*

For answers, please turn to the Answer Key.

INTRODUCTION

At every step in the flow of food through the operation—from receiving through final service—foodhandlers can contaminate food and cause customers to become ill. Good personal hygiene is a critical protective measure against foodborne illness, and customers expect it.

You can minimize the risk of foodborne illness by establishing a personal hygiene program that spells out your specific hygiene policies, provides your employees with training on those policies, and enforces established policies. When employees have the proper knowledge, skills, and attitudes toward personal hygiene, you are one step closer to operating a safe food system.

HOW FOODHANDLERS CAN CONTAMINATE FOOD

In previous chapters, you learned that foodhandlers can cause an illness by transferring microorganisms to food they touch. Many times these microorganisms come from the foodhandlers themselves. Foodhandlers can contaminate food when they

- have a foodborne illness.
- show symptoms of **gastrointestinal illness** (an illness relating to the stomach or intestine).
- have **infected lesions** (infected wounds or injuries).
- live with or are exposed to a person who is ill.
- touch anything that may contaminate their hands.

Even an apparently healthy person may be hosting foodborne pathogens. With some illnesses, such as hepatitis A, an individual is at the most infectious stage of the disease for several weeks before symptoms appear. With other illnesses, the pathogens may remain in a person's system for months after all signs of infection have ceased. Some people are called **carriers** because they might carry pathogens and infect others, yet never become ill themselves.

The next three paragraphs will help illustrate the routes by which employees can contaminate food.

1. A deli foodhandler who was diagnosed with salmonellosis failed to inform his manager that he was ill for fear of losing wages. It was later determined that he was the cause of an outbreak that involved more than two hundred customers through twelve different products.
2. A foodhandler suffering from diarrhea, a symptom of gastrointestinal illness, did not wash his hands and made approximately five thousand people ill when he mixed a vat of buttercream frosting with his bare hands and arms. Another large foodborne-illness outbreak was caused by a foodhandler who scratched an infected facial lesion and then handled a large amount of sliced pepperoni.
3. A foodborne-illness outbreak was traced to a woman who prepared food for a dinner party. The investigation revealed that

Key Point

Simple acts such as rubbing an ear, scratching the scalp, or touching a pimple or sore can contaminate food.

the woman was caring for her infant son, who had diarrhea. The woman could not recall washing her hands after changing the infant's diaper. As a result, twelve of her dinner guests became violently ill with symptoms that included diarrhea and vomiting.

Simple acts such as nose picking, rubbing an ear, scratching the scalp, touching a pimple or an open sore, or running fingers through the hair can contaminate food. Thirty to fifty percent of healthy adults carry *Staphylococcus aureus* in their noses, and about twenty to thirty-five percent carry it on their skin. If these microorganisms contaminate a foodhandler's hands that then touch food, the consequences can be severe. Because of these factors, foodhandlers must pay close attention to what they do with their hands and maintain good personal hygiene.

DISEASES NOT TRANSMITTED THROUGH FOOD

In recent years, the public has expressed growing concern over communicable diseases spread through intimate contact or by direct exchange of bodily fluids. Diseases such as AIDS (Acquired Immune Deficiency Syndrome), hepatitis B and C, and tuberculosis are not spread through food.

Key Point

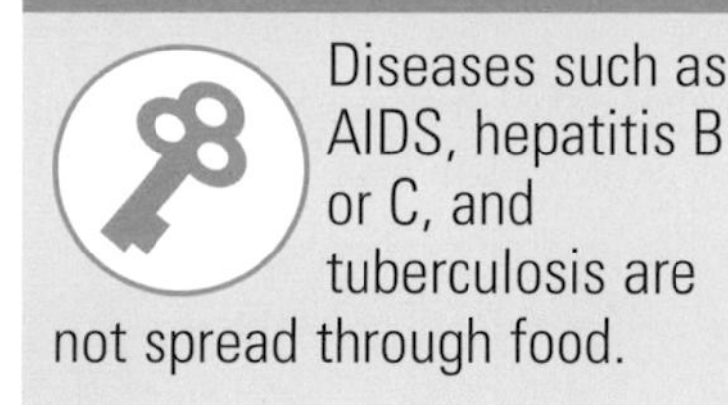

Diseases such as AIDS, hepatitis B or C, and tuberculosis are not spread through food.

Although these diseases are not transmitted through food, as a manager you should be aware of the following laws concerning employees who are HIV-positive (Human Immunodeficiency Virus), have hepatitis B or C, or have tuberculosis.

- The Americans with Disabilities Act (ADA) provides civil-rights protection to individuals who are HIV-positive or have hepatitis B, and thus prohibits employers from firing people or transferring them out of foodhandling duties simply because they have these diseases.
- Employers must maintain the confidentiality of employees who have any nonfoodborne illness.

Personal Hygiene

Managers must train foodhandlers when and how to wash their hands properly, and then must monitor handwashing frequency.

COMPONENTS OF A GOOD PERSONAL HYGIENE PROGRAM

Good personal hygiene is key to the prevention of foodborne illness. Good personal hygiene includes:

- Following hygienic hand practices
- Maintaining personal cleanliness
- Wearing clean and appropriate uniforms and following dress codes
- Avoiding unsanitary habits and actions
- Maintaining good health
- Reporting illnesses

Key Point

There are products on the market that help teach proper handwashing techniques. These products are applied to the hands, which are then washed normally. The employees' hands are then illuminated to reveal how well they were washed.

Picture provided courtesy of Brevis Corp
225 West 2855 South, Salt Lake City, UT 84115
800.383.3377

Hygienic Hand Practices

Handwashing

While it may appear fundamental, many foodhandlers fail to wash their hands properly and as often as needed. As a manager, it is your responsibility to train your foodhandlers and then monitor them. Never take this simple action for granted.

To ensure proper handwashing in your establishment, train your foodhandlers to follow these steps. (See *Exhibit 4a* on the next page.)

- **Step 1: Wet your hands with running water as hot as you can comfortably stand (at least 100°F [38°C]).**
- **Step 2: Apply soap.** Apply enough soap to build up a good lather.
- **Step 3: Vigorously scrub hands and arms for at least twenty seconds.** Lather well beyond the wrists, including the exposed portions of the arms.
- **Step 4: Clean under fingernails and between fingers.** A nail brush might be helpful.
- **Step 5: Rinse thoroughly under running water.** Turn off the faucet using a **single-use paper towel** if available.
- **Step 6: Dry hands and arms.** Use single-use paper towels or a warm-air hand dryer. Never use aprons or wiping cloths to dry hands after washing.

Proper Handwashing Procedure

❶ **Wet your hands with running water as hot as you can comfortably stand (at least 100°F/38°C).**

❷ **Apply soap.**

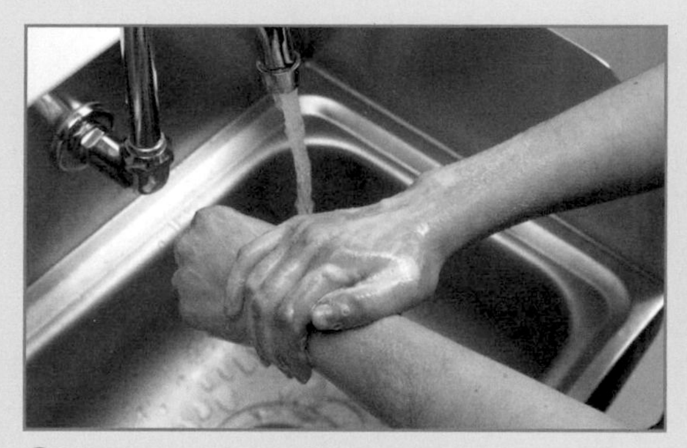

❸ **Vigorously scrub hands and arms for at least twenty seconds.**

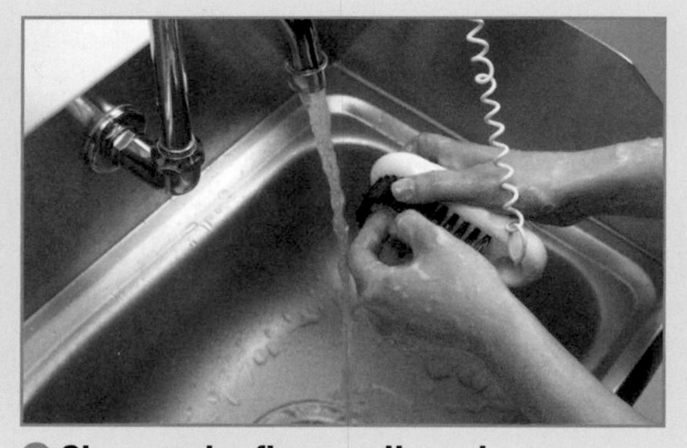

❹ **Clean under fingernails and between fingers.**

❺ **Rinse thoroughly under running water.**

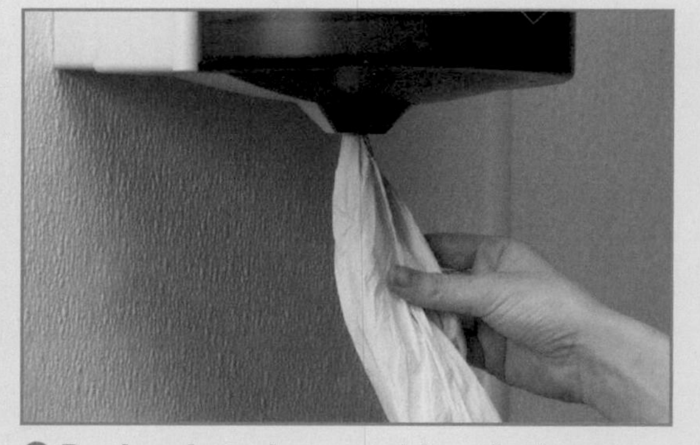

❻ **Dry hands and arms with a single-use paper towel or warm-air hand dryer.**

Personal Hygiene

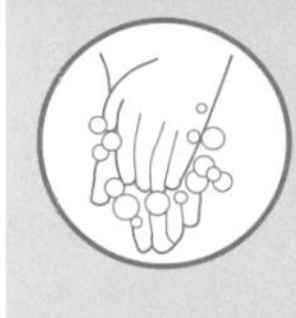

Hand sanitizers should never be used in place of proper handwashing.

Hand sanitizers (a liquid used to lower the number of microorganisms on the surface of the skin) or hand dips may be used after washing, but should never be used in place of proper handwashing. If hand sanitizers are used, foodhandlers should not touch food or food-preparation equipment until the hand sanitizer has dried. Establishments must only use hand sanitizers that have been approved by the FDA.

Foodhandlers must wash their hands before they start work and after the following activities:

- Using the restroom
- Handling raw food (before *and* after)
- Touching the hair, face, or body
- Sneezing, coughing, or using a handkerchief or tissue
- Smoking, eating, drinking, or chewing gum or tobacco
- Handling chemicals that might affect the safety of food
- Taking out garbage or trash
- Clearing tables or busing dirty dishes
- Touching clothing or aprons
- Touching anything else that may contaminate hands, such as unsanitized equipment, work surfaces, or washcloths

Bare-Hand Contact with Ready-To-Eat Food

Proper handwashing minimizes the risk of contamination associated with bare-hand contact with ready-to-eat food. For those jurisdictions that allow bare-hand contact with this food, establishments must have a verifiable written policy on handwashing procedures. Check with your regulatory agency for requirements in your jurisdiction.

Personal Hygiene

Foodhandlers must keep fingernails short and clean, and should not wear false nails or nail polish.

Hand Maintenance

In addition to proper washing, hands need other regular care to ensure that they will not transfer microorganisms to food. To keep food safe, make sure foodhandlers follow these guidelines:

- **Keep fingernails short and clean.** Long fingernails, false fingernails, and acrylic nails should not be worn while handling food since they may be difficult to keep clean and can break off into food. Some jurisdictions allow false nails if single-use gloves are worn. Check your local requirements.

- **Do not wear nail polish.** It can disguise dirt under nails and may flake off into food. Some jusrisdictions allow nail polish if single-use gloves are worn.
- **Cover all hand cuts and sores with clean bandages.** If hands are bandaged, clean gloves or **finger cots,** a protective covering, should be worn at all times to protect the bandage and to prevent it from falling off into food. You may need to move the foodhandler to another job, where he or she will not handle food or touch food-contact surfaces, until the injury heals.

Key Point

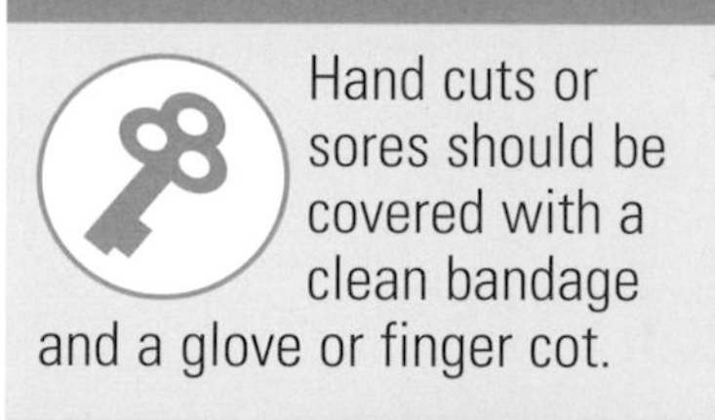

Hand cuts or sores should be covered with a clean bandage and a glove or finger cot.

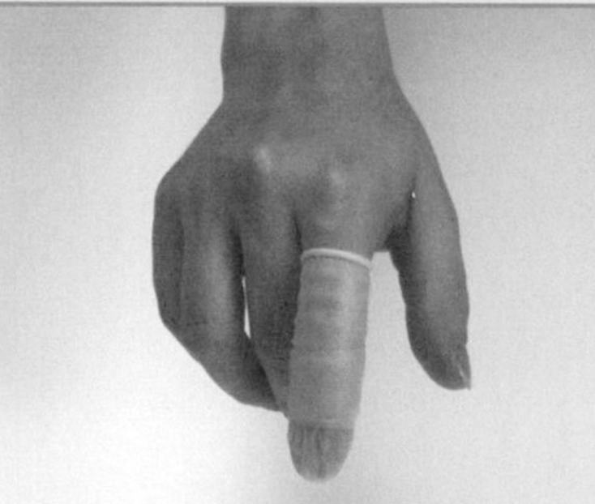

Use of Gloves

Gloves can help keep food safe by creating a barrier between hands and food. (See *Exhibit 4b.*) When purchasing gloves for handling food, managers should:

- **Buy the right glove for the task.** Long gloves, for example, should be used for hand-mixing salads. Colored gloves can also be used to help prevent cross-contamination.
- **Provide a variety of glove sizes.** Gloves that are too big will not stay on the hand and those that are too small will tear or rip easily.

Exhibit 4b

Gloves

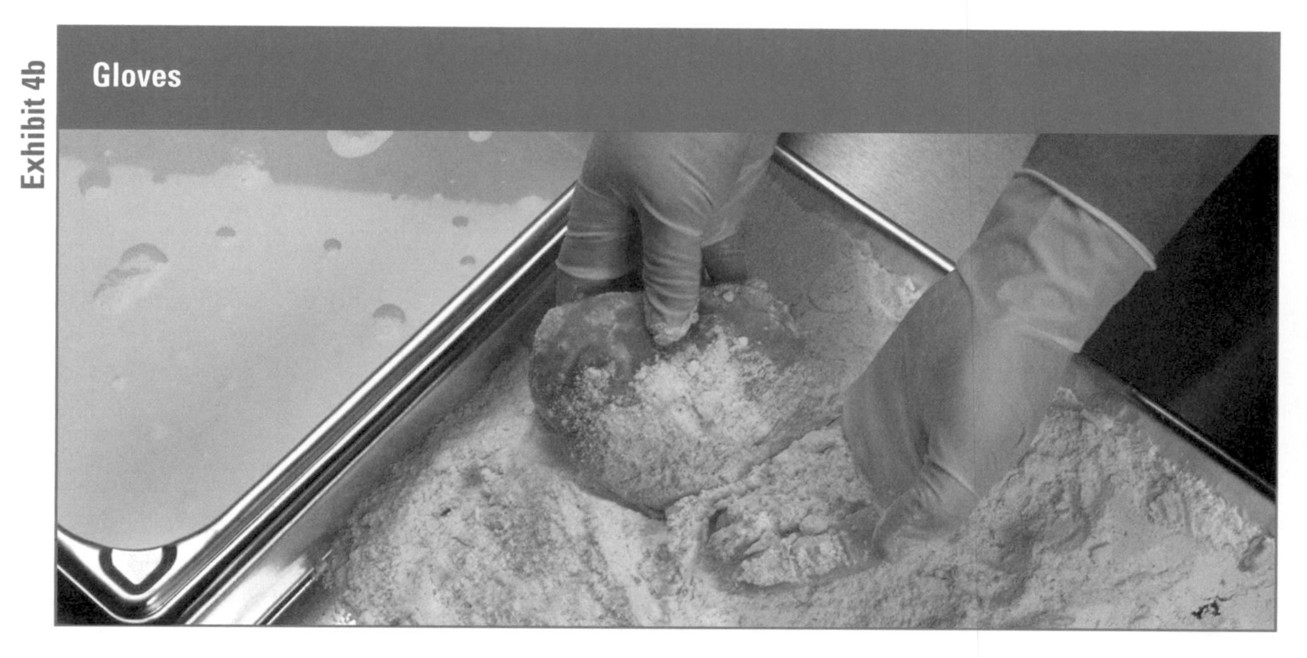

Gloves can help keep food safe by creating a barrier between hands and food.

- **Consider latex alternatives for employees who are sensitive to the material.**
- **Focus on safety, durability, and cleanliness.** Make sure you purchase gloves specifically formulated for food contact, which include gloves bearing the NSF certification mark.

Key Point

Gloves must never be used in place of handwashing.

Gloves must never be used in place of handwashing. Hands must be washed before putting on gloves and when changing to a new pair. Gloves used to handle food are for single use only and should never be washed and re-used. They should be removed by grasping them at the cuff and peeling them off inside out over the fingers while avoiding contact with the palm and fingers. Foodhandlers should change their gloves

- as soon as they become soiled or torn.
- before beginning a different task.
- at least every four hours during continual use, and more often when necessary.
- after handling raw meat and before handling cooked or ready-to-eat food.

Often, foodhandlers consider gloves more sanitary than bare hands. Because of this false sense of security, they might not change gloves as often as necessary. Managers must reinforce the habit of proper hand sanitation with foodhandlers. An effective handwashing education and compliance program will establish a habit that will help prevent foodborne illness.

Other Good Personal Hygiene Practices

Personal hygiene can be a sensitive subject for some people, but because personal cleanliness is vital to food safety, as a manager, you must address the subject with every foodhandler.

Key Point

Foodhandlers must keep their hair clean, since oily hair can harbor pathogens.

General Personal Cleanliness

In addition to following proper hand-hygiene practices, your foodhandlers must maintain personal cleanliness. Foodhandlers should bathe or shower before work. They must also keep their hair clean, since oily, dirty hair can harbor pathogens.

Proper Work Attire

A foodhandler's attire plays an important role in the prevention of foodborne illness. Dirty clothes may harbor pathogens and give customers a bad impression of your establishment. Therefore, managers should make sure foodhandlers observe strict dress standards. (See *Exhibit 4c.*) Foodhandlers should:

- **Wear a clean hat or other hair restraint.** A hair restraint will keep hair away from food and keep the foodhandler from touching it. Foodhandlers with facial hair should also wear beard restraints.
- **Wear clean clothing daily.** The type of clothing chosen should minimize contact with food and equipment, and should reduce the need for adjustments. If possible, foodhandlers should put on work clothes at the establishment.
- **Remove aprons when leaving food-preparation areas.** For example, aprons should be removed and properly stored prior to taking out garbage or using the restroom.
- **Wear appropriate shoes.** Wear clean, closed-toe shoes with a sensible, nonslip sole.
- **Remove jewelry prior to preparing or serving food or while around food-preparation areas.** Jewelry can harbor microorganisms, often tempts foodhandlers to touch it, and may pose a safety hazard around equipment. Remove rings (except for a plain band), bracelets (including medical information jewelry), watches, earrings, necklaces, and facial jewelry (such as nose rings, etc.).

Key Point

Foodhandlers must always wear clean clothes, since dirty clothes can harbor pathogens.

Key Point

Foodhandlers must remove jewelry prior to preparing or serving food since it can harbor pathogens.

Check with your local regulatory agency regarding requirements. These requirements should be reflected in written policies that are consistently monitored and enforced. All potential employees should be made aware of these policies prior to employment.

Policies Regarding Eating, Drinking, Chewing Gum, and Tobacco

Small droplets of saliva can contain thousands of disease-causing microorganisms. In the process of eating, drinking, chewing gum, or smoking, this saliva can be transferred to the foodhandler's hands or directly to the food they are handling. For this reason, foodhandlers must not smoke, chew gum or tobacco,

Exhibit 4c

Proper and Improper Attire

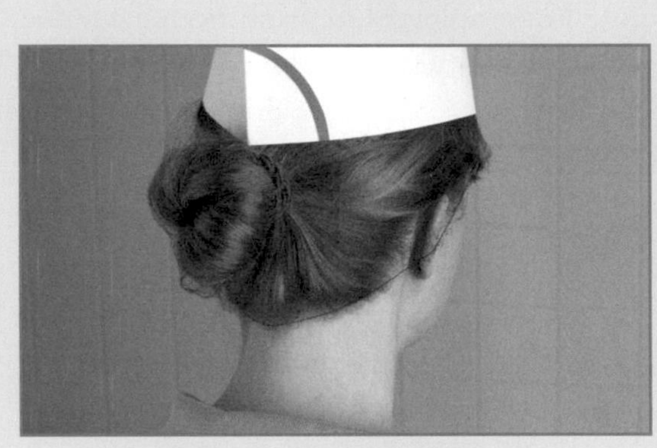

Hair properly restrained

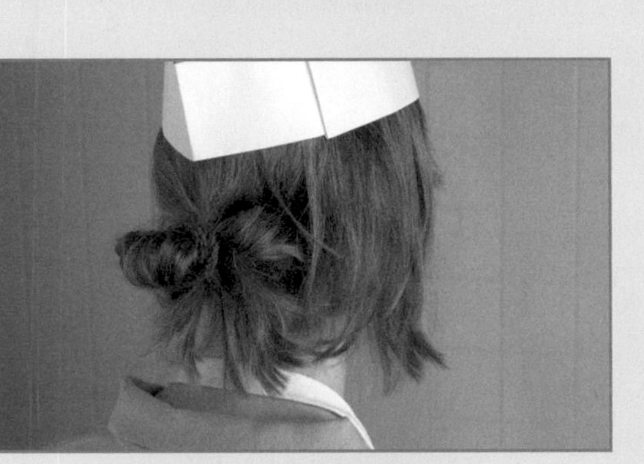

Hair improperly restrained

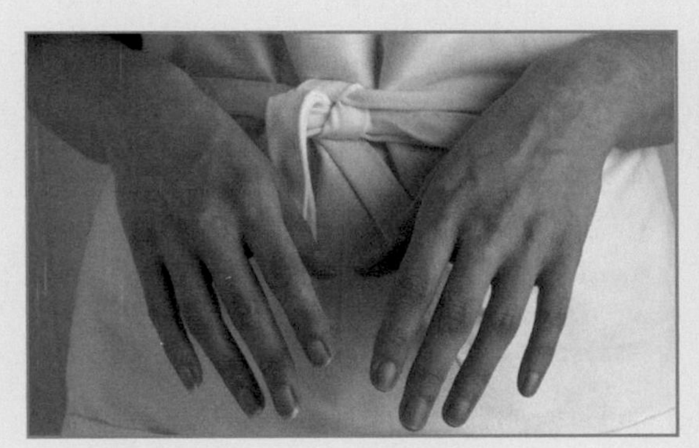

Proper hand hygiene:
clean, short fingernails; no jewelry or nail polish

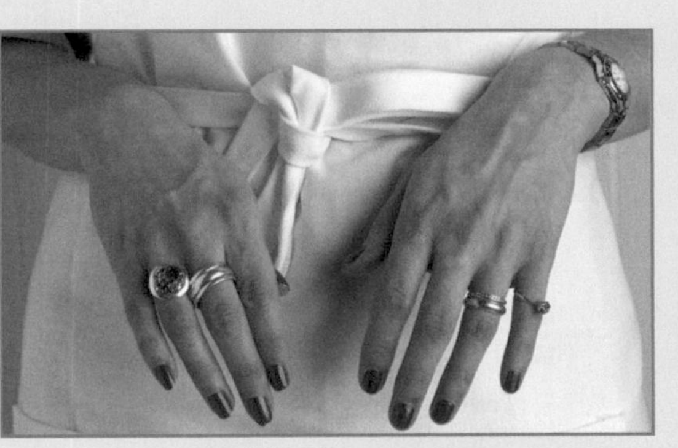

Improper hand hygiene:
long fingernails, jewelry, nail polish

Proper apron: clean

Improper apron: dirty and stained

Exhibit 4d

Safe and Unsafe Food-Tasting Practices

Safe

Safe

Unsafe

or eat or drink while preparing or serving food, while in food preparation areas, or in areas used to clean utensils and equipment.

Some jurisdictions allow employees to drink from a covered container with a straw while in these areas. Check with your local regulatory agency. Foodhandlers should eat, drink, chew gum, or use tobacco products only in designated areas, such as an employee break room. They should never be allowed to spit in the establishment.

If food must be tasted during preparation, it must be placed in a separate dish and tasted with a clean utensil. (See *Exhibit 4d.*) The dish and utensil should then be removed from the food-preparation area for cleaning and sanitizing.

Policies for Reporting Illness and Injury

Foodhandlers must report health problems to the manager of the establishment before working with food. If they become ill while working, they must immediately report their condition, and if food or equipment could become contaminated, the foodhandler must stop working and see a doctor. There are several instances when a foodhandler must either be *restricted* from working with or around food or *excluded* from working within the establishment. (See *Exhibit 4e.*)

If the foodhandler must refrigerate personal medication while working, it must be stored inside a covered, leak-proof container that is clearly labeled.

Any cuts, burns, boils, sores, skin infections, or infected wounds should be covered with a bandage when the foodhandler is working with or around food or food-contact surfaces. Bandages should be clean, dry, and must prevent leakage from the wound. As previously mentioned, waterproof, disposable gloves or finger cots should be worn over bandages on hands. Foodhandlers wearing bandages may need to be temporarily reassigned to duties not involving contact with food or food-contact surfaces.

Vaccination for Hepatitis A

Hepatitis A is a disease-causing inflammation of the liver. It is transmitted to food by poor personal hygiene or contact with contaminated water. It infects many people each year, resulting in community-wide outbreaks. Of all foodborne illnesses facing the foodservice industry, hepatitis A is the only one that can be prevented by vaccine. While effective handwashing is a critical practice to prevent contamination, vaccinating your foodhandlers for hepatitis A can provide an additional barrier. This may be especially recommended in areas where hepatitis A outbreaks are highly prevalent.

Exhibit 4e

Handling Employee Illnesses

If	Then
The foodhandler has one of the following symptoms: ▶ Fever ▶ Diarrhea ▶ Vomiting ▶ Sore throat with fever ▶ **Jaundice** (a yellowing of the skin and eyes)	Restrict them from working with or around food. Exclude them from the establishment if you primarily serve a high-risk population.
The foodhandler has been diagnosed with a foodborne illness.	Exclude them from the establishment and notify the local regulatory agency. Managers must report employee illnesses resulting from the following pathogens to the local health department: ▶ *Salmonella typhi* ▶ *Shigella* spp. ▶ Shiga toxin-producing *E. coli* ▶ Hepatitis A virus The manager must work with the local regulatory agency to determine when the foodhandler can safely return to work.

Personal Hygiene

Management should model proper personal hygiene practices at all times.

MANAGEMENT'S ROLE IN A PERSONAL HYGIENE PROGRAM

Management plays a critical role in the effectiveness of a personal hygiene program. Responsibilities include:

- Establishing proper personal hygiene policies
- Revising policies when laws and regulations change, as well as when changes are recognized in the science of food safety, and retraining foodhandlers as necessary
- Training foodhandlers on personal hygiene policies
- Modeling proper behavior for foodhandlers at all times
- Supervising sanitary practices continuously, and retraining foodhandlers as necessary

Job Assignments

Cross-Contamination

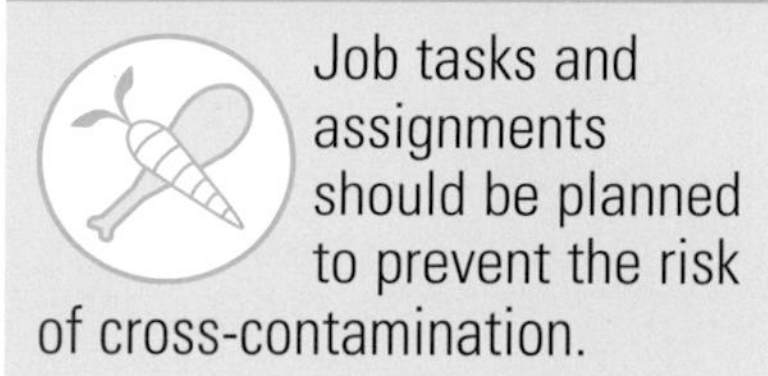

Job tasks and assignments should be planned to prevent the risk of cross-contamination.

When job descriptions are developed and responsibilities assigned, consider the risk of cross-contamination and plan tasks to prevent it. The risk may be higher if foodhandlers are required to perform several different duties than if specific foodhandlers are assigned to a single duty. For example, an employee expected to prepare and wrap food, clear off tables, and then return to food-preparation duties could more easily contaminate food than an employee who is assigned to just one of these tasks. By planning tasks to prevent cross-contamination, you will minimize the amount of time needed for supervision and enable employees to follow sanitation rules more easily.

SUMMARY

Foodhandlers can contaminate food at every step in its flow through the establishment. Good personal hygiene is a critical protective measure against contamination and foodborne illness. A successful personal hygiene program depends on trained foodhandlers who possess the knowledge, skills, and attitude necessary to maintain a safe food system.

Foodhandlers have the potential to contaminate food when they have been diagnosed with a foodborne illness, when they show symptoms of a gastrointestinal illness, when they have infected lesions, or when they touch anything that might contaminate their hands. Foodhandlers must pay close attention to what they do with their hands since simple acts such as nose picking or running fingers through the hair can contaminate food. Proper handwashing must also be practiced. This is especially important before starting work and after using the restroom; after sneezing, coughing, smoking, eating, or drinking; and before and after handling raw food. It is up to the manager to monitor handwashing to make sure it is thorough and frequent. In addition, hands need other care to ensure they will not transfer contaminants to food. Fingernails should be kept short and clean. Cuts and sores should be covered with clean bandages. Hand cuts should also be covered with gloves or finger cots.

Gloves can create a barrier between hands and food; however, they should never be used in place of handwashing. Hands must be washed before putting on gloves and when changing to a new pair. Gloves used to handle food are for single use and should never be washed or reused. They must be changed whenever contamination occurs.

Personal hygiene can be a sensitive subject for some people, but, because it is vital to food safety, it must be addressed with every employee. All employees must maintain personal cleanliness. They should bathe or shower before work and keep their hair clean.

Prior to handling food, foodhandlers must put on a clean hair restraint, put on clean clothing, remove jewelry, and put on appropriate shoes. Aprons should always be removed and properly stored when the employee leaves food-preparation areas.

Establishments should implement strict policies regarding eating, drinking, smoking, and chewing gum and tobacco. These activities should not be allowed when the foodhandler is preparing or serving food, while in food-preparation areas or in areas used to clean utensils and equipment.

Foodhandlers must be encouraged to report health problems to management before working with food. If their condition could contaminate food or equipment, they must stop working and see a doctor. Managers must not allow foodhandlers diagnosed with a foodborne illness to work, and must report illnesses resulting from *Salmonella typhi, Shigella* spp., shiga toxin-producing *E. coli,* and the hepatitis A virus to the local regulatory agency. Managers must restrict foodhandlers from working with or around food if they have symptoms that include fever, diarrhea, vomiting, sore throat with fever, or jaundice. If a foodhandler has any one of these symptoms and the establishment primarily serves a high-risk population, the foodhandler must be excluded from the establishment.

Management plays a critical role in the effectiveness of a personal hygiene program. By establishing a program that includes specific policies, and by training and enforcing those policies, managers can minimize the risk of causing a foodborne illness. Most importantly, managers must set a good example by modeling proper personal hygiene practices.

Apply Your Knowledge

1. Does this situation represent a threat to food safety?
2. Explain what Chris did right.
3. Explain what Chris did wrong.

For answers, please turn to the Answer Key.

A Case in Point 1

Chris works at a quick-service restaurant. She is suffering from seasonal allergies, so she carries a small pack of tissues with her. Her assigned responsibility is to make salads. She washes her hands properly and puts on single-use gloves before she starts her shift. When Chris needs to sneeze, she steps away from the food-preparation area, pulls a clean tissue out of her pocket, sneezes into it, then discards it. Because her medication gives her a dry mouth, Chris keeps a glass of water at her station. Her clearly labeled bottle of allergy medication has to be kept cold, so she stores it in the walk-in refrigerator.

Apply Your Knowledge

1. Explain how Marty might have caused an outbreak of shigellosis.
2. What measures should have been taken to prevent it?

For answers, please turn to the Answer Key.

A Case in Point 2

Marty works for a catering company. A few days ago, he was serving hot food from chafing dishes at an outdoor music festival. He did not wear gloves because he used spoons and tongs to serve the food. His manager noticed that Marty made multiple trips to the bathroom during his four-hour shift. These trips did not interrupt service to customers because there were plenty of staff members on hand and Marty hurried to and from the restroom.

The nearest restroom had soap, separate hot and cold water faucets, and a working hot-air dryer, but no paper towels. Each time Marty used the restroom, he washed his hands quickly and then dried them on his apron. Throughout the following week, the manager of the catering company received several telephone calls from people who had attended the music festival and had eaten their food. They each complained of diarrhea, fever, and chills. One call was from a mother of a young boy who was hospitalized for dehydration. The doctor reported that the boy had shigellosis.

Apply Your Knowledge

Use these questions to review the concepts presented in this chapter.

Discussion Questions

1. What requirements must be met by employees regarding work attire?
2. What personal behaviors can contaminate food?
3. What is the proper procedure for covering cuts, wounds, and sores?
4. What procedures must foodhandlers follow when using gloves?
5. What employee health problems are a possible threat to food safety? What are the appropriate actions that should be taken?

For answers, please turn to the Answer Key.

Apply Your Knowledge

Use these questions to test your knowledge of the concepts presented in this chapter.

Multiple-Choice Study Questions

1. **Which of the following personal behaviors can contaminate food?**
   A. Touching a pimple
   B. Touching hair
   C. Nose picking
   D. All of the above

2. **After you have washed your hands, which of the following items should be used to dry them?**
   A. Your apron
   B. A wiping cloth
   C. A common cloth
   D. Single-use paper towels

3. **A deli worker stops making sandwiches to use the restroom. She must first**
   A. wash her hands.
   B. take off her hat.
   C. take off her apron and properly store it.
   D. change her uniform.

4. **Which of the following items can contaminate food?**
   A. Rings
   B. A watch
   C. Earrings
   D. All of the above

5. **Which of the following is the proper procedure for washing your hands?**
   A. Run hot water (at least 100°F [38°C]), moisten hands and apply soap, vigorously scrub hands and arms, apply sanitizer, dry hands.
   B. Run hot water (at least 100°F [38°C]), moisten hands and apply soap, vigorously scrub hands and arms, rinse hands, dry hands.
   C. Run cold water (at least 41°F [5°C]), moisten hands and apply soap, vigorously scrub hands and arms, rinse hands, dry hands.
   D. Run cold water (at least 41°F [5°C]), moisten hands and apply soap, vigorously scrub hands and arms, apply sanitizer, dry hands.

Continued on next page...

Apply Your Knowledge — Multiple-Choice Study Questions *continued*

6. **Establishments must only use hand sanitizers that**
   A. dry quickly.
   B. can be dispensed in a liquid.
   C. have been approved by the FDA.
   D. can be applied before handwashing.

7. **Which foodhandler is least likely to contaminate the food she will handle?**
   A. A foodhandler who keeps her fingernails long
   B. A foodhandler who keeps her fingernails short
   C. A foodhandler who wears false fingernails
   D. A foodhandler who wears nail polish

8. **Kim wore disposable gloves while she formed raw ground beef into patties. When she was finished, she continued to wear the gloves while she sliced hamburger buns. What mistake did Kim make?**
   A. She failed to wash her hands and put on new gloves after handling raw meat and before handling the ready-to-eat buns.
   B. She failed to wash her hands before wearing the same gloves to slice the buns.
   C. She failed to wash and sanitize her gloves before handling the buns.
   D. She failed to wear reusable gloves.

9. **A foodhandler who has been diagnosed with shigellosis should be**
   A. told to stay home.
   B. told to wear gloves while working with food.
   C. told to wash his hands every fifteen minutes.
   D. assigned to a nonfoodhandling position until he is feeling better.

10. **Managers must report employee illnesses resulting from this pathogen.**
    A. *Shigella* spp.
    B. *Vibrio vulnificus*
    C. *Clostridium perfringens*
    D. *Clostridium botulinuim*

Continued on next page...

Multiple-Choice Study Questions *continued*

11. Some jurisdictions will allow bare-hand contact with cooked and ready-to-eat food if

A. employees double-wash their hands.

B. employees keep their fingernails short and clean.

C. employees use hand sanitizers after properly washing their hands.

D. the establishment has a verifiable written policy on handwashing procedures.

12. Foodhandlers should be restricted from working with or around food if they are experiencing which of the following symptoms?

A. Soreness, itching, fatigue

B. Fever, vomiting, diarrhea

C. Headache, irritability, thirst

D. Muscle cramps, insomnia, sweating

13. Which of the following policies should be implemented at establishments?

A. Employees must not smoke while preparing or serving food.

B. Employees must not eat while in food-preparation areas.

C. Employees must not chew gum or tobacco while preparing or serving food.

D. All of the above

14. Stephanie has a small cut on her finger and is about to prepare chicken salad. How should Stephanie's manager respond to the situation?

A. Send Stephanie home immediately.

B. Cover the hand with a glove or finger cot.

C. Cover the cut with a clean bandage and a glove or finger cot.

D. Cover the cut with a clean bandage.

15. Hands should be washed after which of the following activities?

A. Touching your hair

B. Eating

C. Using a handkerchief

D. All of the above

Continued on next page...

Apply Your Knowledge — Multiple-Choice Study Questions *continued*

16. Al, the prep cook at the Great Lakes Senior Citizen Home, called his manager and told her that he had a bad headache, upset stomach, and a sore throat with fever. What is the manager required to do with Al?
    A. Tell him to rest for a couple hours and then come in.
    B. Tell him to go to the doctor and then immediately come to work.
    C. Tell him that he cannot come to work and that he should see a doctor.
    D. Tell him that he can come in for a couple of hours and then go home.

For answers, please turn to the Answer Key.

ADDITIONAL RESOURCES

Books and Periodicals

Cliver, D. O. 1997. Virus transmission via food. *Food Technology.* 51 (4):71–78.

Don't just rinse 'em, wash 'em. 2001. *Food Safety Illustrated.* 1 (1):3.

Hernandez, J. 1998. Food safety: Preparation and cooking. *Food Management.* 33 (5):90.

Look, Ma, no hands: Touch-free devices gain popularity among the handwashing crowd. 2002. *Food Safety Illustrated.* 2 (1):18.

Outfitted for food safety. 2001. *Food Safety Illustrated.* 1 (3):3.

Proper use of gloves. 2003. *Food Safety Illustrated.* 3 (1):13.

When to wash your hands. 2003. *Food Safety Illustrated.* 3 (3):3.

Web Sites

Centers for Disease Control and Prevention (CDC)

www.cdc.gov

The mission of the CDC is to promote health and quality of life by preventing and controlling disease, injury, and disability. To prevent and control foodborne illness, the CDC collect data on outbreaks. This Web site provides general information on foodborne illnesses and their prevention.

Colgate-Palmolive Company

www.colpalipd.com

A manufacturer of powerful, performance-tested cleaners and sanitizers, Colgate-Palmolive provides a Web site that contains useful information on its foodservice products and on food safety issues.

FDA Food Code

http://vm.cfsan.fda.gov/~dms/foodcode.html

As the basis for many local sanitation codes, as well as the basis for information in this textbook, the FDA Food Code, available at this Web address, is a useful resource for information relating to food safety for the restaurant and foodservice industry.

FoodHandler Inc.
www.foodhandler.com
FoodHandler Inc. is a manufacturer of foodservice safety equipment, such as gloves, aprons, and food-storage systems. The site contains information on products, as well as issues related to food safety.

Hepatitis Control Report
www.hepatitiscontrolreport.com
This site is an online quarterly newsletter devoted to news on the public-health control of viral hepatitis. The mission of the site is to provide accurate and balanced reporting of developments in hepatitis epidemiology, control programs, and public policy.

National Center for Infectious Diseases (NCID)
www.cdc.gov/ncidod
NCID is one of the centers of the CDC. Its mission is to prevent illness, disability, and death caused by infectious diseases around the world. NCID accomplishes this mission by conducting surveillance, epidemic investigations, epidemiologic and laboratory research, and training. It also sponsors public education programs to develop, evaluate, and promote prevention and control strategies for infectious diseases. This Web site serves as another great resource for information on foodborne illness.

Apply Your Knowledge

Notes

Cooper
0
20
40
60
80
100
120
140
160
180
200
220
CT220
°F

Unit 2

The Flow of Food Through the Operation

5

The Flow of Food: An Introduction

Inside this chapter:

- Preventing Cross-Contamination
- Time and Temperature Control
- Monitoring Time and Temperature

After completing this chapter, you should be able to:

- Identify methods for preventing cross-contamination.
- Identify methods for preventing time-temperature abuse.
- Identify different types of temperature-measuring devices and their uses.
- Calibrate and maintain different temperature-measuring devices.
- Properly measure the temperature of food at each point in the flow of food.

Key Terms

Flow of food
Bimetallic stemmed thermometer
Time-temperature indicator (TTI)
Calibration

Apply Your Knowledge

Check to see how much you know about the concepts in this chapter. Use the page references provided to explore the topic in each question.

Test Your Food Safety Knowledge

1. **True or False:** The longer food stays at 85°F (29°C), the more time microorganisms have to multiply. *(See page 5-4.)*
2. **True or False:** The flow of food begins with purchasing and ends with cooking. *(See page 5-2.)*
3. **True or False:** When checking the temperature of a roast, insert the thermometer stem into the thinnest part of the product. *(See page 5-10.)*
4. **True or False:** When calibrating a bimetallic stemmed thermometer, it should be set to 32°F (0°C) prior to placing it into ice water. *(See page 5-12.)*
5. **True or False:** Washing and rinsing a cutting board will prevent it from cross-contaminating the next product it touches. *(See page 5-3.)*

For answers, please turn to the Answer Key.

Exhibit 5a

The Flow of Food

Purchasing
Receiving
Storing
Preparing
Cooking
Holding
Cooling
Reheating
Serving
The Flow of Food

INTRODUCTION

Your responsibility for the safety of the food in your establishment starts long before you serve meals. Many things can happen to a product on its path through the establishment, from purchasing and receiving, through storing, preparing, cooking, holding, cooling, reheating, and serving—known as the **flow of food.** (See *Exhibit 5a.*) A frozen product that leaves the processor's plant in good condition, for example, may thaw on its way to the distributor's warehouse and go unnoticed during receiving. Once in your establishment, the product might not be stored properly or cooked to the correct internal temperature, potentially causing a foodborne illness.

The safety of the food you serve at your establishment will depend largely on your understanding of food safety concepts throughout the flow of food, especially the prevention of cross-contamination and time and temperature control. It also depends on your ability to develop a system that prioritizes, monitors, and verifies the most important food safety practices, which will be discussed in Chapter 10.

PREVENTING CROSS-CONTAMINATION

A major hazard to the flow of food in your operation is cross-contamination, which is the transfer of microorganisms from one food or surface to another. Microorganisms move around easily in a kitchen. They can be transferred from food or unwashed hands to prep tables, equipment, utensils, cutting boards, dish towels, sponges, or other food.

Cross-contamination can occur at almost any point in an operation. When you know where and how microorganisms can be transferred, cross-contamination is fairly simple to prevent. Prevention starts with the creation of barriers between food products. These barriers can be physical or procedural.

Physical Barriers for Preventing Cross-Contamination

- **Assign specific equipment to each type of food product.** For example, use one set of cutting boards, utensils, and containers for poultry, another set for meat, and a third set for produce. Some manufacturers make colored cutting boards and utensils with colored handles. Color coding can tell employees which equipment to use with what products, such as green for produce, yellow for chicken, and red for meat. Although color-coding is helpful, it does not eliminate the need to follow proper practices (i.e., cleaning and sanitizing, minimizing cross-contamination, etc.). Color coding helps to minimize the risk from actually occurring.
- **Clean and sanitize all work surfaces, equipment, and utensils after each task.** After cutting up raw chicken, for example, it is not enough to simply rinse the cutting board. Wash, rinse, and sanitize cutting boards and utensils in a three-compartment sink, or run them through a warewashing machine. Make sure employees know which cleaners and sanitizers to use for each job. Sanitizers used on food-contact surfaces must meet local or state department codes conforming to the Code of Federal Regulations (21CFR178.1010). *(See Chapter 12 for more information on cleaning and sanitizing.)*

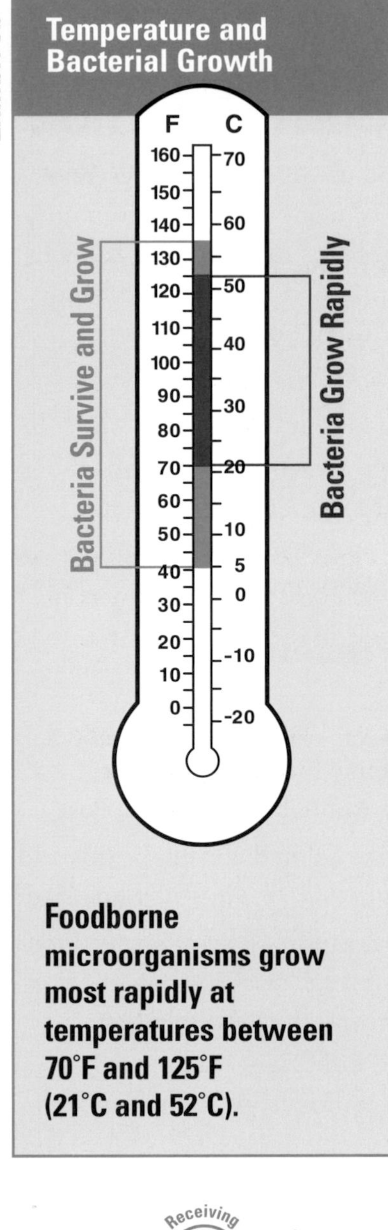

Foodborne microorganisms grow most rapidly at temperatures between 70°F and 125°F (21°C and 52°C).

Procedural Barriers for Preventing Cross-Contamination

- **When using the same prep table, prepare raw and ready-to-eat food at different times.** For example, establishments with limited prep space can prepare lunch salads in the morning, clean and sanitize the utensils and surfaces, and then debone chicken for dinner entrées in the same space in the afternoon.
- **Purchase ingredients that require minimal preparation.** For example, an establishment can switch from buying raw chicken breasts to purchasing precooked breasts.

TIME AND TEMPERATURE CONTROL

One of the biggest factors in foodborne-illness outbreaks is time-temperature abuse. Disease-causing microorganisms grow and multiply at temperatures between 41°F and 135°F (5°C and 57°C), which is why this range is known as the temperature danger zone. Microorganisms grow much faster in the middle of the zone, at temperatures between 70°F and 125°F (21°C and 52°C). (See *Exhibit 5b.*) Whenever food is held in the temperature danger zone, it is being abused.

As you learned in Chapter 2, time also plays a critical role in food safety. Microorganisms need both time and temperature to grow. The longer food stays in the temperature danger zone, the more time microorganisms have to multiply and make food unsafe. To keep food safe throughout the flow of food, you must minimize the amount of time it spends in the temperature danger zone. It is recommended that food does not remain in the zone for more than four hours.

Common opportunities for time-temperature abuse throughout the flow of food include:

- Not cooking food to its required minimum internal temperature
- Not cooling food properly
- Failing to reheat food to 165°F (74°C) for fifteen seconds within two hours

- Failing to hold food at a minimum internal temperature of 135°F (57°C) or higher or 41°F (5°C) or lower

The best way to avoid time-temperature abuse is to establish procedures employees must follow and then monitor them. Make time and temperature control part of every employee's job. Some suggestions include:

- **Decide the best way to monitor time and temperature in your establishment.** Determine which foods should be monitored, how often, and who should check them. Then assign responsibilities to employees in each area. Make sure employees understand exactly what you want them to do, how to do it, and why it is important.
- **Make sure the establishment has the right kind of thermometers available in the right places.** Give employees their own calibrated thermometers. Have them use timers in prep areas to monitor how long food is being kept in the temperature danger zone.
- **Regularly record temperatures and the times they are taken.** (See *Exhibit 5c.*) Print simple forms employees can use to record temperatures and times throughout the shift. Post

Exhibit 5c

Time and Temperature Control

Print simple forms employees can use to record temperatures.

these forms on clipboards outside of refrigerators and freezers, near prep tables, and next to cooking and holding equipment.

- **Incorporate time and temperature controls into standard operating procedures for employees.** These might include:
  - Removing from the refrigerator only the amount of food that can be prepared in a short period of time
  - Refrigerating ingredients and utensils before preparing certain recipes, such as tuna or chicken salad
  - Cooking potentially hazardous food to required minimum internal temperatures
- **Develop a set of corrective actions.** Decide what action should be taken if time and temperature standards are not met. For example, an establishment that holds egg rolls on a steam table might throw them out if their internal temperature falls below 135°F (57°C) for more than four hours, or they might reheat the egg rolls to 165°F (74°C) for at least fifteen seconds within two hours.

MONITORING TIME AND TEMPERATURE

To manage both time and temperature, you need to monitor and control them. The thermometer may be the single most important tool you have to protect your food.

Choosing the Right Thermometer

There are many types of thermometers used in an establishment. Each is designed for a specific purpose. Some are used to measure the temperature of refrigerated or frozen storage areas. Others measure the temperature of equipment, such as ovens, hot-holding cabinets, and warewashing machines. Perhaps the most important types are thermometers that measure the temperature of food. The most common types used in establishments are the bimetallic stemmed thermometer, the thermocouple, and the thermistor. Infrared thermometers are also becoming increasingly popular.

Exhibit 5d

Bimetallic Stemmed Thermometer

Courtesy of Cooper-Atkins Corporation

Bimetallic Stemmed Thermometer

The most common and versatile type of thermometer used in the restaurant and foodservice industry is the **bimetallic stemmed thermometer.** (See *Exhibit 5d*.) This type of thermometer measures temperature through a metal probe with a sensor in the end. Bimetallic stemmed thermometers often have scales measuring temperatures from 0°F to 220°F (–18°C to 104°C). This makes them useful for measuring the temperatures of everything from incoming shipments to the internal temperature of food in hot-holding units. When you select this type of thermometer, it should have (see *Exhibit 5e*)

- an adjustable calibration nut to keep it accurate.
- easy-to-read, numbered temperature markings.
- a dimple to mark the end of the sensing area (which begins at the tip).
- accuracy to within ±2°F (±1°C).

Exhibit 5e

Components of a Bimetallic Stemmed Thermometer

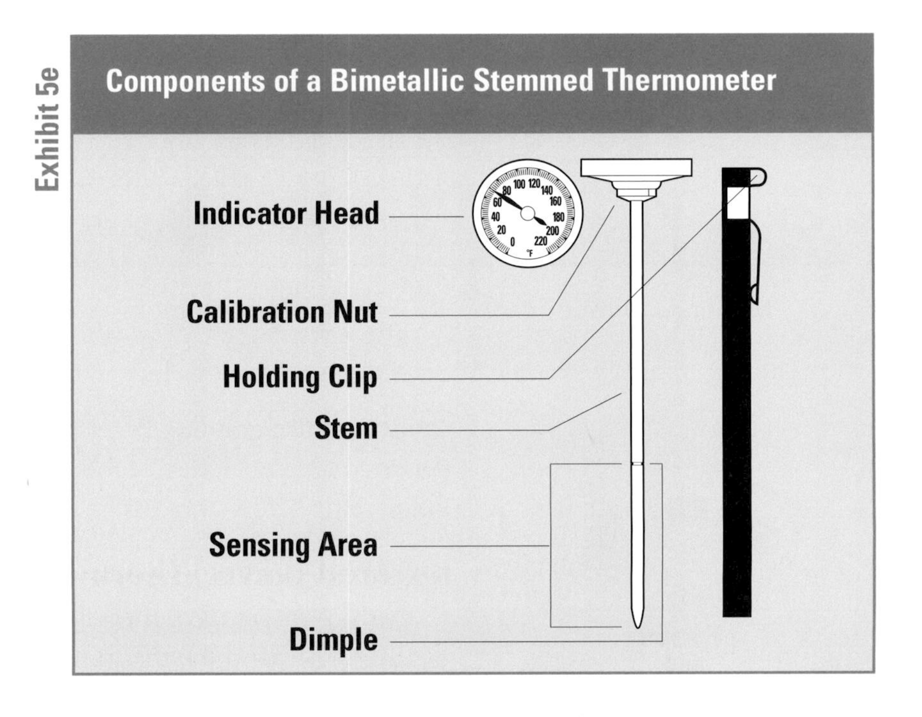

Exhibit 5f

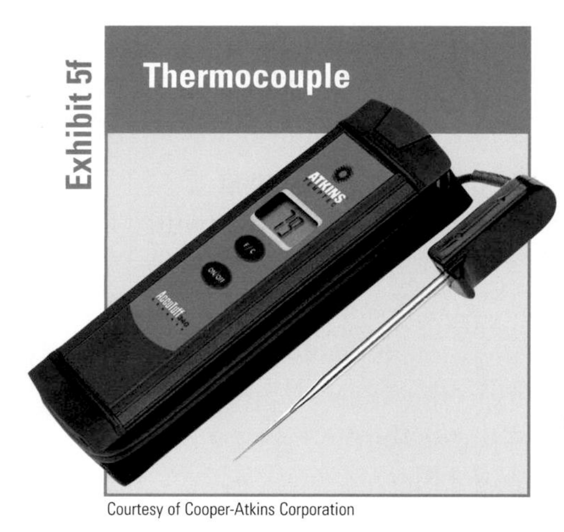

Courtesy of Cooper-Atkins Corporation

Thermocouples and Thermistors

Thermocouples and thermistors measure temperatures through a metal probe or sensing area and display results on a digital readout. They come in a wide variety of styles and sizes, from small pocket models to panel-mounted displays. (See *Exhibit 5f.*) Many come with interchangeable temperature probes designed to measure the temperature of equipment and food.

Basic types of probes include immersion, surface, penetration, and air probes. (See *Exhibit 5g.*) Immersion probes are designed to measure temperatures of liquids, such as soups, sauces, or frying oil. Surface probes measure temperatures of flat cooking equipment like griddles. Penetration probes are used to measure the internal temperature of food. Air probes measure temperatures inside refrigerators or ovens.

Exhibit 5g

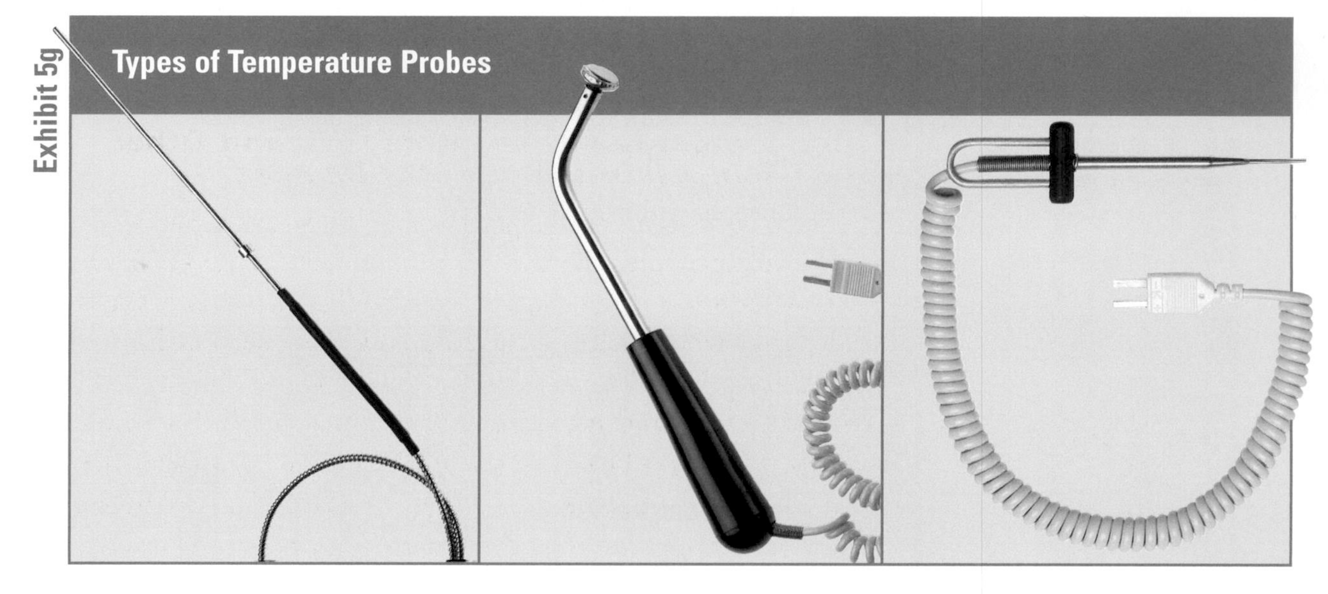

Courtesy of Cooper-Atkins Corporation

Infrared (Laser) Thermometers

Infrared thermometers use infrared technology to produce accurate temperature readings of food and equipment surfaces. They are quick and easy to use. Infrared thermometers are noncontact thermometers that, when used properly, can reduce the risk of cross-contamination and damage to food products. (See *Exhibit 5h.*)

Exhibit 5h

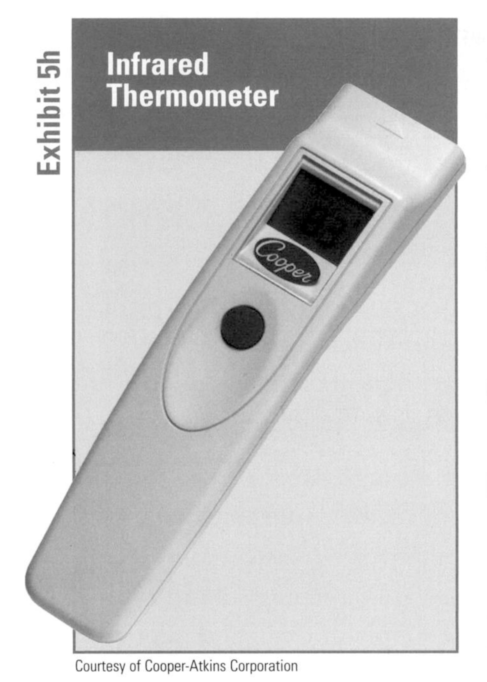

Courtesy of Cooper-Atkins Corporation

To gain the most accurate thermometer reading, remove any barriers between the thermometer and the product being measured and hold the thermometer as close as possible to the product without touching it.

When using infrared thermometers, remember the following:

- Infrared thermometers should not be used to measure air temperature or the internal temperature of food. They are designed to measure surface temperature.
- Do not take temperature measurements through glass or shiny or polished-metal surfaces, such as stainless steel or aluminum.
- Always follow the manufacturer's guidelines for tips on obtaining the most accurate temperature reading with the infrared thermometer you are using.

Time-Temperature Indicators (TTI) and Other Time-Temperature Recording Devices

Some instruments are designed to monitor both time and product temperature. The **time-temperature indicator (TTI)** is one example. Some suppliers attach these self-adhesive tags or sticks to a food shipment to determine if the product's temperature has exceeded safe limits during shipment or later storage. If the product's temperature has exceeded these limits, the TTI provides an irreversible record of the incident. A change in color inside the TTI windows notifies the receiver that the product has undergone time-temperature abuse. (See *Exhibit 5i.*)

Exhibit 5i

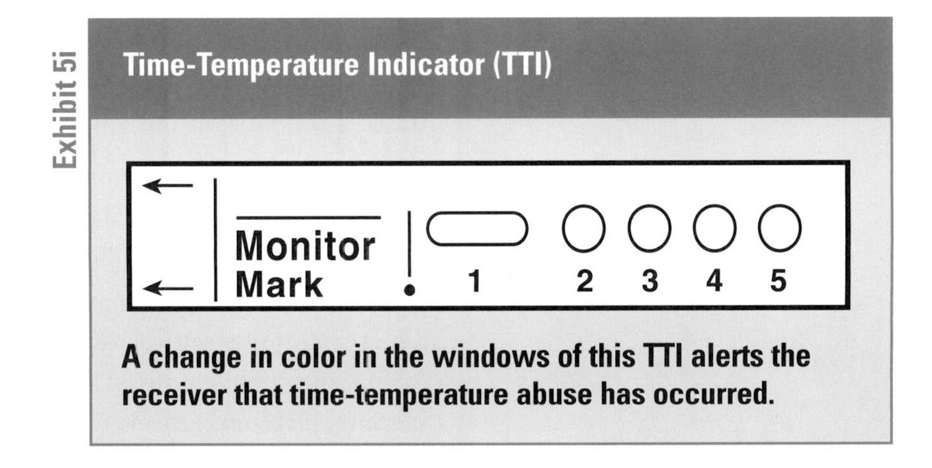

A change in color in the windows of this TTI alerts the receiver that time-temperature abuse has occurred.

More suppliers are using recording devices that continuously monitor temperatures in their delivery trucks. When delivered products appear to have suffered time-temperature abuse, the recording device can be checked to see if the temperature in the delivery truck changed at any time during transit.

Cross-Contamination

Wash, rinse, sanitize, and air-dry thermometers before and after each use to prevent cross-contamination. Use an approved food-contact surface sanitizing solution to sanitize all thermometers.

General Thermometer Guidelines

It is important to know how to use and care for each type of thermometer found in your operation. When it comes to maintenance, always follow manufacturers' recommendations. Here are a few simple guidelines for using thermometers.

- **Keep thermometers and their storage cases clean.** Thermometers should be washed, rinsed, sanitized, and air-dried before and after each use to prevent cross-contamination. Use an approved food-contact surface sanitizing solution to sanitize them. Have an adequate supply of clean and sanitized thermometers on hand.
- **Calibrate thermometers regularly to ensure accuracy.** This should be done before each shift or before each day's deliveries. Thermometers should also be recalibrated any time they suffer a severe shock (for example, after being dropped or after an extreme change in temperature). Thermometers that hang or sit in refrigerators or freezers can be damaged easily. To make sure these thermometers are accurate, use a thermocouple with an air probe to check the temperature. Hanging thermometers usually cannot be recalibrated and must be replaced if they are not accurate.
- **Never use glass thermometers filled with mercury or spirits to monitor the temperature of food.** They can break and pose a serious danger to employees and customers.
- **Measure internal temperatures of food by inserting the thermometer stem or probe into the thickest part of the product (usually the center).** It is a good practice to take at least two readings in different locations because product temperatures may vary across the food portion. When checking the internal temperature of food using a bimetallic stemmed thermometer, insert the stem into the product so that it is at least immersed from the tip to the end of the sensing area.

When measuring the internal temperature of thin food, such as meat or fish patties, small diameter probes should be used.

- **Wait for the thermometer reading to steady before recording the temperature of a food item.** Wait at least fifteen seconds from the time the thermometer stem or probe is inserted into the food.

How to Calibrate Thermometers

Calibration is the process of ensuring that a thermometer gives accurate recordings by adjusting it to a known standard. Most thermometers can be easily calibrated. Two accepted methods of calibration are the boiling-point method (see *Exhibit 5j*) and the ice-point method—which is the most commonly used. (See *Exhibit 5k* on the next page.) To calibrate your thermometers properly, follow one of these methods.

Exhibit 5j

Boiling-Point Method for Calibrating a Thermometer

Step	Process	Notes
1	Bring clean tap water to a boil in a deep pan.	
2	Put the thermometer stem or probe into the boiling water so the sensing area is completely submerged. Wait thirty seconds, or until the indicator stops moving.	Do not let the stem or probe touch the pan's bottom or sides. The thermometer stem or probe must remain in the boiling water.
3	Hold the calibration nut securely with a wrench or other tool and rotate the head of the thermometer until it reads 212°F (100°C) or the appropriate boiling-point temperature for your elevation.	The boiling point of water is about 1°F (about 0.5°C) lower for every 550 feet (168 m) you are above sea level. On some thermocouples or thermistors, it may be possible to press a reset button to adjust the readout.

Ice-Point Method for Calibrating a Thermometer

Step	Process	Notes
1	Fill a large container with crushed ice. Add clean tap water until the container is full.	Stir the mixture well.
2	Put the thermometer stem or probe into the ice water so the sensing area is completely submerged. Wait thirty seconds, or until the indicator stops moving.	Do not let the stem or probe touch the container's bottom or sides. The thermometer stem or probe must remain in the ice water.
3	Hold the calibration nut securely with a wrench or other tool and rotate the head of the thermometer until it reads 32°F (0°C).	On some thermocouples or thermistors, it may be possible to press a reset button to adjust the readout.

SUMMARY

The flow of food is the path that food takes through your establishment from purchasing and receiving through storing, preparing, cooking, holding, cooling, reheating, and serving. Many things can happen to food as it flows through the establishment, but hazards that the manager has most control over are cross-contamination and time-temperature abuse.

Cross-contamination is the transfer of microorganisms from one food or surface to another. Prevention starts with the creation of physical or procedural barriers between food products. Physical barriers include assigning specific equipment to each type of food product, and cleaning and sanitizing all work surfaces, equipment, and utensils after each task. Procedural barriers include purchasing ingredients that require minimal preparation and preparing raw and ready-to-eat food at different times.

Food has been time-temperature abused any time it has been allowed to remain at temperatures between 41°F and 135°F (5°C and 57°C). This temperature range is known as the temperature danger zone. To keep food safe, you must minimize the amount of time food spends in this dangerous range. To prevent time-temperature abuse in your establishment, you should incorporate time and temperature controls into your standard operating procedures, make thermometers available to your employees, and regularly record temperatures and the times they were taken.

Thermometers are the most important tools managers have to prevent time-temperature abuse. Managers should make sure employees know what different thermometers are used for and how to calibrate and use them properly. Thermometers should be washed, rinsed, sanitized, and air-dried before and after each use to prevent cross-contamination. They should also be calibrated regularly to ensure accuracy. When measuring the internal temperature of food, the thermometer stem or probe should be inserted in the thickest part of the product (usually the center). When using a bimetallic stemmed thermometer, the stem should be immersed in the product from the tip to the end of the sensing area. When measuring the internal temperature of thin food, use a small diameter probe. Always wait for the thermometer reading to steady before recording the temperature of the food. Never use glass thermometers filled with mercury to measure food temperatures.

Thermometers can be calibrated by using either the ice-point or boiling-point methods. Using the ice-point method, the thermometer is submerged in ice water and adjusted to 32°F (0°C), while the boiling-point method requires the thermometer to be adjusted to 212°F (100°C) after the stem or probe is placed in boiling water.

Apply Your Knowledge

Use these questions to review the concepts presented in this chapter.

Discussion Questions

1. What are some ways that food can be time-temperature abused?
2. How can cross-contamination be prevented in the establishment?
3. How is a thermometer calibrated using the ice-point method?

For answers, please turn to the Answer Key.

Apply Your Knowledge

Use these questions to test your knowledge of the concepts presented in this chapter.

Multiple-Choice Study Questions

1. **An employee has just trimmed raw chicken on a cutting board and must now use the board to prepare vegetables. What should the employee do with the board prior to preparing the vegetables?**
   A. Wash, rinse, and sanitize the cutting board.
   B. Dry it with a paper towel.
   C. Rinse it under very hot water.
   D. Turn it over and use the reverse side.

2. **All of the following practices can help prevent cross-contamination during food preparation *except***
   A. preparing meat separately from ready-to-eat food.
   B. assigning specific equipment for preparing specific food.
   C. rinsing cutting boards between preparing raw food and ready-to-eat food.
   D. using specific storage containers for specific food.

3. **Infrared thermometers should be used to measure the**
   A. air temperature of a refrigerator.
   B. internal temperature of a cooked turkey.
   C. surface temperature of a steak.
   D. internal temperature of a batch of soup.

4. **All of the following practices can help prevent time-temperature abuse *except***
   A. storing milk at 41°F (5°C).
   B. holding chicken noodle soup at 120°F (49°C).
   C. reheating chili to 165°F (74°C) for fifteen seconds within two hours.
   D. holding the ingredients for tuna salad at 39°F (4°C).

5. **You have a thermocouple with several different types of probes. Which probe should you use to check the temperature of a large stockpot of soup?**
   A. Penetration probe
   B. Surface probe
   C. Air probe
   D. Immersion probe

Continued on next page...

Apply Your Knowledge — Multiple-Choice Study Questions *continued*

6. **Which step for calibrating a thermometer using the ice-point method is *incorrect*?**
   A. First, insert the thermometer stem or probe into a container of ice water.
   B. Second, wait thirty seconds or until the temperature indicator stops moving.
   C. Third, remove the thermometer stem or probe from the ice water.
   D. Finally, adjust the thermometer until it reads 32°F (0°C).

7. **Your manager has asked you to purchase a new thermometer for the restaurant. Which type would *not* be a proper choice?**
   A. Bimetallic stemmed thermometer accurate to ±2°F (±1°C)
   B. Thermistor
   C. Thermocouple
   D. Mercury-filled glass thermometer

For answers, please turn to the Answer Key.

ADDITIONAL RESOURCES

Books and Periodicals

Cross out cloth contamination. 2001. *Food Safety Illustrated.* 1 (2):13.

Golden rules for ground beef. 2001. *Food Safety Illustrated.* 1 (2):14.

How to calibrate thermometers. 2002. *Food Safety Illustrated.* 2 (3):3.

How to organize your prep area. 2002. *Food Safety Illustrated.* 2 (1):14.

Sanitize your thermometers. 2001. *Food Safety Illustrated.* 1 (2):3.

Thermometers: Probing for answers. 2001. *Food Safety Illustrated.* 1 (1):14.

Web Sites

Cooper-Atkins Corporation

www.cooperinstrument.com

Cooper-Atkins Corporation is a supplier of temperature, time, and humidity instruments for a variety of global markets. This Web site contains information on how to calibrate thermometers properly and maintain food temperatures during cooking, holding, cooling, etc.

FDA Food Code

http://vm.cfsan.fda.gov/~dms/foodcode.html

As the basis for many local sanitation codes, as well as the basis for information in this textbook, the FDA Food Code, available at this Web address, is a useful resource for information relating to food safety for the restaurant and foodservice industry.

KatchAll/San Jamar

www.sanjamar.com

KatchAll is an industry leader in innovative food safety products for the restaurant and foodservice industry. The Web site contains information on all KatchAll products that control time and temperature and help avoid cross-contamination.

National Restaurant Association
www.restaurant.org
The National Restaurant Association is the leading business association for the restaurant industry. Together with the National Restaurant Association Educational Foundation, the Association's mission is to represent, educate, and promote the rapidly growing restaurant and foodservice industry. This Web site should be your starting place for all issues and concerns related to your establishment. This Web site has it all, from tips for running your establishment to vital data on your customers' spending habits.

Vitsab TTI
www.vitsab.com
Vitsab time temperature indicator (TTI) labels are inexpensive monitoring devices that can be placed on temperature-sensitive containers to track their possible exposure to out-of-range, high-temperature conditions. The indicator tracks time, as well as temperature, so the result is a true indication of temperature exposure. This Web site provides technical information on TTIs and Vitsab.

Apply Your Knowledge

Notes

ENRICO'S
ITALIAN DINING

The Flow of Food: Purchasing and Receiving

Inside this chapter:

- Choosing a Supplier
- Inspection Procedures
- Receiving and Inspecting Specific Food

After completing this chapter, you should be able to:

- Identify an approved food source.
- Identify accept and reject criteria for:
  - Meat and poultry
  - Seafood
  - Milk and dairy products
  - Eggs
  - Fruit and vegetables
  - Canned goods and other dry food
  - Ready-to-eat food
  - Frozen food
  - Bakery goods

Key Terms

Shellstock identification tags
Modified atmosphere packaging (MAP)
Vacuum-packed food
Sous vide food
Ultra-high temperature (UHT) pasteurization
Aseptically packaged food

Apply Your Knowledge

Check to see how much you know about the concepts in this chapter. Use the page references provided to explore the topic in each question.

Test Your Food Safety Knowledge

1. **True or False:** Upon arrival, a delivery of fresh fish should be received at an internal temperature of 41°F (5°C) or lower. *(See page 6-9.)*
2. **True or False:** Turkey should be rejected if the texture is firm and springs back when touched. *(See page 6-8.)*
3. **True or False:** You should reject a delivery of frozen steaks covered in large ice crystals since the steaks have probably been thawed and refrozen. *(See page 6-18.)*
4. **True or False:** If a sack of flour is dry upon delivery, the contents may still be contaminated. *(See page 6-20.)*
5. **True or False:** A supplier that is in compliance with local, state, and federal law can be considered an approved source. *(See page 6-3.)*

For answers, please turn to the Answer Key.

INTRODUCTION

To be sure the food you serve is safe, you must first control the quality and safety of food that comes in your back door.

The final responsibility for the safety of food entering your establishment rests with you. You can avoid many potential food safety hazards by making sure products are properly received. Using approved suppliers and inspecting products when they are delivered are the first steps in the process.

CHOOSING A SUPPLIER

A number of factors go into selecting the right suppliers. While the level of service needs to be considered, choosing a supplier who can deliver safe food is the ultimate goal. Make sure your suppliers meet or exceed your quality standards.

Quality Standards

- **Make sure suppliers are getting their products from approved sources.** An approved food source is one that has been inspected and is in compliance with applicable local, state, and federal law. Before you accept any deliveries, it is your responsibility to ensure that food you purchase comes from suppliers (distributors) and sources (points of origin) that have been approved.
- **Make sure suppliers are reputable.** Ask other operators what their experience has been with a particular supplier.
- **Inspect your supplier's warehouse or plant from time to time, if possible.** See if it is clean and well run.
- **Ask your suppliers if they have a HACCP program in place.** (If they supply fresh produce, ask if they have a Good Agricultural Practices Plan.) If not, ask what precautions or procedures they take to ensure product safety.
- **Find out if your supplier's employees are trained in food safety.**
- **Check the condition of the supplier's delivery trucks.** Are they clean and well maintained? Do they hold refrigerated or frozen products at the proper temperatures? Are raw products separated from processed food and fresh produce?
- **Check your supplier's shipments for consistent product quality.** Inspect deliveries for unsafe packaging. Broken boxes, leaky packages, or dented cans are signs of careless handling.
- **Request that suppliers deliver products when your staff has time to receive them properly.**

Rejecting Shipments

Remember that you have the right to refuse any delivery that does not meet your standards. You should have a company policy about returns, and your suppliers should be aware of it and agree to it. When food does not meet your standards, your employees should know what to do. To reject a product or shipment:

- **Set the rejected product aside.** Keep it separate from other food and supplies.

- **Tell the delivery person exactly what is wrong with the rejected product.** Use your purchase agreement and company standards to back up your decision to reject the product.
- **Get a signed adjustment or credit slip from the delivery person before throwing the product away or letting the delivery person remove it.**
- **Log the incident on the invoice or receiving document.** Note the food involved, including lot number and expiration date if appropriate, the standard that was not met, and the corrective action you took.

Once you have established a relationship with a supplier, continue to be a smart customer. Always inspect deliveries. Randomly check weights and product temperatures, break down cases, and take counts. Do not take anything for granted. You are buying more than products; you are buying service and food safety.

INSPECTION PROCEDURES

If you establish procedures for inspecting products, you can reduce hazards before they enter your establishment.

Here are some general guidelines that can help you improve the way you receive deliveries.

- **Train employees to inspect deliveries properly.** Ideally, you should assign the responsibility for inspecting and receiving deliveries to specific employees. These employees should be trained to judge product quality, check products for proper temperatures, identify code dates, identify food that has been thawed and refrozen, spot damage or insect infestation, and so on. They also should be authorized to accept, reject, and sign for deliveries.
- **Plan ahead for shipments.** Have clean hand trucks, carts, dollies, and containers available in the receiving area. Make sure enough space is available in walk-ins and storerooms prior to receiving a shipment. Some operations use a refrigerator and freezer in the receiving area for temporary storage. If products need to be washed or broken down and rewrapped, make work space available as close to the receiving area as possible. This will prevent dirt and pests from being brought into storage areas or the kitchen.

- **Schedule deliveries during off-peak hours.** Arrange it so products are delivered during times when employees have adequate time to inspect them.
- **Plan a backup menu in case you have to return food items.** If food is not safe or does not meet your standards, you may have to take an item off the menu, substitute another menu item, or try to arrange delivery from another supplier.
- **If possible, receive only one delivery at a time.** Inspect and store each delivery before accepting another one to avoid potential confusion and product abuse in the receiving area.
- **Have the right information available.** Receivers should have a purchase order or order sheet ready to check against a supplier's invoice. The sheet should list quantities, quality specifications, and agreed-upon prices. The sheet should also have room to record the date and time of delivery, product temperatures, and other notes.
- **Inspect deliveries immediately.** Do a thorough visual inspection to count quantities, check for damaged products, and look for items that might have been repacked or mishandled. Spot-check weights and take sample temperatures of all refrigerated food. (See *Exhibit 6a.*) There is always a

Exhibit 6a

Checking Temperatures During Receiving

Take sample temperatures of refrigerated food.

possibility that food—even government-inspected products—may have been mishandled during shipment.

- **Correct mistakes immediately.** If any products are damaged, not at the correct temperature, or have not been delivered to specifications, do not accept them.
- **Put products away as quickly as possible, especially products requiring refrigeration.**
- **Keep the receiving area clean and well lighted to discourage pests.**

How to Check Temperatures of Deliveries

Use the following guidelines when checking the temperature of deliveries. (See *Exhibit 6b*.)

- **Check the temperature of meat, poultry, and fish by inserting the thermometer stem or probe into the thickest part of the product (usually the center).** Additionally, you may want to check the surface temperature using an appropriate thermometer since it can be a better indicator of potential temperature abuse.
- **Check the temperature of MAP, vacuum-packed, or *sous vide* food *(see page 6-18)* by inserting the thermometer stem or probe between two packages.** Be careful not to puncture the product wrapping. Wait until the indicator stops moving, and record the temperature.
- **Check the temperature of liquids or other packaged food by opening the package and inserting the thermometer stem or probe into the food until the sensing area is immersed.** Do not let the thermometer stem or probe touch the container's sides. Wait until the indicator stops moving and record the temperature.
- **Check the temperature of bulk liquids by folding the bag or pouch around the thermometer stem or probe.** Be careful not to puncture the bag or pouch.

Exhibit 6b

Meat, Poultry, Fish

MAP, Vacuum-Packed, and *Sous Vide* Food

Liquids or Other Packaged Food

Bulk Liquids

- **Check the temperature of live, molluscan shellfish by inserting the stem or probe into the middle of the carton or case, between the shellfish, for an ambient reading.** Check the temperature of shucked shellfish by inserting the stem or probe into the container until the sensing area is immersed.
- **When receiving eggs, check the ambient (air) temperature of the delivery truck, as well as the truck's temperature chart recorder for extreme temperature fluctuations during transport.** Temperature fluctuations, high humidity, and warm temperatures may result in the growth of harmful microorganisms.
- **Always be sure to use a clean, sanitized thermometer each time you check a temperature.** If you do not have extra thermometers at the back door, clean and sanitize your thermometer after each use. Keep a bucket of sanitizing solution on the dock, or use approved sanitizing wipes.

RECEIVING AND INSPECTING SPECIFIC FOOD

Every food product delivered to your establishment should be inspected carefully for damage, poor quality, or potential contamination. Internal temperatures of products should be checked and recorded. While receiving temperatures for fresh food are product-specific, frozen food should always be received frozen. In addition, note each food's appearance, texture, smell, and, in some cases, taste. (See *Exhibit 6c* on the next page.)

Seafood

Fish and shellfish are very sensitive to rough treatment and time-temperature abuse. Both fresh and frozen seafood deteriorate quickly if improperly handled. Time-temperature abuse can result in the rapid growth of microorganisms, which can lead to foodborne illness.

Exhibit 6c

Accept and Reject Criteria for Receiving Seafood, Meat, Poultry, and Eggs

Food	Accept	Reject
Fresh Fish Receive at an internal temperature of 41°F (5°C) or lower	**Color:** bright red gills; bright, shiny skin **Odor:** mild ocean or seaweed smell **Eyes:** bright, clear, and full **Texture:** firm flesh that springs back when touched	**Color:** dull, gray gills; dull, dry skin **Odor:** strong fishy or ammonia smell **Eyes:** cloudy, red-rimmed, sunken **Texture:** soft, leaves an imprint when pressed
Fresh Shellfish (clams, mussels, oysters, scallops) **Live:** Receive on ice or at an ambient temperature of 45°F (7°C) or lower **Shucked:** Receive at an internal temperature of 45°F (7°C) or lower	**Odor:** mild ocean or seaweed smell **Shells:** closed and unbroken (indicates shellfish are alive) **Condition:** received alive; identified by shellstock identification tag. Retain tags for 90 days after product is used.	**Odor:** strong fishy smell **Shells:** open shells that do not close when tapped (indicates shellfish are dead); broken shells **Condition:** dead on arrival **Texture:** slimy, sticky, or dry
Fresh Crustaceans (lobster, shrimp, and crab) **Live:** Must be received alive **Processed:** Receive at an internal temperature of 41°F (5°C) or lower	**Odor:** mild ocean or seaweed smell **Shells:** hard and heavy for lobsters and crabs **Condition:** shipped alive; packed with seaweed and kept moist	**Odor:** strong fishy smell **Shells:** soft **Condition:** dead on arrival, tail fails to curl when lobster is picked up
Fresh Meat Receive at an internal temperature of 41°F (5°C) or lower	**Beef Color:** bright, cherry red **Lamb Color:** light red **Pork Color:** pink lean meat, white fat **Texture:** firm and springs back when touched	**Color:** brown or greenish; brown, green, or purple blotches; black, white, or green spots **Texture:** slimy, sticky, or dry **Packaging:** broken cartons, dirty wrappers, or torn packaging **Odor:** sour odor
Fresh Poultry Receive at an internal temperature of 41°F (5°C) or lower	**Color:** no discoloration **Texture:** firm and springs back when touched **Packaging:** poultry should be surrounded by crushed, self-draining ice	**Color:** purple or green discoloration around the neck; dark wing tips (red wing tips are acceptable) **Texture:** stickiness under the wings or around joints **Odor:** abnormal, unpleasant odor
Fresh Eggs (Shell) Receive at an air temperature of 45°F (7°C) or lower	**Odor:** none **Shells:** clean and unbroken	**Odor:** sulfur smell or off odor **Shells:** dirty or cracked

Fish

Fresh fish should be packed in self-draining, crushed or flaked ice. Upon delivery, it should be received at a temperature of 41°F (5°C) or lower. Fresh fish in good condition should meet the following standards:

- Clear eyes
- Firm flesh
- Pleasant, mild scent of ocean or seaweed
- Bright red and moist gills
- Bright skin

Carelessly handled fish is not appetizing, let alone safe to eat. The following conditions are grounds for rejecting a shipment of fish:

- Strong fishy or ammonia smell
- Cloudy, red-rimmed, sunken eyes
- Dull, gray gills
- Dry skin
- Soft skin that leaves an imprint when touched
- Tumors, abscesses, and cysts on the skin

See *Exhibit 6d* for examples of acceptable and unacceptable fish.

Exhibit 6d

Acceptable vs. Unacceptable Fish

Acceptable | Unacceptable

Frozen fish should be received frozen. If there is any indication it has been allowed to thaw, do not accept it. Fish that has thawed and then been refrozen before reaching your establishment may have a sour odor and be off color. Fillets often turn brown at the edges when they have been refrozen. Other signs include large amounts of ice or liquid in the bottom of the shipping box and moist, discolored, or slimy wrapping paper.

Shellfish

Shellfish include molluscan bivalves (mollusks), such as clams, oysters, and mussels. They can be shipped live, frozen, in the shell, or shucked.

Interstate shipping of molluscan shellfish is monitored by the FDA and the Interstate Shellfish Sanitation Conference. Shellfish must be bought only from suppliers listed in the *National Shellfish Sanitation Program Guide for the Control of Molluscan Shellfish,* or from sources included in the Interstate Certified Shellfish Shippers List. *(The list is available on the FDA Seafood Information and Resources Web site; see page 6-31.)* Shucked shellfish must be packaged in nonreturnable containers clearly labeled with the name, address, and certification number of the packer. Packages containing less than one-half gallon must have a sell-by date. Packages containing more than one-half gallon should list the date the shellfish was shucked.

Live, molluscan shellfish must be received on ice or at an ambient temperature of 45°F (7°C) or lower, while shucked product must be received at an internal temperature of 45°F (7°C) or lower. When shipped live, shellfish must be delivered alive in nonreturnable containers. The FDA requires that live, molluscan shellfish carry **shellstock identification tags**. (See *Exhibit 6e.*) Restaurant and foodservice operators must write the date of delivery on the tags. The tag should remain attached to the container the shellfish came in until the container is empty. Operators then must keep the tags on file for ninety days after the last shellfish has been used. Never mix shellfish from one shipment with another.

Shells of clams, mussels, and oysters will be closed if alive. Partly open shells may mean they are dead. To find out, tap on

Exhibit 6e

Shellstock Identification Tags

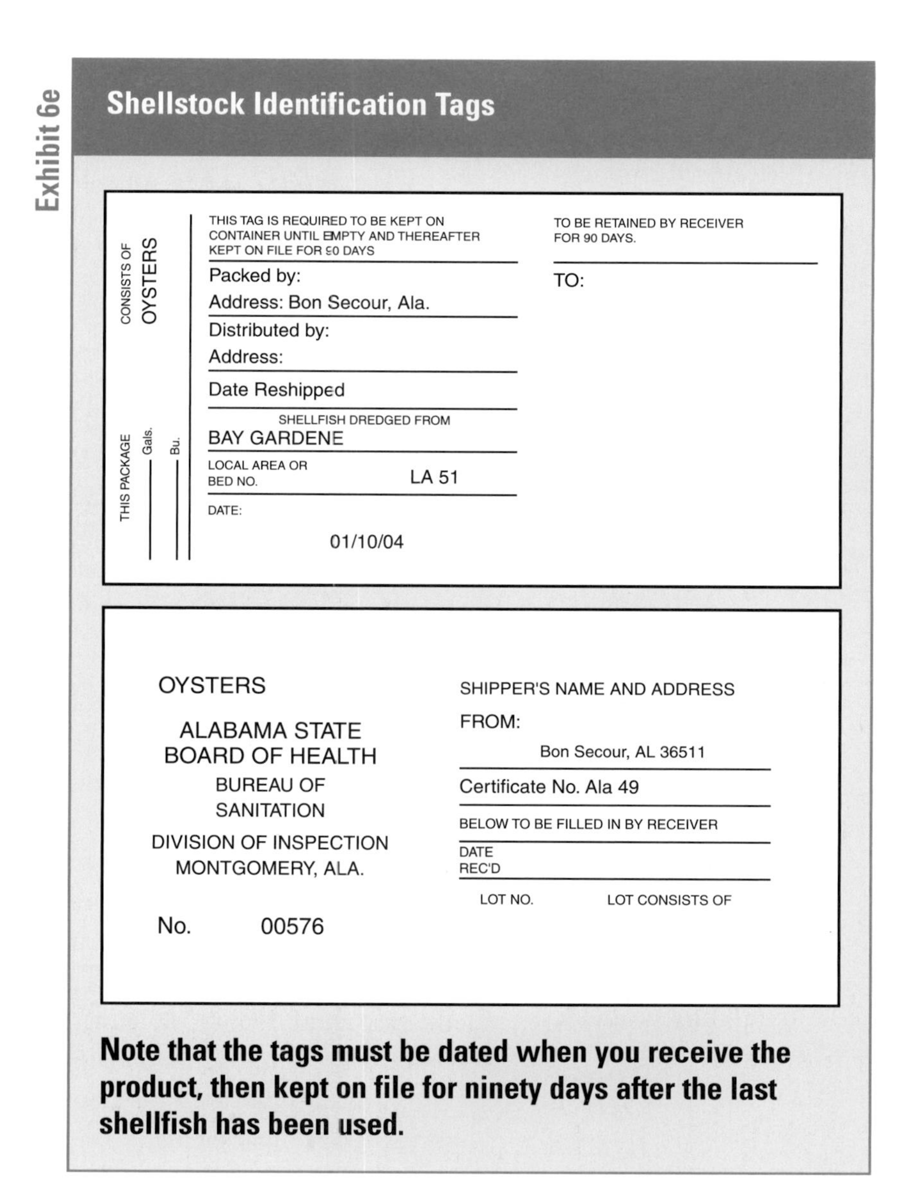
CONSISTS OF OYSTERS

THIS PACKAGE Gals. Bu.

THIS TAG IS REQUIRED TO BE KEPT ON CONTAINER UNTIL EMPTY AND THEREAFTER KEPT ON FILE FOR 90 DAYS

Packed by:
Address: Bon Secour, Ala.
Distributed by:
Address:
Date Reshipped

SHELLFISH DREDGED FROM
BAY GARDENE

LOCAL AREA OR BED NO. LA 51

DATE: 01/10/04

TO BE RETAINED BY RECEIVER FOR 90 DAYS.

TO:

OYSTERS

ALABAMA STATE
BOARD OF HEALTH
BUREAU OF
SANITATION
DIVISION OF INSPECTION
MONTGOMERY, ALA.

No. 00576

SHIPPER'S NAME AND ADDRESS
FROM:
Bon Secour, AL 36511
Certificate No. Ala 49
BELOW TO BE FILLED IN BY RECEIVER
DATE REC'D
LOT NO. LOT CONSISTS OF

Note that the tags must be dated when you receive the product, then kept on file for ninety days after the last shellfish has been used.

the shells. If they close, the mollusks are still alive. If the shells do not close or are badly cracked or broken, they should be discarded.

Crustacea

Crustacea include shrimp, crab, and lobster. Fresh lobsters or crabs in good condition are easy to tell from those that have been poorly handled. A fresh lobster or crab will meet the following standards:

- Show signs of movement
- Have a hard and heavy shell

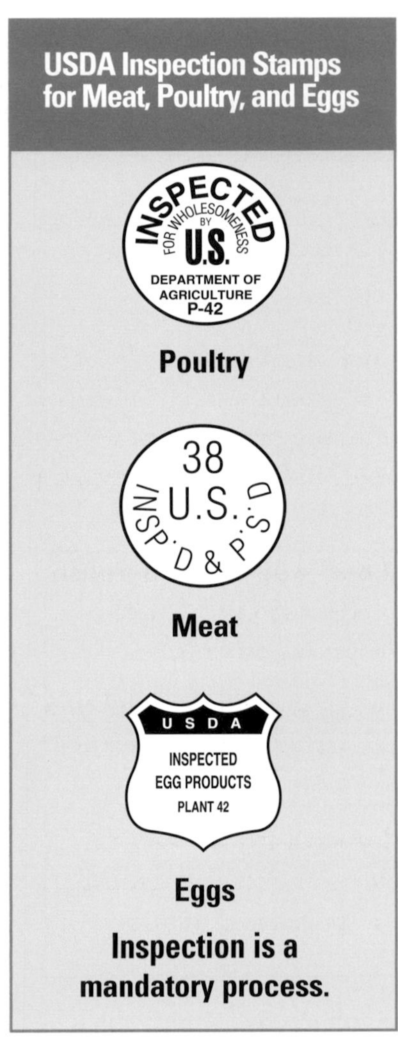

- React when its eyes are pinched
- Curl its tail under when turned on its back (lobsters)

Live lobsters or crabs must be received alive. Those showing weak signs of life should be cooked right away, while dead ones must be discarded or returned to the vendor for credit. All processed crustacean must be received at an internal temperature of 41°F (5°C) or lower.

Fresh Meat and Poultry

We often hear news stories of foodborne illnesses caused by *Salmonella* spp. and *Campylobacter* spp. from poultry or shiga toxin-producing *E. coli* from beef or other meat. While fresh meat and poultry are carefully inspected for wholesomeness by government agencies, it may be impossible to prevent microorganisms from contaminating the product during processing. Almost all cases of foodborne illness can be prevented, however, if food is properly handled and prepared.

Fresh meat and poultry should be delivered at 41°F (5°C) or lower. When it arrives, inspect it closely, checking temperature, color, odor, texture, and packaging. Meat and poultry must be purchased from plants inspected by the USDA or state department of agriculture. Meat products that have been inspected will be stamped with abbreviations for "inspected and passed" by the inspecting agency, along with the number identifying the processing plant. (See *Exhibit 6f.*) Stamps will not appear on every cut of meat, but one should be present on every inspected carcass and on packaging.

Meat and poultry inspection is mandatory. During the inspection process, products are checked for wholesomeness. USDA inspectors examine the carcass and viscera of each animal for possible signs of illness and check processing plants for sanitary conditions. The USDA inspection stamp means that both product and processing plant have met certain standards. It does not mean the product is free of microorganisms that can cause foodborne illness. Operators are still responsible for properly handling and preparing this food to be sure it is safe for consumers to eat.

USDA Grading Stamps for Meat, Poultry, and Eggs

Grading is a voluntary service that processors and packers pay for.

Most meat and poultry also carry a stamp indicating its "grade," or palatability, and level of quality. Grading is a voluntary service offered by the USDA and is paid for by processors and packers. USDA grades are printed inside a shield-shaped stamp. (See *Exhibit 6g.*)

When receiving meat and poultry, consider the following factors:

Meat

- **Beef should be bright, cherry red.** Aged beef may be darker in color. Any beef turning brown or green should be rejected. Beef usually spoils on or near the surface first. A slimy texture or sour or off odor are signs the meat has begun to deteriorate. (See *Exhibit 6h.*)
  - **Vacuum-packed refrigerated beef will appear purplish in color while packaged.** The product should be rejected if the seal has been broken or the package is torn.
- **Lamb is light red when fresh and properly exposed to air.** Do not accept any fresh lamb that is brown or has a whitish surface covering the lean meat.
- **Fresh pork is light pink in color with firm, white fat portions.** An excessively dark color, soft or rancid fat, and a sour odor all indicate the meat is spoiled and should be rejected.

Exhibit 6h

Acceptable vs. Unacceptable Beef

Acceptable Unacceptable

Poultry

Fresh poultry should be shipped in self-draining, crushed ice and delivered at a temperature of 41°F (5°C) or lower, or chill-packed. Poultry shipped and stored at temperatures of 28°F (–2°C) will likely have a significantly longer shelf life.

Mishandled poultry is easy to spot by its appearance. (See *Exhibit 6i*.) Poultry that has started to spoil may have the following characteristics:

- Purplish or greenish color
- Abnormal odor
- Stickiness under the wings and around joints
- Dark wing tips (red tips are acceptable)

Poultry is inspected by federal or state agriculture agencies in much the same way as meat. As with meat, the grading system is voluntary and paid for by processors.

Eggs

Purchase fresh eggs from approved, government-inspected suppliers. The USDA inspection stamp on egg cartons indicates federal regulations are enforced to maintain quality and reduce contamination. As with meat and poultry, grading is voluntary and is provided by the USDA. The official grade stamp certifies the eggs have been graded for quality under federal and/or state supervision. Cases and cartons of shell eggs for direct sale to the consumer must display safe-handling instructions on them.

Exhibit 6i

Acceptable vs. Unacceptable Poultry

Exhibit 6j

Acceptable Eggs

Eggs should be clean, dry, and free of cracks.

Choose suppliers who can deliver eggs within a few days of the packing date. Eggs must be delivered in refrigerated trucks. These trucks should be capable of documenting air temperature during transportation. When the eggs arrive, the truck's air temperature should be 45°F (7°C) or lower, and the eggs must be stored immediately in refrigeration units that will hold them at an ambient (air) temperature of 45°F (7°C) or lower.

Shells should be clean, dry, and free of cracks. (See *Exhibit 6j*.) Fresh eggs should have no odor. Refrigerate fresh eggs in their original containers.

Liquid, frozen, and dehydrated eggs must be pasteurized as required by law, and bear the USDA inspection mark. When delivered, they should be refrigerated or frozen at the proper temperature. Check packages for damage or indication of refreezing and note use-by dates to be sure the product is in good condition and still usable.

Dairy Products

Key Point

Purchase only pasteurized dairy products.

Purchase only pasteurized dairy products. Unpasteurized milk and other dairy products are potential sources of microorganisms such as *Salmonella* spp., *Campylobacter jejuni*, and *Listeria monocytogenes*, all of which can cause serious illness. All milk and milk products should be labeled "Grade A." This means they meet standards for quality and sanitary processing methods set by the FDA and U.S. Public Health Service. Dairy products with the Grade A label—such as cream, dried milk, cottage cheese, cream cheese, butter, cheese, and frozen products like ice cream—are made with pasteurized milk.

Like other refrigerated products, milk and dairy products should be received at 41°F (5°C) or lower, unless the law governing their distribution specifies a different temperature. Check with the proper regulatory agency for temperature requirements.

Fresh milk has a sweetish taste. Any milk that tastes sour, bitter, or moldy should be rejected. Milk has a sell-by date stamped on the container. Milk delivered after that date, as well as milk with any off odor, should be rejected.

Butter should have a sweet flavor, uniform color, and firm texture. Check for signs of mold, specks, or other foreign matter. Make sure containers are clean and undamaged. Do not accept any butter that is rancid or has absorbed odors.

In the U.S., all cheeses must meet certain standards of identity. In order for the product to be called "Cheddar" or "mozzarella," for example, the government specifies ingredients that must be used, maximum moisture content, minimum fat content, and general characteristics. When cheeses are delivered, check to see each type has its typical flavor, texture, and color. If the cheese has a rind, it should be clean and unbroken. Cheese should have no signs of unnatural mold or off odors. As always, check for proper temperature and clean, undamaged packaging.

Key Point

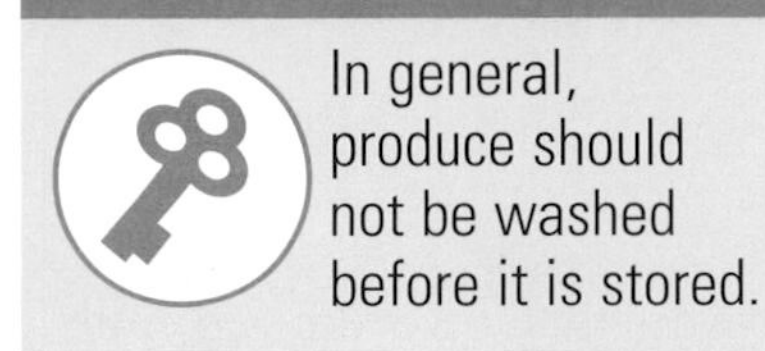

Fresh Produce

Fresh fruit and vegetables have different temperature requirements for transportation and storage, so they may be held at different temperatures. No specific temperature is mandated by regulation for the transportation and storage of fresh produce, with a few exceptions. Cut melons, a potentially hazardous food, must be held at 41°F (5°C) or lower. Fresh-cut produce is best held at 33°F to 41°F (1°C to 5°C) to maintain quality.

Fresh fruit and vegetables are highly perishable and should be put into storage quickly. Most produce should not sit out at room temperature, let alone on a warm dock. Have a system in place to get fresh-cut and highly perishable items into a cooler.

In general, produce should not be washed before it is stored. While washing would not hurt leafy green items, many other products are likely to decay faster if washed before storage. This is

Exhibit 6k

Inspecting the Quality of Fruit

Spoilage will appear in a variety of ways, including mold, blemishes, mushiness, discoloration, wilting, or dull appearance.

especially true for mushrooms and berries. Wash produce just before preparing and serving it.

All fruit and vegetables should be handled with care. If they are pinched, squeezed, or roughly handled, they will bruise and spoil more quickly.

Check products being delivered for signs of mishandling and insect infestation, including insect eggs and egg cases. Visually inspect produce for quality. (See *Exhibit 6k*.) Spoilage will appear in a variety of ways, including mold, blemishes, cuts, mushiness, discoloration, wilting, or dull appearance. What applies to one fruit or vegetable may not apply to another. For example, peaches with cuts in them could be considered poor quality, but potatoes and carrots with cuts would be considered acceptable. Discoloration in produce may vary as well. For example, oranges may actually revert to a green color without affecting the quality of the orange or its juice.

Use smell and taste to help determine product quality. Unpleasant odors will tell you when a product is not acceptable. With fruit, sometimes taste is the best test. Outer peels or skins can be blemished without affecting flavor or quality. Be sure to wash or peel fruit and vegetables carefully before tasting them.

Since there are so many ways produce can show signs of spoilage, employees need to learn not only how to identify obviously unacceptable produce, but also produce that will spoil quickly in storage.

Refrigerated and Frozen Processed Food

More and more establishments are using prepared food that is either refrigerated or frozen. This includes precut meats, Individually Quick Frozen (IQF) poultry, frozen or refrigerated entrées that only require heating, and fresh-cut fruit and vegetables (including salads). Processed food can save time and money, but only if it is treated with the same care given to other food products. Mishandled products that end up causing a foodborne illness can put an establishment out of business.

The temperature of refrigerated, processed food should be 41°F (5°C) or lower when delivered. While these products are usually fully cooked or ready to eat, they still require careful handling. Inspect packaging for tears or holes, and check use-by dates.

Key Point

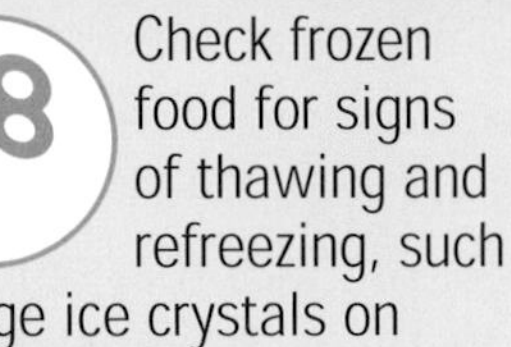

Check frozen food for signs of thawing and refreezing, such as large ice crystals on the product.

All frozen food should be delivered frozen, with the exception of ice cream. Ice cream may be delivered and stored at temperatures of 6°F to 10°F (–14°C to –12°C) without affecting product safety or quality.

Frozen food should be checked for signs of thawing and refreezing. Simply because a product is frozen upon receipt does not mean it has not thawed and been refrozen during prior handling. Obvious signs are blocks of ice or liquid at the bottom of the case or large ice crystals on the product itself. Other signs include product discoloration or dryness and stains on the outer packaging. Food should be wrapped in airtight packaging, and boxes and outer cartons should be clean and undamaged.

MAP, Vacuum-Packed, and *Sous Vide* Food

MAP Food

MAP stands for **modified atmosphere packaging**. By this method, air is removed from a food package and replaced with gases, such as carbon dioxide and nitrogen. These gases help extend the shelf life of the product. Many fresh-cut produce items are packaged using MAP methods.

Key Point

Many fresh-cut produce items are packaged using MAP methods.

Vacuum-Packed Food

Vacuum-packed food is processed by removing the air around a food product sealed in a package. Bacon is an example of food packaged this way. (See *Exhibit 6I.*)

Exhibit 6I

Vacuum-Packed and *Sous Vide* Products

Sous Vide Food

Sous vide (pronounced *soo-veed*) is a French term meaning "under vacuum." Food processed by this method is vacuum-packed in individual pouches, partially or fully cooked, and then chilled. This food is then heated for service in the establishment.

Sous vide products may be received either refrigerated or frozen, depending on the manufacturer. (See *Exhibit 6I.*) Some frozen, precooked meals are packaged using this method.

Receiving MAP, Vacuum-Packed, and *Sous Vide* Food

Removing oxygen from packaged food can reduce or prevent the growth of some microorganisms that need oxygen to grow. However, the same conditions can promote the growth of microorganisms that thrive without oxygen, such as those that produce the botulism toxin.

For this reason, the FDA does not allow establishments to package food on-site using a MAP method except under certain conditions. To prepare *sous vide* or MAP food, operators must have a HACCP plan in place and limit the foods being packaged to those that cannot support the growth of *Clostridium botulinum.*

When you receive MAP, vacuum-packed, and *sous vide* food, use the following guidelines:

- Make sure the supplier has a HACCP plan in place.
- Refrigerated products must be delivered and stored at 41°F (5°C) or lower unless otherwise specified by the manufacturer.
- Frozen products should be frozen when they arrive.
- Reject packages with leaks.

- Reject products that appear slimy or have bubbles.
- Reject products that are an unacceptable color.
- Reject packages that contain product with an expired code date.

Dry and Canned Products

Dry and canned products seem to pose little threat to consumers. Most have a fairly long shelf life and are usually used long before they have a chance to spoil. However, canned products provide a good environment for the microorganisms that cause botulism. Dry food can be contaminated by a variety of sources that can cause foodborne illnesses.

Key Point

Reject dry food if there are moisture stains on the packaging.

Dry food must be kept dry. Most microorganisms need moisture to grow and multiply, which is why dry food has a much longer shelf life than fresh food. Check both outer cases and inner packaging for dampness or moisture. Reject the shipment if it is damp or shows signs of prior wetness (moisture stains).

Dry food often attracts pests. Because it can be stored at room temperature, dry food might not be sealed and stored as securely as fresh, refrigerated, or frozen food. Insects and rodents have an easier time getting into dry-food packages. Carefully inspect packages for holes, tears, or punctures. Check products themselves for signs of infestation. You can spot insects or insect eggs in cereal or flour by sprinkling some of the product on brown paper. Rodents leave signs such as chewed packages and droppings in and around cartons.

Key Point

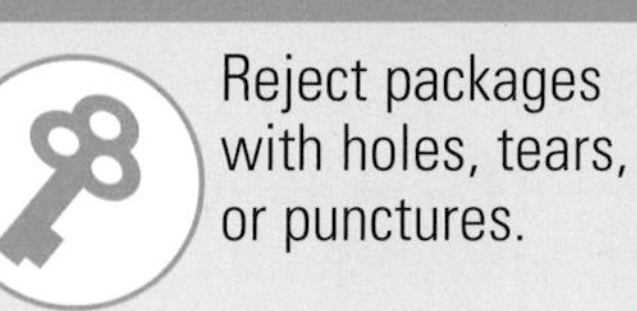

Reject packages with holes, tears, or punctures.

Use sight and smell to inspect dry food too. Off colors and odors, spots of mold, or a slimy appearance are signs of spoilage.

Canned food must also be checked carefully for damage. (See *Exhibit 6m.*) Check can exteriors first, looking for the following signs of contamination:

- **Swollen ends.** One or both ends of a can may bulge from gas produced by the presence of chemicals or the growth of foodborne bacteria inside. If one end bulges out when the other is pressed, discard it, because the can has not gone through the proper heat-treating process to eliminate foodborne microorganisms.

Exhibit 6m

Conditions for Rejecting Damaged Cans

Reject cans with swollen ends, rust, or dents.

- **Leaks and flawed seals.** If there are any leaks, flaws, or irregularities along the top or side seals, reject the can.
- **Rust.** If a can is rusted, the contents may be too old or the rust may have eaten holes in the can, which can lead to contamination.
- **Dents.** Do not accept cans with dents along side or top seams or dents large enough to make it impossible to open the can with a can opener. The seams may be broken. (Check with your regulatory agency regarding dented cans; some do not allow any dents.)

Health Alert

Never taste canned foods you are unsure of.

Any cans received without labels should be rejected. Once the exteriors of the cans have been checked, spot-check the contents. Any food that does not have a normal color, texture, or odor, that is foamy, or that contains a milky-colored liquid should be thrown out immediately. Never taste canned food you are unsure of. Botulism, a foodborne illness associated with canned food, is so dangerous that people have died from just tasting and spitting out contaminated food.

Ultra-High Temperature (UHT) Pasteurized and Aseptically Packaged Food

Some foods are heat-treated to kill microorganisms at very high temperatures in a process called **ultra-high temperature (UHT) pasteurization.** These foods are often also

aseptically packaged, sealed under sterile conditions to keep them from being contaminated. Examples include some puddings, juices, and creamers and milk products.

Once food has been UHT-pasteurized and aseptically packaged, it can be received and stored at room temperature. Once opened, however, it should be refrigerated at 41°F (5°C) or lower. Always follow manufacturers' directions for receiving and storing these products, and check packaging and seals to make sure they are intact. Reject the product if packaging is punctured or seals are broken.

Bakery Goods

It is common for establishments to regularly receive shipments of various types of bakery items. While safe handling depends upon the items received, it is always important to follow manufacturers' recommendations, especially regarding time and temperature control. Additionally, these items should be rejected if they have passed their expiration date or show signs of mold or pest damage.

Potentially Hazardous Hot Food

Occasionally, operators may receive a shipment of hot food. Hot food must be properly cooked as required by local or federal codes. *(See Chapter 8 for guidelines.)* If you purchase hot food, make sure the supplier has a HACCP plan or other means of documenting proper cooking methods and temperatures. Potentially hazardous hot food must be delivered at a temperature of 135°F (57°C) or higher. This food should be delivered in appropriate containers that can maintain these temperatures.

SUMMARY

Even though federal and state agencies regulate and monitor the production and transportation of food such as meats, poultry, seafood, eggs, dairy products, and canned goods, it is your responsibility to check the quality and safety of food that comes into your establishment.

Make sure suppliers are getting their products from approved sources—those that have been inspected and are in compliance with local, state, and federal law. Take steps to ensure that the suppliers you have chosen are reputable by asking other operators what their experiences have been with those suppliers.

Operators must plan delivery schedules so products can be handled promptly and correctly. Employees assigned to receive deliveries should be trained to inspect food properly, as well as to distinguish between products that are acceptable and those that are not. They should also be authorized to reject products that do not meet company standards and to sign for products that do.

All products arriving at the establishment should meet agreed-upon standards. Packaging should be clean and undamaged. Use-by dates should be current. Food should show no signs of mishandling.

Products must be delivered at the proper temperature. All products—especially meat, poultry, and fish—should be checked for proper color, texture, and odor. Live, molluscan shellfish and crustacea must be delivered alive. Eggs should be inspected for freshness and for dirty and cracked shells. Dairy products must be checked for freshness. Produce should be fresh and wholesome. Frozen food should be inspected for signs of thawing and refreezing. MAP, vacuum-packed, and *sous vide* food should not bubble or appear slimy, its packaging should be intact, and code dates should not be expired. Canned food must be carefully examined for signs of damage. Dry food should be inspected for pest infestation and moisture. Bakery goods should not be moldy, show signs of pest damage, and should not have passed their expiration date. Potentially hazardous hot food must be delivered at 135°F (57°C) or higher.

Apply Your Knowledge

1. What was done incorrectly?

For answers, please turn to the Answer Key.

A Case in Point 1

On Monday, a large food delivery arrived at the Sunnydale Nursing Home during the busy lunch hour. It included: cases of frozen ground beef patties, canned vegetables, frozen shrimp, fresh tomatoes, a case of potatoes, and fresh chicken.

Betty, the new assistant manager, thought the best thing to do was to put everything away and check it later, since she was very busy. She told Ed, in charge of receiving, to sign for the delivery and put the food into storage. Ed asked her if it would be better to ask the delivery driver to come back later. Since she needed the chicken for dinner, Betty asked Ed to accept the delivery now and went back to the front of the house.

Ed put the frozen shrimp and ground beef patties in the freezer and the fresh chicken in the refrigerator. Then he put the fresh tomatoes, potatoes, and canned vegetables in dry storage. When he was finished, he went back to work in the kitchen.

Apply Your Knowledge

❶ What was done wrong?

For answers, please turn to the Answer Key.

A Case in Point 2

ABC Seafood makes its usual Thursday afternoon delivery to The Fish House. John, a prep cook, is the only person in the kitchen when the driver rings the bell at the back door. The kitchen manager, who is in charge of receiving, is in a managers' meeting. The chef is out on an errand, and the rest of the kitchen staff is on break, though some are still in the restaurant.

John follows the driver onto the dock, where the driver unloads two crates of ice-packed fresh fish, bags of live mussels, two buckets of live oysters, a case of shucked oysters in plastic containers, and a case each of frozen shrimp and frozen lobster tails. John goes back into the kitchen to get a bimetallic stemmed thermometer and remembers to look in the chef's office for a copy of the order form. He takes both out to the dock and begins to inspect the shipment.

John checks the products against both the order sheet and the invoice, then begins to check product temperatures. First he checks the temperature of the shucked oysters by taking the cover off one container and inserting the thermometer stem into it. The thermometer reads 45°F (7°C), which worries him. He remembers being taught that refrigerated products should be 41°F (5°C) or lower. He decides not to say anything.

After wiping the thermometer stem on his apron, John checks the internal temperature of a whole fish packed in ice. Finally, he reaches into one of the buckets of live oysters with his hand to see if it feels cold. He notices a few mussels and oysters with open and broken shells. He removes those with broken shells, knowing the chef will not use them.

John records all his findings on the order sheet he took from the chef's office, signs for the delivery, and starts putting the products away. He first puts away the live shellfish, dumping the few still left in the refrigerator into the new containers to make room. Then he puts the shucked oysters and fresh fish in the refrigerator and the shrimp and lobster into the freezer.

Apply Your Knowledge

Discussion Questions

Use these questions to review the concepts presented in this chapter.

1. What are some general guidelines for receiving food safely?
2. What are proper methods for checking the temperature of fresh poultry delivered on ice and bulk milk? What should the temperature be for each?
3. What are three conditions that would result in rejecting a shipment of fresh poultry?
4. What four types of external damage to cans are cause for rejection?

For answers, please turn to the Answer Key.

Apply Your Knowledge

Use these questions to test your knowledge of the concepts presented in this chapter.

Multiple-Choice Study Questions

1. **Which is the most important factor in choosing a food supplier?**
   A. Its prices are the lowest.
   B. Its warehouse is close to your establishment.
   C. It offers a convenient delivery schedule.
   D. It has been inspected and is compliant with local, state, and federal law.

2. **Which of the following shipments should be rejected?**
   A. Beef that is bright, cherry red in color
   B. Lamb with flesh that is firm and springs back when touched
   C. Fish that arrives with sunken eyes
   D. Chicken received at 41˚F (5˚C)

3. **How should whole, fresh salmon be packaged for delivery and storage?**
   A. Layered with salt
   B. Vacuum-sealed
   C. Wrapped in dry, clean cloth
   D. Packed in self-draining, crushed ice

4. **Which condition would cause you to reject a shipment of live oysters?**
   A. Most of the oysters have closed shells.
   B. The oysters have a mild seaweed scent.
   C. Most of the oysters shells are open and do not close when they are tapped.
   D. The oysters are delivered with shellstock identification tags.

Continued on next page...

Apply Your Knowledge

Multiple-Choice Study Questions *continued*

5. A box of sirloin steaks carries both a USDA inspection stamp and a USDA-choice grading stamp. What do these stamps tell you?

A. The farm that supplied the beef uses only USDA-certified animal feed.

B. The meat and processing plant have met USDA standards, and the meat quality is acceptable.

C. The meat wholesaler meets USDA quality-grading standards.

D. The steaks are free of disease-causing microorganisms.

6. Which condition would cause you to reject a shipment of fresh chicken?

A. The flesh is firm and springs back when touched.

B. The wing tips are brown.

C. The grading stamp is missing.

D. The chicken is odorless.

7. Statements from a dairy supplier's sales brochure are listed below. Which statement should tell you not to hire this supplier?

A. We make our cheese with only the freshest unpasteurized milk.

B. From farm to you, our milk is kept at temperatures below 41°F (5°C).

C. Money-back guarantee—our prices are the lowest in the area.

D. We deliver according to your schedule and needs.

8. A shipment of eggs should be rejected for all of the following reasons *except*

A. the shells are cracked.

B. they have a sulfur smell.

C. they lack an inspection stamp.

D. the air temperature of the delivery truck was 45°F (7°C).

Continued on next page...

Apply Your Knowledge

Multiple-Choice Study Questions *continued*

9. Which of the following food does not have to be received at 41°F (5°C) or lower?
   - A. Beef
   - B. Live shellfish
   - C. Pork
   - D. Fish

10. All of the following would be grounds for rejecting a case of frozen food *except*
    - A. there are large ice crystals on the frozen food inside the case.
    - B. the outside of the case is water-stained.
    - C. the food in the box is frozen solid.
    - D. there is frozen liquid at the bottom of the case.

11. Which delivery should be rejected?
    - A. Several cans in a case of peaches have torn labels.
    - B. Several cans in a case of tomato soup have swollen ends.
    - C. A bag of oatmeal is delivered at 60°F (16°C).
    - D. A case of rice is missing a USDA inspection stamp.

12. You have just received a shipment of pumpkin pies. What would be grounds for rejecting the shipment?
    - A. The pies have been received frozen.
    - B. The pies have passed their expiration date.
    - C. The pies contain preservatives.
    - D. The pies have been received at 41°F (5°C).

For answers, please turn to the Answer Key.

ADDITIONAL RESOURCES

Books and Periodicals

First in, first out! The FIFO rule of storage. 2003. *Food Safety Illustrated.* 3 (2):3.

National Restaurant Association Educational Foundation. 1998. *A practical approach to HACCP: Coursebook.* Chicago: National Restaurant Association Educational Foundation.

Partnering with suppliers. 2002. *Food Safety Illustrated.* 2 (1):8–10.

Receiving foods. Caution is the key. 2003. *Food Safety Illustrated.* 3 (1):14.

Seafood safety. 2002. *Food Safety Illustrated.* 2 (3):14–15.

Web Sites

American Egg Board (AEB)

www.aeb.org

As the egg industry's promotional arm, the AEB is the U.S. egg producer's link to the consumer for communicating the value of the egg. This Web site offers a wealth of information about eggs and egg safety.

American Meat Institute

www.meatami.com

The American Meat Institute is a membership trade association representing the interests of the U.S. meat and poultry industry to the federal government, the media, and the customer. This Web site provides up-to-date information about what is happening in the meat and poultry processing industry.

Bunzl Distribution, Inc.

www.bunzldistribution.com

Bunzl Distribution is a supplier of disposable paper and plastic packaging supplies, as well as specialty items for service and packaging for the restaurant and foodservice industry.

Cooper-Atkins Corporation

www.cooperinstrument.com

Cooper-Atkins Corporation is a supplier of temperature, time, and humidity instruments for a variety of global markets. This Web site contains information on how to calibrate thermometers properly and maintain food temperatures during cooking, holding, cooling, etc.

FDA Center for Food Safety and Applied Nutrition (CFSAN)

www.cfsan.fda.gov

As the center within the FDA responsible for food safety and nutrition, CFSAN promotes and protects public health by researching and implementing guidelines, policies, and standards to ensure food is safe, nutritious, wholesome, and properly labeled. This Web site provides a wealth of information on food safety and sanitation, including corresponding guidelines, policies, and standards.

FDA Food Code

http://vm.cfsan.fda.gov/~dms/foodcode.html

As the basis for many local sanitation codes, as well as the basis for information in this coursebook, the FDA Food Code, available at this Web address, is a useful resource for information relating to food safety for the restaurant and foodservice industry.

FDA Seafood Information and Resources

http://vm.cfsan.fda.gov/seafood1.html

The FDA operates an oversight compliance program for fishery products under which responsibility for product safety, wholesomeness, identity, and economic integrity rests with the processor or importer, who must comply with regulations under the Federal Food, Drug, and Cosmetic (FD&C) Act. This Web site houses information on the seafood program, foodborne pathogens, and contaminants associated with seafood, and HACCP compliance.

Food Marketing Institute (FMI)

www.fmi.org

The FMI is a nonprofit association conducting programs in research, education, industry relations, and public affairs on behalf of its members, which include large multistore chains, small regional firms, and independent supermarkets.

International Fresh-Cut Produce Association (IFPA)

www.fresh-cuts.org

IFPA serves commercial producers, suppliers, and distributors of fresh-cut produce. It provides members with a forum for representation, education, technical support, networking, information, and resources. This site contains information about recent industry happenings and regulations, and provides guidance documents for purchasing, receiving, and storing fresh-cut produce.

The Mushroom Council
www.mushroomcouncil.com
This interesting and useful Web site from the Mushroom Council contains recipes and information about how to handle and store mushrooms to ensure safety and quality.

National Cattlemen's Beef Association (NCBA)
www.beef.org
NCBA is a consumer-focused, producer-directed organization representing the largest segment of the nation's food and fiber industry. This Web site, as well as a companion site for foodservice, www.beeffoodservice.com, provides excellent information on purchasing, cooking, and serving beef.

National Chicken Council
www.eatchicken.com
This Web site is brought to you by the National Chicken Council and the U.S. Poultry & Egg Association, which represent a wide spectrum of companies and individuals in the chicken business. This fun and interesting Web site contains chicken safety tips and nutritional and statistical information.

National Food Processors Association (NFPA)
www.nfpa-food.org
The NFPA is the voice of the food processing industry on scientific and public policy issues involving food safety, nutrition, technical and regulatory matters, and consumer affairs. Its Web site offers many resources on food safety and food processing for the entire food industry.

National Frozen Food Association (NFFA)
www.nffa.org
NFFA's mission is to promote the sales and consumption of frozen food through education training, research, sales planning, and menu development, and to provide a forum for industry dialogue. Publications and information on frozen foods and how to market them to consumers are available on this Web site.

National Pork Producers Council

www.nppc.org

Billing itself as the place for pork lovers, pork producers, and anyone who wants to learn more about pigs and pork, the National Pork Producers Council's Web site provides information on pork safety and quality, nutrition and industry information, and publications.

National Restaurant Association

www.restaurant.org

The National Restaurant Association is the leading business association for the restaurant industry. Together with the National Restaurant Association Educational Foundation, the Association's mission is to represent, educate, and promote the rapidly growing restaurant and foodservice industry. This Web site should be your starting place for all issues and concerns related to your establishment. This Web site has it all, from tips for running your establishment to vital data on your customers' spending habits.

National Turkey Federation

www.eatturkey.org

The National Turkey Federation advocates for all segments of the turkey industry, providing services and conducting activities which increase demand for its members' products by protecting and enhancing their ability to profitably provide wholesome, high-quality, nutritious products. Its Web site contains an area specifically for the restaurant and foodservice industry, including recipe and menu ideas, cooking demonstrations, and a reference on purchasing, preparing, and promoting turkey.

Produce Marketing Association

www.pma.com

The Produce Marketing Association is a not-for-profit trade association serving members who market fresh fruit, vegetables, and floral products worldwide. Its members are involved in the production, distribution, retail, and foodservice sectors of the industry. This Web site provides promotional ideas and information about hot topics in the produce industry.

SYSCO Corporation
www.sysco.com
Sysco is North America's leading marketer and distributor of food and foodservice products. In addition to providing information about Sysco brands and products, this Web site contains a recipe database, food safety information, and information about building your business.

Tyson Foods, Inc.
www.tyson.com
Tyson Foods is the world's largest poultry producer. This Web site is a great resource about chicken products for the restaurant and foodservice industry. It includes recipes and menu ideas, information on new products, and information about Tyson Foods' efficient distribution and transportation system.

USDA–FSIS
www.fsis.usda.gov
The Food Safety and Inspection Service (FSIS) is the public health agency of the USDA responsible for ensuring that the nation's commercial supply of meat, poultry, and egg products is safe, wholesome, and correctly labeled and packaged. This Web site contains a wealth of information on food safety relating to meat, poultry, and eggs.

U.S. Foodservice, Inc.
www.usfoodservice.com
U.S. Foodservice is one of the largest broadline foodservice distributors in the U.S., distributing food and related products to restaurants and institutional foodservice establishments across the country.

Apply Your Knowledge

Notes

Whole Berry
CRANBERRY SAUCE
SWEET RELISH
Nutrition Facts

7

The Flow of Food: Storage

Inside this chapter:

- General Storage Guidelines
- Types of Storage
- Storage Techniques
- Storing Specific Food

After completing this chapter, you should be able to:

- Properly label and date-mark refrigerated, frozen, and dry food prior to storage.
- Properly store refrigerated, frozen, dry, and canned food.
- Apply first in, first out (FIFC) practices as they relate to refrigerated, frozen, and dry storage areas.
- Properly store raw food to prevent cross-contamination.
- Identify temperature requirements for refrigerated and dry storage areas.
- Identify proper storage containers for refrigerated, frozen, and dry food.

Key Terms

First in, first out (FIFO)
Refrigerated storage
Frozen storage
Dry storage
Shelf life
Hygrometer

Apply Your Knowledge

Check to see how much you know about the concepts in this chapter. Use the page references provided to explore the topic in each question.

Test Your Food Safety Knowledge

1. **True or False:** Potato salad that has been prepared in-house and stored at 41°F (5°C) must be discarded after three days. *(See page 7-3)*
2. **True or False:** Food can be stored near chemicals as long as the chemicals are stored in sturdy, clearly labeled containers. *(See page 7-4.)*
3. **True or False:** Storing cans of stewed tomatoes at 65°F (18°C) is acceptable. *(See page 7-7.)*
4. **True or False:** Raw chicken must be stored below ready-to-eat food, such as pumpkin pie, if it is stored in the same walk-in refrigerator. *(See page 7-6.)*
5. **True or False:** If stored food has passed its expiration date, you should cook and serve it at once. *(See page 7-3.)*

For answers, please turn to the Answer Key.

INTRODUCTION

When food is stored improperly and not used in a timely manner, quality and safety suffer. Poor storage practices can cause food to spoil quickly with potentially serious results.

GENERAL STORAGE GUIDELINES

Every facility has a wide variety of products that need to be stored. A few general rules can be applied to most storage situations.

- **Label food.** All potentially hazardous, ready-to-eat food prepared on-site that has been held for longer than twenty-four hours must be labeled with either the date it was prepared, or the date it should be sold, consumed, or discarded. If an item has been previously cooked and stored and is later mixed with another food item to make a new dish, the label on the new dish must indicate the preparation or discard date for the previously cooked item. For example, if ground beef has been cooked and stored at 41°F (5°C) or lower, and later is used to make meat sauce, the meat sauce must be labeled with either the preparation or discard date of the ground beef.

Key Point

Rotate products in storage to ensure that the oldest are used first.

- **Rotate products to ensure that the oldest inventory is used first.** The **first in, first out (FIFO)** method is commonly used to ensure that refrigerated, frozen, and dry products are properly rotated during storage. By this method, a product's use-by, expiration, or preparation date is first identified. The products are then stored to ensure that the oldest are used first. One way to do this is to train employees to store products with the earliest use-by or expiration dates in front of products with later dates. Once shelved, those stored in front are used first. (See *Exhibit 7a*).

Exhibit 7a

Follow FIFO When Storing Food

One way to follow FIFO is to store products with the earliest use-by or expiration dates in front of products with later dates and use those first.

- **Discard food that has passed the manufacturer's expiration date.** All potentially hazardous, ready-to-eat food that has been prepared in-house can be stored for a maximum of seven days at 41°F (5°C) or lower before it must be discarded.
- **Establish a schedule to ensure that stored product is depleted on a regular basis.** If the product has not been sold or consumed by a pre-determined date, discard it.
- **Transfer food between containers properly.** If you take food out of its original package, put it in a clean, sanitized container and cover it. The new container must be labeled with the name of the food being stored and its original use-by or expiration date. Never use empty food containers to store chemicals or put food in empty chemical containers.
- **Keep potentially hazardous food out of the temperature danger zone.** Store deliveries as soon as they have been inspected. Take out only as much food as you can prepare at one time, and put prepared food away until needed. Properly cool and store cooked food as soon as it is no longer needed. *(See Chapter 8 for more information on cooling cooked food.)*
- **Check temperatures of stored food and storage areas.** Temperatures should be checked at the beginning of the shift. Many establishments use a preshift checklist to guide employees through this process.

Exhibit 7b

Improper Storage

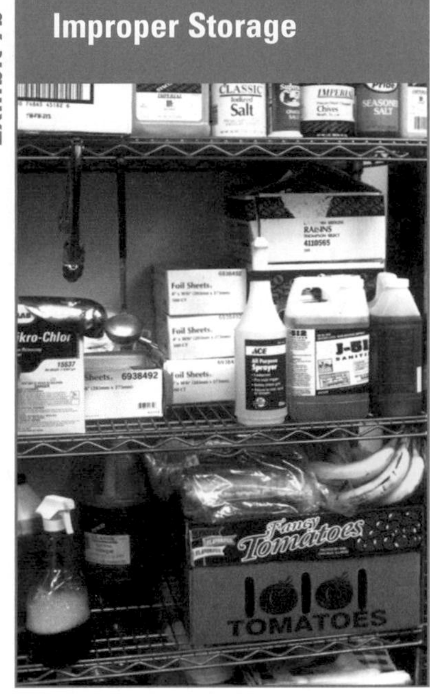

Never store food near chemicals or cleaning supplies.

- **Store food only in designated storage areas.** Do not store food products near chemicals or cleaning supplies (see *Exhibit 7b*); in restrooms, locker rooms, janitor closets, furnace rooms or vestibules; or under stairways or pipes of any kind. Food can easily be contaminated in any of these areas.
- **Keep all storage areas clean and dry.** Floors, walls, and shelving in refrigerators, freezers, dry storerooms, and heated holding cabinets should be properly cleaned on a regular basis. Clean up spills and leaks right away to keep them from contaminating other food.
- **Clean dollies, carts, transporters, and trays often.**

TYPES OF STORAGE

Most restaurants and foodservice establishments have several types of storage areas in their facilities. The most common include:

- **Refrigerated storage.** These areas are typically used to hold potentially hazardous food at 41°F (5°C) or lower. Refrigeration slows the growth of microorganisms and helps keep them from multiplying to levels high enough to cause illness. (Some jurisdictions allow food to be held at an internal temperature of 45°F [7°C] or lower. Check with the local regulatory agency for specific regulations.)
- **Frozen storage.** These areas are typically used to hold frozen food at 0°F (–18°C) or lower. Freezing does not kill all microorganisms, but it does slow their growth substantially.
- **Dry storage.** These areas are used to hold dry and canned food. To maintain the quality of this food, dry-storage areas should be kept at the appropriate temperature and humidity levels. Storerooms should be clean, well ventilated, and well lighted.

Managers should monitor the use of storage areas since improper storage practices can affect the safety of food. For example, an overstocked refrigerator may not be able to hold the proper temperature and may not allow stock to be rotated properly.

Key Point

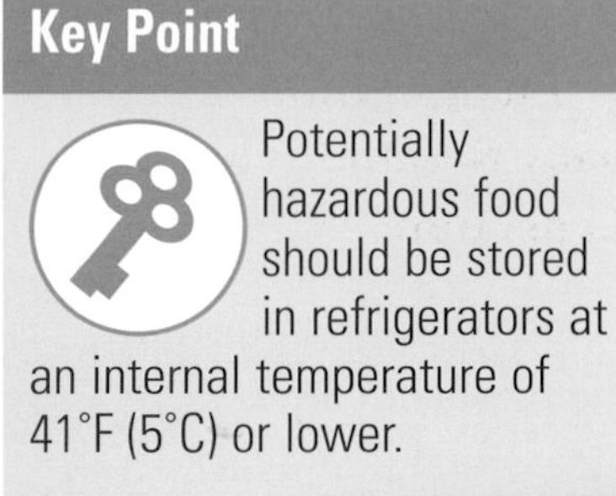

Potentially hazardous food should be stored in refrigerators at an internal temperature of 41°F (5°C) or lower.

Storage spaces should also be located to ease the flow of food through the operation and to prevent food contamination. They must be accessible to receiving, food-preparation, and cooking areas, but must be located so that food is stored away from warewashing and garbage areas.

STORAGE TECHNIQUES

A few commonsense rules apply to each of these storage areas. Make sure employees follow these rules to keep food safe.

Cross-Contamination

Store raw meat, poultry, and fish separately from cooked and ready-to-eat food whenever possible.

Refrigerated Storage

In general, the colder food is, the safer it is. Keeping food as cold as possible without freezing it also extends its **shelf life.** Ideal storage temperatures will vary depending on the food. Fruit and vegetables will freeze if stored at temperatures ideal for fish. Meat and poultry will have a shorter shelf life if stored at temperatures better suited for produce. If possible, store food such as meat and poultry in separate refrigerators to hold them at optimal temperatures. If this is impractical, store meat, poultry, fish, and dairy products in the coldest part of the unit, away from the door.

Exhibit 7c

Monitor Refrigerated Food Temperatures Regularly

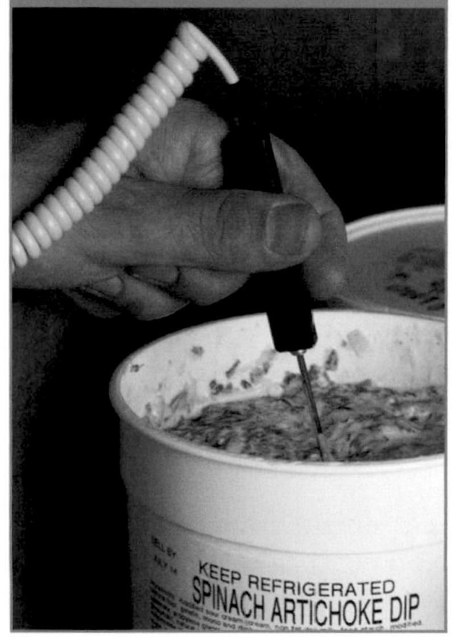

Randomly sample the internal temperature using a calibrated thermometer.

While there are many types of refrigeration equipment available to operators, from walk-in refrigerators to refrigerated drawers, some general guidelines apply when using all of them.

- **To hold food at a specific internal temperature, refrigerator air temperature usually must be at least 2°F (1°C) lower than the desired temperature**. For example, to hold poultry at an internal temperature of 41°F (5°C), the air temperature in the refrigerator should be at least 39°F (4°C). Use hanging thermometers in the warmest part of the refrigerator. Some units have a readout panel outside to check the temperature without opening the door. These should also be checked for accuracy. At least once during each shift, check the temperature of the unit.
- **Monitor food temperature regularly.** Randomly sample the internal temperature of food stored inside using a calibrated thermometer. (See *Exhibit 7c.*) You may also want to monitor food temperatures by using a product-mimicking device.

Exhibit 7d

Refrigerator Storage

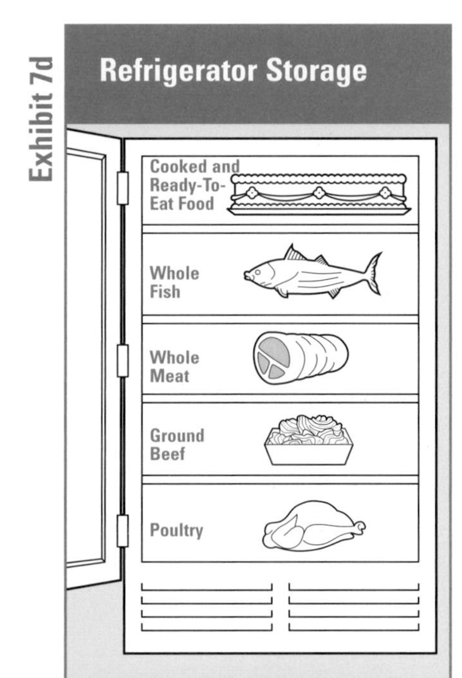

Recommended top-to-bottom storage of different raw food in the same refrigerator

Exhibit 7e

Improper Storage of Different Raw Products in the Same Refrigerator

Never store raw products this way.

- **Never place hot food in the refrigerator.** This can warm the interior enough to put other food in the temperature danger zone.
- **Do not overload the unit.** Storing too many products prevents good airflow and makes the unit work harder to stay cold.
- **Use open shelving.** Lining shelves with aluminum foil or paper restricts circulation of cold air in the unit.
- **Keep the refrigerator door closed as much as possible.** Frequent opening lets warm air inside, which can affect food safety and make the refrigerator work harder. Consider using cold curtains to help maintain walk-in refrigerator temperatures.
- **Wrap all food properly.** Leaving food uncovered can lead to cross-contamination.
- **Store raw meat, poultry, and fish separately from cooked and ready-to-eat food to prevent cross-contamination.**
- **Store cooked or ready-to-eat food above raw meat, poultry, and fish if these items are stored in the same unit.** This will prevent raw-product juices from dripping onto the prepared food and causing a foodborne illness. It is also recommended that raw meat, poultry, and fish be stored in the following top-to-bottom order in the refrigerator: whole fish, whole cuts of beef and pork, ground meats and fish, whole and ground poultry. This order is based upon the required minimum internal cooking temperature of each food. (See *Exhibits 7d* and *7e.*)

Frozen Storage

Some general guidelines for using freezers include:

- **Check unit temperatures regularly.** Use a calibrated thermometer to check the accuracy of hanging thermometers or unit readout.
- **Keep freezer temperatures at 0°F (–18°C) or lower unless the food you are storing requires a different temperature.**
- **Place frozen food deliveries in the freezer as soon as they have been inspected.** Clearly label the food, identifying the

package's contents, date of delivery, and use-by date, if there is one. Never hold frozen food at room temperature.

- **Use caution when placing food into a freezer.** Warm food can raise the temperature in the unit and partially thaw the food inside. Preparing several small batches of food is a better idea than freezing leftovers. Store food to allow good air circulation. Overloading a freezer makes it work harder and makes it more difficult to find and rotate food properly.
- **Defrost freezer units on a regular basis.** They will operate more efficiently when free of frost. Move contents to another freezer when defrosting, or use them immediately.
- **Open the unit as infrequently as possible.** Use cold curtains to help maintain temperatures.
- **Use frozen food as quickly as possible.**

Dry Storage

Dry food remains safe and retains quality if held in the right conditions. Moisture and heat are the biggest problems. The storeroom temperature should be between 50°F and 70°F (10°C and 21°C). Keep relative humidity around fifty to sixty percent. If high humidity is a problem, consider using a dehumidifier. When storing dry food, follow these guidelines:

- **Monitor temperature and humidity regularly.** Use a hanging thermometer to record temperatures and a **hygrometer** to measure relative humidity. Units are available that combine both instruments.
- **Store dry food away from walls and at least six inches off the floor.**
- **Keep dry food out of direct sunlight.**
- **Keep the area clean and dry.**
- **Make sure storerooms are well ventilated.** This will help keep temperature and humidity constant throughout the storage area.

STORING SPECIFIC FOOD

The general storage guidelines previously discussed apply to most food. However, certain types of food have special requirements.

Meat

- Store meat immediately after delivery and inspection in its own storage unit or in the coldest part of the refrigerator. Fresh meat must be held at an internal temperature of 41°F (5°C) or lower. Frozen meat should be stored at a temperature that will keep it frozen.
- Wrap raw cuts of meat, especially ground beef, so they are airtight. Meat will turn brown when exposed to air. Frozen meat should be wrapped in airtight, moisture-proof material or placed in containers to prevent freezer burn.
- Primal cuts, quarters, sides of raw meat, and slab bacon can be hung on clean, sanitized hooks or can be placed on sanitized racks. To prevent cross-contamination, do not store meat above any other food.
- Throw out any meat showing signs of spoilage.

Poultry

- Store raw, fresh poultry at an internal temperature of 41°F (5°C) or lower. Frozen poultry should be stored at temperatures that will keep it frozen. If it has been removed from its original packaging, place it in airtight containers or wrap it in airtight material.
- Ice-packed poultry can be stored in a refrigerator as is. Containers must be self-draining. Change the ice and sanitize the container often.

Fish

- Store fresh fish at an internal temperature of 41°F (5°C) or lower. Keep fillets and steaks in original packaging or wrap them in moisture-proof materials. Fresh, whole fish can be packed in flaked or crushed ice, and ice beds must be self-draining. Change the ice and clean and sanitize the container regularly.

Key Point

Fresh raw meat, poultry, and fish must be stored at an internal temperature of 41°F (5°C) or lower.

- Store frozen fish at temperatures that will keep it frozen. Wrap fish in moisture-proof packaging.
- Fish that will be served raw or partially cooked should be frozen by the processor, as follows, prior to shipment to kill any parasites that may be present.
  - –4°F (–20°C) or lower for seven days (168 hours) in a storage freezer; or
  - –31°F (–35°C) or lower for fifteen hours in a blast freezer.

The only exceptions are shellfish and certain species of tuna not susceptible to parasites. Ask your suppliers whether the fish you are purchasing meet these conditions.

Shellfish

- Store live shellfish in their containers at an ambient (air) temperature of 45°F (7°C) or lower. You can store live molluscan shellfish (clams, oysters, mussels, scallops) in a display tank under one of two conditions:
  - The tanks carry a sign stating that the shellfish are for display only.
  - You obtain a **variance** from the health department to serve the shellfish on display. A variance is a document from the regulatory agency authorizing a modification or waiver of a requirement. To obtain a variance, you must submit a HACCP-based plan showing the following:
    - The water in the tank will not come into contact with any other fish.
    - Using the tank will not affect product quality or safety.
    - You retain shellstock identification tags as required.

Time & Temperature

Fish to be served raw or partially cooked must be frozen by the processor to specific temperatures prior to shipment.

Eggs

- Eggs received at an air temperature of 45°F (7°C), in compliance with laws governing their shipment from suppliers, must be placed immediately upon receipt in refrigeration equipment capable of maintaining an ambient temperature of 45°F (7°C) or lower. Maintain constant temperature and humidity levels in refrigerators used to store eggs. Do not wash eggs before storing them since they are washed and sanitized at the packing facility.
- Use the FIFO method of stock rotation. Plan to use all eggs within four to five weeks of the packing date.
- Keep shell eggs in cold storage right up until the time they are used. Take out only as many eggs as are needed for immediate use.
- Do not combine cracked eggs in a bowl (pool them) unless you intend to use them right away or will hold them at 41°F (5°C) or lower.
- Store frozen egg products at temperatures that will keep them frozen.
- Store liquid eggs according to the manufacturer's recommendations.
- Dried egg products can be stored in a cool, dry storeroom away from light. Once they are reconstituted (mixed with water), store them in the refrigerator at 41°F (5°C) or lower.

Key Point

Store dairy products at 41°F (5°C) or lower.

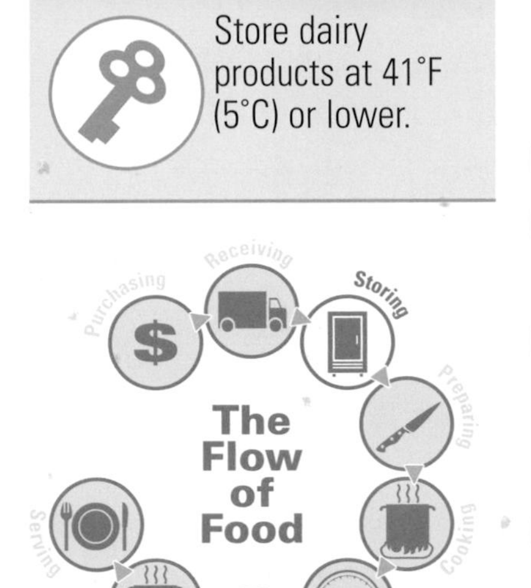

Dairy Products

- Store dairy products at 41°F (5°C) or lower.
- Frozen dairy products such as ice cream and frozen yogurt can be stored at 6°F to 10°F (–14°C to –12°C).
- Always use the FIFO method of stock rotation. Discard products if they have passed their use-by or expiration dates.

Fresh Produce

- Fruit and vegetables have various temperature requirements for storage. While many whole, raw fruit and vegetables can be stored at 41°F (5°C) or lower, not all will be stored at these temperatures. Whole, raw produce and raw, cut vegetables—

Key Point

While many whole, raw fruit and vegetables can be stored at 41°F (5°C) or lower, not all will be stored at these temperatures.

such as celery, carrots, and radishes—delivered packed in ice can be stored that way. The containers must be self-draining, and ice should be changed regularly.

- Fruit and vegetables kept in the refrigerator can dry out quickly. Keep the relative humidity at eighty-five to ninety-five percent.
- Although most produce can be stored in the refrigerator, avocados, bananas, pears, and tomatoes ripen best at room temperature.
- Most produce should not be washed before storage. Moisture promotes the growth of mold in many instances. Instead, wash produce before preparing or serving it.
- Store whole citrus fruit, hard-rind squash, eggplant, and root vegetables—such as potatoes, sweet potatoes, rutabagas, and onions—in a cool, dry storeroom. Temperatures of 60°F to 70°F (16°C to 21°C) are best. Make sure containers are well ventilated. Store onions away from other vegetables that might absorb odor.

MAP, Vacuum-Packed, and *Sous Vide* Food

- Always store MAP, vacuum-packed, and *sous vide* food at temperatures recommended by the manufacturer. Most should be stored at 41°F (5°C) or lower. Frozen MAP, vacuum-packed, and *sous vide* food should be stored at temperatures that will keep it frozen. Store and handle these products carefully.
- Vacuum packaging will not stop the growth of microorganisms that do not require oxygen to grow. MAP fish, for example, is especially susceptible to the growth of *Clostridium botulinum*. Discard product if the package is torn or slimy, if it contains excessive liquid, or if the product bubbles, indicating the possible growth of *Clostridium botulinum*.
- Always check the expiration date before using MAP, vacuum-packed, and *sous vide* products. Labels should clearly list contents, storage temperature, preparation instructions, and a use-by date.

- Operators who produce MAP food on-site must follow specific labeling rules. The FDA rules for processing MAP food on-site are very strict.

UHT and Aseptically Packaged Food

- Food that has been pasteurized at ultra-high temperatures (UHT) and aseptically packaged (the packaging is free of microorganisms) can be stored at room temperature. Since much of this food, such as milk and pudding, is served cold, you might want to store it in the refrigerator.
- Once opened, store UHT, aseptically packaged food in the refrigerator at 41°F (5°C) or lower.
- UHT products not aseptically packaged must be stored at an internal temperature of 41°F (5°C) or lower.

Canned Goods

Key Point

Storage temperatures higher than 70°F (21°C) may shorten the shelf life of canned goods.

- Store canned goods and other dry food at a temperature between 50°F and 70°F (10°C to 21°C). Even canned food spoils over time. Higher storage temperatures may shorten shelf life. Acidic food, such as canned tomatoes, does not last as long as food low in acid. The acid can also form pinholes in the metal over time.
- Keep storerooms dry. Too much moisture will cause cans to rust.
- Wipe cans clean with a sanitized cloth before opening them to help prevent dirt from falling into the contents of the can.

Dry Food

- Keep flour, cereal, and grain products such as pasta or crackers in airtight containers. They can quickly become stale in a humid room and can become moldy if there is too much moisture.
- Before using dry food, check containers or packages for damage from insects or rodents. Cereal and grain products are favorite targets for these pests.
- Salt and sugar, if stored in the right conditions, can be held almost indefinitely.

SUMMARY

When food is stored improperly, quality and safety will suffer. Although different food has different storage needs, some common rules apply. Food should be stored in designated areas and rotated to ensure that the oldest product is used first. It should also be stored in its original packaging. If food must be removed from its original packaging, wrap it in clean, moisture-proof materials or place it in clean and sanitized containers with tight-fitting lids. Make sure all packaging and containers are labeled with the name of the food being stored. All potentially hazardous, ready-to-eat food prepared on-site that has been held for longer than twenty-four hours must be labeled with either the date it was prepared, or the date it should be sold, consumed, or discarded. It can be stored for a maximum of seven days at 41°F (5°C) or lower before it must be discarded. Discard all food that has passed its manufacturer's expiration date. Check the temperatures of stored food and the storage area regularly, and keep these areas clean and dry to prevent contamination.

Hold potentially hazardous food in storage areas refrigerated at 41°F (5°C) or lower—temperatures that slow the growth of microorganisms. To hold food at the proper temperature, keep the air temperature of the unit at least 2°F (1°C) lower than the desired food temperature. Never place hot food in the refrigerator, which could raise the temperature inside. Do not line refrigerator shelves, overload the unit, or open the door too often. These practices make the unit work harder to keep food cold. If possible, raw meat, poultry, and fish should be stored separately from cooked and ready-to-eat food to prevent cross-contamination. If not, store these items below cooked or ready-to-eat food.

While freezing does not kill microorganisms, it does slow their growth substantially. Freezers should be kept at 0°F (–18°C) or lower unless the food being stored requires a different temperature. Unit temperatures should be checked often.

Dry storage areas should be kept at the appropriate temperature and humidity levels and should be clean and well ventilated to maintain food quality. Food in dry storage should be stored away from walls and at least six inches off the floor. Do not store food products near chemicals or cleaning supplies since food can easily become contaminated. Empty food containers should never be used to store chemicals.

Fresh meat, poultry, fish, and dairy products should be stored at 41°F (5°C) or lower. Fish and poultry can be stored under refrigeration in crushed ice as long as the containers are self-draining, the ice is changed, and the container is sanitized regularly. Eggs should be refrigerated at an ambient (air) temperature of 45°F (7°C) or lower right up until they are used.

Live, molluscan shellfish should be stored in their original containers at an ambient temperature of 45°F (7°C). Fresh produce has various temperature requirements for storage. Produce should not be washed before storage because it can promote mold growth. MAP, vacuum-packed, and *sous vide* food should be stored at temperatures recommended by the manufacturer. Packages should be checked for signs of contamination, including bubbling, excessive liquid, and tears and slime. Once opened, UHT and aseptically packaged foods should be stored in the refrigerator at 41°F (5°C) or lower. Dry and canned food should be stored at temperatures between 50°F and 70°F (10°C to 21°C).

Apply Your Knowledge

1. What storage errors were made?
2. What food items are at risk?

For answers, please turn to the Answer Key.

A Case in Point 1

On Monday afternoon, the kitchen staff at the Sunnydale Nursing Home was busy cleaning up after lunch and preparing for dinner. Pete, a kitchen assistant, put a large stockpot of hot, leftover vegetable soup in the refrigerator to cool. Angie, a cook, began deboning the chicken breasts that were stored earlier. When she finished, she put the chicken on an uncovered sheet pan and stored the pan in the refrigerator. She carefully placed the raw chicken on the top shelf, away from the hot soup. Next, Angie iced a carrot cake she had baked that morning. She put the carrot cake in the refrigerator on the shelf directly below the chicken breasts.

Apply Your Knowledge

A Case in Point 2

Anticipating a slow dinner service, Ed, the owner of Big City Diner, gave his kitchen manager the afternoon off. Some time later, Acme Distributors delivered an order of canned and dry food and supplies, including canned tomato sauce, canned soup, crackers, pasta, paper napkins, warewashing detergent and sanitizer.

After checking the order, Ed stacked the canned and dry food onto a dolly and wheeled it into the storeroom. There was an open case of tomato sauce on the shelf with one can left in it and a full case behind it. Ed removed the open case and replaced it with the case on the dolly. He put the single can in an open spot on another shelf. By shifting some boxes around, Ed was able to put the cases of soup, napkins, and pasta on shelves in front of cases already stored there. While moving cases, he accidentally knocked over a container of flour, spilling some on the floor. There was no room on the shelves for the crackers, so he left the case on the floor. He was sweating from exertion and wondered if the storeroom was too hot. He checked the hanging thermometer, which read 70°F (21°C).

Ed went back to the dock to get the detergent and sanitizer. He wheeled them into the storeroom and stacked them against the wall next to several cases of pasta. When he returned the dolly to the back door, a truck from Mike's Produce pulled up to deliver cases of potatoes and several bags of onions. Ed knew the produce room was overstocked, so he stored the produce in a space sometimes used for storage underneath the dining room stairs.

1. What did Ed do wrong?

For answers, please turn to the Answer Key.

Apply Your Knowledge

Use these questions to review the concepts presented in this chapter.

Discussion Questions

1. What is the recommended top-to-bottom order for storing the following food in the same refrigerator: raw trout, an uncooked beef roast, raw chicken, and raw ground beef?
2. How should food that has been taken out of its original package be stored?
3. What is the proper storage temperature for frozen dairy products?
4. What temperature should the air inside a walk-in refrigerator be to hold products at an internal temperature of 41°F (5°C)?
5. Explain the FIFO method of stock rotation.

For answers, please turn to the Answer Key.

Apply Your Knowledge

Use these questions to test your knowledge of the concepts presented in this chapter.

Multiple-Choice Study Questions

1. **At what internal temperature would stored ground beef most likely become unsafe to use?**
   A. 0°F (–17°C)
   B. 30°F (–1°C)
   C. 41°F (5°C)
   D. 60°F (16°C)

2. **Dry-storage rooms should be kept at**
   A. 35°F to 41°F (2°C to 5°C).
   B. 45°F to 50°F (7°C to 10°C).
   C. 50°F to 70°F (10°C to 21°C).
   D. 70°F to 80°F (21°C to 27°C).

3. **Under which condition could you use a tank to display live mussels you will be serving to customers?**
   A. You have obtained a variance from the health department.
   B. You clean the display tank at least once a month.
   C. You will mix them with other mussels that have not been on display.
   D. You have removed the shellstock identification tags as required by law.

4. **Which of the following is *not* a good storage practice?**
   A. Shelving food based on its expiration date
   B. Storing raw poultry at temperatures between 41°F and 135°F (5°C and 57°C)
   C. Storing live shellfish at an ambient temperature of 45°F (7°C) or lower
   D. Storing raw meat below ready-to-eat food.

5. **The first in, first out (FIFO) method helps ensure all of the following during storage *except***
   A. products are properly rotated.
   B. the oldest products are used first.
   C. prepared items are used before they expire.
   D. items that have passed their expiration date are used first.

Continued on next page...

Multiple-Choice Study Questions *continued*

6. When storing products using the FIFO method, the products with the earliest use-by dates should be
   A. stored in front of products with later use-by dates.
   B. stored behind products with later use-by dates.
   C. stored alongside products with later use-by dates.
   D. stored away from products with later use-by dates.

7. A restaurant that has prepared tuna salad can store it at 41°F (5°C) for a maximum of
   A. 1 day.
   B. 3 days.
   C. 7 days.
   D. 14 days.

8. To keep refrigerated food at an internal temperature of 41°F (5°C), the air temperature in the refrigerator should be at least
   A. 0°F (–18°C).
   B. 26°F (–3°C).
   C. 32°F (0°C).
   D. 39°F (4°C).

9. All of the following are incorrect storage practices *except*
   A. lining refrigerator shelving with aluminum foil.
   B. cooling hot food in a refrigerator.
   C. storing products with the earliest expiration dates in front of products with older dates.
   D. storing fresh poultry over ready-to-eat food.

10. Which storage practice is incorrect?
    A. Storing fresh lamb at 41°F (5°C)
    B. Storing eggs at room temperature
    C. Storing raw clams in their shipping crate at an ambient temperature of 45°F (7°C)
    D. Storing raw ground beef in its original packaging

For answers, please turn to the Answer Key.

ADDITIONAL RESOURCES

Books and Periodicals

Electronic monitoring systems. 2002. *Food Safety Illustrated.* 2 (3):18–19.

Organize your walk-in cooler. 2001. *Food Safety Illustrated.* 1 (4):13.

Seafood safety. 2002. *Food Safety Illustrated.* 2 (3):14–15.

Web Sites

American Egg Board

www.aeb.org

As the egg industry's promotional arm, the American Egg Board is the U.S. egg producer's link to the consumer for communicating the value of the egg. Its Web site offers a wealth of information about eggs and egg safety.

American Meat Institute

www.meatami.com

The American Meat Institute is a membership trade association representing the interests of the U.S. meat and poultry industry to the federal government, the media, and the customer. This Web site keeps you up-to-date on what is happening in the meat and poultry processing industry.

Cooper-Atkins Corporation

www.cooperinstrument.com

Cooper-Atkins Corporation is a supplier of temperature, time, and humidity instruments for a variety of global markets. This Web site contains information on how to calibrate thermometers properly and maintain food temperatures during cooking, holding, cooling, etc.

FDA Center for Food Safety and Applied Nutrition (CFSAN)

http://vm.cfsan.fda.gov

As the center within the FDA responsible for food safety and nutrition, CFSAN promotes and protects public health by researching and implementing guidelines, policies, and standards to ensure that food is safe, nutritious, wholesome, and properly labeled. This Web site provides a wealth of information on food safety and sanitation, including corresponding guidelines, policies, and standards.

FDA Food Code

http://vm.cfsan.fda.gov/~dms/foodcode.html

As the basis for many local sanitation codes, as well as the basis for information in this textbook, the FDA Food Code, available at this Web address, is a useful resource for information related to food safety for the restaurant and foodservice industry.

FDA Seafood Information and Resources

http://vm.cfsan.fda.gov/seafood1.html

The FDA operates an oversight compliance program for fishery products under which responsibility for product safety, wholesomeness, identity, and economic integrity rests with the processor or importer, who must comply with regulations under the Federal Food, Drug, and Cosmetic (FD&C) Act. This Web site houses information on the seafood program, foodborne pathogens and contaminants associated with seafood, and HACCP compliance.

International Fresh-Cut Produce Association (IFPA)

www.fresh-cuts.org

IFPA serves commercial producers, suppliers, and distributors of fresh-cut produce. It provides members with a forum for representation, education, technical support, networking, information, and resources. This site contains information about recent industry happenings and regulations, and provides guidance documents for purchasing, receiving, and storing fresh-cut produce.

The Mushroom Council

www.mushroomcouncil.com

This useful Web site from the Mushroom Council contains recipes and information about how to handle and store mushrooms to ensure safety and quality.

National Cattlemen's Beef Association (NCBA)

www.beef.org

NCBA is a consumer-focused, producer-directed organization representing the largest segment of the nation's food and fiber industry. This Web site, as well as its companion site for foodservice, **www.beeffoodservice.com,** provides excellent information on purchasing, cooking, and serving beef. Both sites also provide important food safety information.

National Chicken Council

www.eatchicken.com

This Web site is brought to you by the National Chicken Council and the U.S. Poultry & Egg Association, representing a wide spectrum of companies and individuals in the chicken business. This fun and interesting site contains chicken safety tips and nutritional and statistical information.

National Frozen Food Association (NFFA)

www.nffa.org

NFFA's mission is to promote the sales and consumption of frozen food through education, training, research, sales planning, and menu development, as well as to provide a forum for industry dialogue. Publications and information on frozen food and how to market it to consumers are available at this Web site.

National Pork Producers Council

www.nppc.org

Billing itself as the place for pork lovers, pork producers, and anyone who wants to learn more about pigs and pork, the National Pork Producers Council Web site provides information on pork safety and quality, nutrition and industry information, and publications.

National Restaurant Association

www.restaurant.org

The National Restaurant Association is the leading business association for the restaurant industry. Together with the National Restaurant Association Educational Foundation, the Association's mission is to represent, educate, and promote the rapidly growing restaurant and foodservice industry. This Web site should be your starting place for all issues and concerns related to your restaurant. This Web site has it all, from tips for running your establishment to vital data on your customers' spending habits.

National Turkey Federation

www.eatturkey.com

The National Turkey Federation advocates for all segments of the turkey industry, providing services and conducting activities that increase demand for its members' products by protecting and enhancing their ability to profitably provide wholesome, high-quality, nutritious products. Its Web site contains an area specifically for the restaurant and foodservice industry, including recipe and menu ideas, cooking demonstrations, and a reference on purchasing, preparing, and promoting turkey.

Produce Marketing Association

www.pma.com

The Produce Marketing Association is a not-for-profit trade association serving members who market fresh fruit, vegetables, and floral products worldwide. Its members are involved in the production, distribution, retail, and foodservice sectors of the industry. This Web site provides information on hot topics in the produce industry and promotional ideas.

Tyson Foods, Inc.

www.tyson.com

Tyson Foods is the world's largest poultry producer. This Web site is a great resource about chicken products for the restaurant and foodservice industry. It includes recipes and menu ideas, information on new products, and information about Tyson Foods' efficient distribution and transportation system.

USDA–FSIS

www.fsis.usda.gov

The Food Safety and Inspection Service (FSIS) is the public health agency of the USDA responsible for ensuring that the nation's commercial supply of meat, poultry, and egg products is safe, wholesome, and correctly labeled and packaged. This Web site contains a wealth of information on food safety relating to meat, poultry, and eggs.

8

The Flow of Food: Preparation

Inside this chapter:

- Preparation
- Preparing Specific Food
- Cooking Food
- Cooking Requirements for Specific Food
- Cooling Food
- Reheating Potentially Hazardous Food

After completing this chapter, you should be able to:

- Identify proper methods for thawing foods.
- Identify the minimum internal cooking time and temperatures for potentially hazardous foods.
- Identify the proper procedure for cooking potentially hazardous food in a microwave.
- Identify methods and time and temperature requirements for cooling cooked food.
- Identify time and temperature requirements for reheating cooked, potentially hazardous food.
- Identify methods for preventing contamination and time and temperature abuse when preparing food.
- Recognize the importance of informing consumers of risks when serving raw or undercooked food.

Key Terms

Slacking
Minimum internal temperature
Two-stage cooling
Ice-water bath
Cold paddle

Apply Your Knowledge

Check to see how much you know about the concepts in this chapter. Use the page references provided to explore the topic in each question.

Test Your Food Safety Knowledge

1. **True or False:** Ground beef should be cooked to a minimum internal temperature of 140°F (60°C). *(See page 8-14.)*
2. **True or False:** Fish cooked in a microwave must be heated to145°F (63°C). *(See page 8-18.)*
3. **True or False:** Cooked potentially hazardous food must be cooled from 135°F to 70°F (57°C to 21°C) within four hours and from 70°F to 41°F (21°C to 5°C) or lower within an additional two hours. *(See page 8-18.)*
4. **True or False:** When potentially hazardous food is reheated for hot-holding, it must be heated to 155°F (68°C) for fifteen seconds within two hours. *(See page 8-21.)*
5. **True or False:** It is acceptable to thaw a beef roast at room temperature. *(See page 8-3.)*

For answers, please turn to the Answer Key.

INTRODUCTION

Once food has been received and stored safely, it is essential that it be prepared, cooked, cooled, and reheated with just as much care. It is at these points in the flow of food that the risk of cross-contamination and time-temperature abuse is greatest.

PREPARATION

Controlling time and temperature and preventing cross-contamination will help keep food safe during preparation. For many types of food, this begins with thawing.

Thawing Food Properly

Often the first step in food preparation is thawing what you intend to cook. If frozen food is thawed improperly, foodborne microorganisms can rapidly grow to unsafe levels.

Exhibit 8a

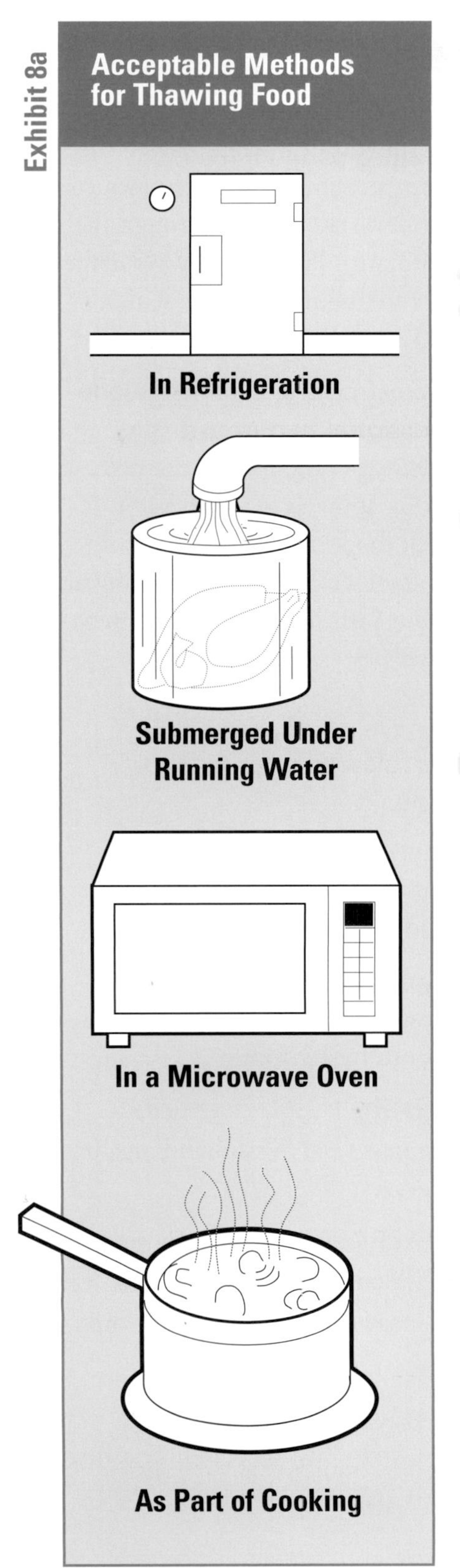

Suppose a cook wants to prepare a twenty-pound frozen turkey. The cook is in a hurry, so instead of putting the turkey in the refrigerator he puts it in a pan on a worktable to thaw overnight. The air temperature in the kitchen is about 72°F (22°C).

There are two problems with this method of thawing. First, when the turkey begins to thaw, the skin and outer layers are exposed to the temperature danger zone even though the turkey's core is still frozen. Microorganisms may multiply to a point where cooking the turkey may not be sufficient to kill them. The cook may end up serving a potentially dangerous dish to customers.

Second, microorganisms on the turkey may contaminate everything in the area: the cook's hands, the worktable, the pan, the cutting board, knives, and other utensils. If the area and equipment are not completely cleaned and sanitized before the cook starts another task, microorganisms could spread throughout the entire kitchen.

There are only four acceptable ways to thaw potentially hazardous, frozen food safely. (See *Exhibit 8a.*)

- **Thaw food in the refrigerator at 41°F (5°C) or lower.** This method requires advance planning. A twenty-pound turkey can take several days to thaw completely in the refrigerator. Employees need to take food out of frozen storage far enough in advance to make sure there is adequate thawed product on hand when it is needed.
- **Submerge the food under running potable water at a temperature of 70°F (21°C) or lower.** The water has to flow fast enough to wash loose food particles into the overflow drain. Make sure the thawed product does not drip water onto other products or food-contact surfaces. Clean and sanitize the sink and work area before and after thawing food this way.
- **Thaw food in a microwave oven if it will be cooked immediately afterward.** Microwave thawing can actually start cooking the product, so do not use this method unless you intend to continue cooking the food immediately. Large items such as roasts or turkeys do not thaw well in the microwave.

- **Thaw food as part of the cooking process as long as the product reaches the required minimum internal cooking temperature.** Frozen hamburger patties, for example, can go straight from the freezer onto a grill without being thawed first. Frozen chicken can go straight into a deep fryer. These products cook quickly enough from the frozen state to pass through the temperature danger zone without harm. However, always make sure you verify the final internal cooking temperature with a calibrated thermometer.

Some frozen food may be slacked before cooking. **Slacking** is the process of gradually thawing frozen food in preparation for deep frying, allowing even heating during cooking. For example, you might allow a large block of frozen spinach to warm from –10°F (–23°C) to 25°F (–4°C). Slack food just before you cook it. Do not let food get any warmer than 41°F (5°C). If slacking at room temperature is permitted in your jurisdiction, a system must be in place to ensure the product does not exceed 41°F (5°C).

PREPARING SPECIFIC FOOD

Meat, Fish, and Poultry

The sources of most cross-contamination in an operation are raw meat, poultry, and seafood. Your staff should follow safe procedures when handling these products.

- **Use clean and sanitized work areas, cutting boards, knives, and utensils.** Prepare raw meat, poultry, and seafood separately or at a different time from fresh produce.
- **Be sure hands are washed properly.** If gloves are used, change them before starting each new task. Wash hands again before putting on a new pair of gloves.
- **Remove from the refrigerator only as much product as can be prepared at one time.** When cubing beef for stew, for example, take out and cube one roast and refrigerate it before taking out another roast.
- **Put raw, prepared meat away or cook it as quickly as possible.** If put back into storage, wrap raw, prepared meat first or put it in covered containers that are clearly labeled. Never

store it above cooked or ready-to-eat food, such as fresh produce, because raw meat juices may drip and cross-contaminate other food.

Salads Containing Potentially Hazardous Food

Chicken, tuna, egg, pasta, and potato salads all have been known to cause foodborne-illness outbreaks. These salads are typically made from food that can easily support the rapid growth of microorganisms. Since they usually will not be cooked after preparation, there is no chance to kill microorganisms that could have been introduced. Therefore, care must be taken to both control time and temperature and prevent cross-contamination when preparing these salads. Handle all ingredients with care and follow these precautions:

- Make sure leftover meat and poultry have been properly cooked, held, cooled, and stored. Leftover meat and poultry should be discarded after seven days if stored in the refrigerator at 41°F (5°C) or lower (four days if stored at 45°F [7°C]). Salads made with these items should be discarded when this storage period expires.
- Make sure foodhandlers wash their hands before preparing food and, if required by local health codes, have them wear clean, single-use gloves.
- Prepare other ingredients away from raw meat, seafood, and poultry. Always prepare raw vegetables using clean, sanitized equipment and utensils.
- Leave food in the refrigerator until all ingredients are ready to be mixed. For example, do not take out chicken and leave it on the prep table while you chop the vegetables.
- Consider chilling all ingredients before making salads. Refrigerate unopened cans of tuna and jars of mayonnaise the day before you make tuna salad, for example. Consider chilling mixing bowls and utensils too. Taking these steps will provide additional barriers that might prevent the growth of harmful microorganisms that could be present.

Exhibit 8b

Salads Containing Potentially Hazardous Food

Do not let large amounts of these types of food sit out at room temperature.

- Prepare food in small batches. This will prevent large amounts of food from sitting out at room temperature for long periods of time. (See *Exhibit 8b.*) You might want to set a working time limit of twenty minutes and make sure you take out only enough ingredients to use within that time period. When time is up, refrigerate what you have made, then get ingredients to make more.

Eggs and Egg Mixtures

Historically, the contents of whole, clean, uncracked shell eggs were considered free of bacteria. Now it is known that, in rare cases, a certain bacteria, *Salmonella* Enteritidis, can be found inside eggs. *Salmonella* Enteritidis can live inside a laying hen and can be deposited in the white of an egg before the shell is formed. Although only a small number of eggs produced in the U.S. are likely to carry this bacteria, all untreated shell eggs are considered to be a potentially hazardous food because they are able to support the rapid growth of microorganisms. As a consequence, eggs should be purchased fresh and should be stored and handled properly. Eggs should be kept at an ambient (air) temperature of 45°F (7°C) or lower until immediately before use. Special handling precautions for eggs are described below.

- Use clean bowls, whisks, blenders, and other equipment when working with eggs and other potentially hazardous food items. To prevent cross-contamination, promptly clean and sanitize all equipment and utensils used with eggs. Place freshly made batters or sauces in clean containers. Never reuse containers that held raw eggs, even to replenish the same recipe.
- Food containing eggs that receive little or no cooking—such as Caesar salad dressing, meringue, mayonnaise, hollandaise sauce, and eggnog—should be prepared with great care. Follow the directions precisely and monitor temperatures using a thermometer. Many jurisdictions now require operators to post a notice to consumers when raw eggs are used as an ingredient in these types of food. The use of pasteurized shell eggs or pasteurized egg products may be a safer alternative when preparing these types of dishes.
- Eggs that are cracked open and combined in a container (pooled eggs) must be handled with special care because

bacteria present in one egg can be spread to the rest. Cross-contamination from one pooled egg mixture to another may also be a hazard if the food is not prepared or handled properly. Pooled eggs, if used, must be cooked promptly after mixing or stored at 41°F (5°C) or lower. Containers that held pooled eggs must be washed and sanitized before being used again. There should be no carryover of egg residue from one food batch to another, even if it is the same recipe.

- Federal public health officials require that operations serving highly susceptible populations—such as those in nursing homes and hospitals—use only pasteurized shell eggs or pasteurized egg products. This is recommended for all types of eggs and egg dishes served to these people and is required when liquid egg mixtures are prepared, when eggs are cooked and held for service, and when egg-containing food will receive little or no cooking.

Batters and Breading

Key Point

Batters and breading can be potentially hazardous and face the same risks of time-temperature abuse and cross-contamination as other food.

If prepared with eggs or milk, batters and breading are at risk from time-temperature abuse and cross-contamination, so they should be handled with care. In some cases, it may be better to buy frozen breaded items that can be taken directly from the freezer and then cooked thoroughly in the oven or fryer. If you prefer to make breaded or battered food from scratch, follow some simple rules:

- Consider making batters with pasteurized shell eggs or egg products instead of raw shell eggs whenever possible.
- Prepare batters in small batches and keep what you do not need in a covered container in the refrigerator. Using small amounts prevents time-temperature abuse of both the batter and the food being coated.
- When making a batch of batter or breading for later use, store it in the refrigerator at 41°F (5°C) or lower. Do not combine it with other batches already stored in the refrigerator.
- If you are breading food to cook later, put the breaded food in the refrigerator as quickly as possible.
- Cook battered and breaded food thoroughly. The coating acts as an insulator, which can prevent food from being thoroughly

cooked. When deep-frying food, make sure the temperature of the oil recovers before loading each batch. Overloading the basket also slows cooking time, which means product could be removed from the fryer before it is thoroughly cooked. Be sure to monitor oil and food temperatures using calibrated thermometers, and watch cooking time.

- Throw out any used batter or breading left over after each shift. Whenever a potentially hazardous food is dipped in batter or dredged in breading, consider the batter or breading contaminated. Never use a batter or breading for more than one product. For example, dipping raw chicken into batter then using the same batter for onion rings may contaminate the onion rings and possibly make someone sick.

Fruit and Vegetables

For the most part, fresh fruit and vegetables are not as likely to carry pathogens as are some other food. But viruses such as hepatitis A, bacteria such as *Listeria* spp., and parasites such as *Cryptosporidium parvum* can survive on produce, especially cut produce. The risk from such microorganisms can be minimized or eliminated by simple preparation safeguards.

- Start with a clean, sanitized work space. Prepare vegetables away from raw meat, poultry, and eggs, as well as from cooked and ready-to-eat food, to prevent cross-contamination.
- Surfaces that come in contact with raw meat—such as hands, gloves, knives, prep tables, or cutting boards—should never make contact with food that will be eaten raw, such as salad or fruit, unless they have been cleaned and sanitized.
- Wash fruit and vegetables thoroughly under running water to remove dirt or other contaminants just before cutting, combining with other ingredients, or cooking. Failure to wash fruit before cutting it may result in contamination of the fruit's interior.
- Pay particular attention to leafy items such as lettuce and spinach because dirt and microorganisms can get into inner leaves. Remove the outer leaves, pull lettuce and spinach completely apart, and rinse thoroughly.
- Cut away bruised or damaged areas when preparing fruit and vegetables.

Cross-Contamination

When preparing raw vegetables, start with a clean, sanitized work space. Prepare vegetables away from raw meat, poultry, and eggs, and from cooked and ready-to-eat food.

Reprinted with permission from Tony Soluri and Charlie Trotter

- Hold cut melons at 41°F (5°C) or lower, since they are potentially hazardous food.
- Do not add sulfites to food products.
- If your establishment primarily serves high-risk populations, do not serve raw seed sprouts.

Fresh Juice

Many establishments prepare fresh fruit and vegetable juices on-site for their customers. However, if the juice is packaged on-site for sale at a later time, the establishment must have a variance from the regulatory agency, and the juice must be treated (i.e. pasteurized) according to an approved HACCP plan or be labeled with the following phrase: *Warning: This product has not been pasteurized and therefore may contain harmful bacteria that can cause serious illness in children, the elderly, and people with weakened immune systems.* Federal public health officials require that establishments serving high-risk populations only serve fresh juice that has been treated to eliminate pathogens that can cause foodborne illness.

Cross-Contamination

Store ice scoops outside of the ice machine in a sanitary, protected location.

Ice

People often forget that ice is also a food. It is used to chill beverages and to chill or dilute food. In Chapter 7, you learned how to store products on ice, such as poultry, fish, fruit, and vegetables. Many establishments also use ice to chill food on display, such as canned beverages or fruit. Any time ice has been used to cool food in this way, you may not reuse it as a food.

Ice used as a food, or used to chill other food, must always be made with potable water (water that is safe to drink). Remember that ice can become contaminated just as easily as other food. When transferring ice from an ice machine to an ice bin or to a display, use a clean, sanitized scoop and container. Store ice scoops outside of the ice machine in a sanitary, protected location. Never hold ice in containers that have been used to store raw meat, poultry, fish, or chemicals.

Preparation Practices that Require a Variance

Some preparation methods require a variance from the regulatory agency. A variance is required whenever an establishment

- smokes or uses food additives as a method of food preservation.
- cures food.
- custom-processes animals for personal use.
- packages food using a reduced-oxygen packaging method (MAP, vacuum-packaging).

COOKING FOOD

Bacterial Growth

While cooking can kill microorganisms, it cannot destroy the spores or toxins they might create. Handling food safely before and after it is cooked will prevent microorganisms from growing or producing spores or toxins.

Whatever cooking method you choose, temperature is key to killing microorganisms that can cause foodborne illness. The food's internal temperature must reach a certain level for a specific amount of time before you can be sure all foodborne microorganisms have been sufficiently reduced in number. The only way to be certain a food has reached the required minimum internal cooking temperature is to check it using a calibrated thermometer.

The **minimum internal temperature** at which foodborne microorganisms are destroyed varies from product to product. Minimum standards have been developed for most cooked food and are included in local and state health codes. (See *Exhibit 8c.*)

It is important to remember that potentially hazardous food—such as meat, eggs, seafood, and poultry—should be cooked to the minimum internal temperatures specified in this chapter unless otherwise ordered by the customer. Potentially hazardous food not cooked to these temperatures—including over-easy eggs, raw oysters, sashimi, and rare hamburgers—generally do not pose an unacceptable risk of foodborne illness to the healthy customer. However, if a customer is in a group at risk for foodborne illness *(see Chapter 1)*, consuming raw or undercooked, potentially hazardous food could possibly increase the risk of illness, sometimes seriously.

Exhibit 8c

Minimum Internal Cooking Temperatures

Product	Temperatures
Poultry (whole or ground) (duck, chicken, turkey) **Stuffing, stuffed meat, and dishes that include previously cooked potentially hazardous ingredients**	165°F (74°C) for 15 seconds
Ground meats (beef, pork, or other meat or fish)	155°F (68°C) for 15 seconds
Injected meats (including brined ham and flavor-injected roasts)	155°F (68°C) for 15 seconds
Pork, beef, veal, lamb	**Steaks/Chops:** Cook to an internal temperature of 145°F (63°C) for 15 seconds **Roasts:** 145°F (63°C) for 4 minutes
Fish	145°F (63°C) for 15 seconds
Fresh shell eggs for immediate service	145°F (63°C) for 15 seconds
Commercially processed ready-to-eat foods held for service	135°F (57°C)
Eggs, poultry, fish, meat cooked in a microwave oven	165°F (74°C); let food stand for 2 minutes after cooking

Adapted from the FDA Food Code

In all cases, high-risk customers should be advised of the potential risk, if they ask about or specifically request undercooked food or any potentially hazardous food (or ingredient) that is raw or not fully cooked. They may want to consult with a physician before regularly consuming these types of food. Additionally, check with your regulatory agency for specific requirements.

Here are general guidelines to follow when cooking:

- **Specify cooking time and required minimum internal cooking temperature in all recipes.**
- **Use properly calibrated thermometers with a suitable size probe, accurate to within ±2°F (±1°C), to measure**

food temperatures. Check internal temperatures in several places, in the thickest part of the food. Clean and sanitize the thermometer after each use.

- **Avoid overloading ovens, fryers, and other cooking equipment.** The equipment or oil temperature might drop, and the food might not cook properly. Crowding food also might lead to cross-contamination.
- **Let the cooking equipment's temperature recover between batches.**
- **Use utensils or gloves to handle food after cooking, and taste food correctly to avoid cross-contamination.** The safest and most sanitary way to taste food is to ladle a small amount into a dish. Taste the food in the dish with a clean utensil. When finished, remove the dish and utensil from the area and have them cleaned and sanitized.

Exhibit 8d

Checking the Temperature of Poultry

Poultry should be cooked to an internal temperature of 165°F (74°C) or higher for fifteen seconds.

COOKING REQUIREMENTS FOR SPECIFIC FOOD

Poultry, Stuffing, and Stuffed Meats

Poultry, stuffing, and stuffed meats should be cooked to an internal temperature of 165°F (74°C) or higher for fifteen seconds. (See *Exhibit 8d.*) Poultry tends to have more types and higher counts of microorganisms than other meat and, therefore, should be cooked more thoroughly.

Stuffing poses a hazard for several reasons. It can be made with potentially hazardous food, such as eggs or oysters, that needs to be fully cooked. It also acts as insulation, preventing heat from reaching the center of meat or poultry. For example, when poultry is stuffed, the thighs, wings, and even the breast can reach temperatures of 165°F (74°C), while the stuffing may still be at an unsafe temperature. If the poultry is taken out of the oven too soon, microorganisms in the stuffing will not be killed, creating a potential hazard.

For this reason, stuffing should be cooked separately, particularly when cooking whole, large birds or large cuts of meat. Smaller cuts of meat, such as pork tenderloins or veal chops, may be stuffed before cooking, but verify that the internal temperature of both the meat and the stuffing has reached 165°F (74°C) for at least fifteen seconds.

Dishes That Include Potentially Hazardous Ingredients

When cooking dishes that include previously cooked, potentially hazardous ingredients, such as the ground beef in a meat sauce, these ingredients must be cooked to 165°F (74°C) for at least fifteen seconds within two hours.

When cooking dishes that include raw, potentially hazardous ingredients, these ingredients must be cooked to their required minimum internal temperature. For example, when cooking jambalaya, you must ensure that the raw shrimp reaches its required minimum internal temperature of 145°F (63°C) for at least fifteen seconds.

Exhibit 8e

Checking the Temperature of Pork

Pork, such as chops or medallions of tenderloin, should be cooked to an internal temperature of 145°F (63°C) or higher for fifteen seconds.

Pork

Cook pork, such as chops or medallions of tenderloin, to an internal temperature of 145°F (63°C) or higher for fifteen seconds. (See *Exhibit 8e.*) This temperature is high enough to destroy any *Trichinella* spp. larvae that might have contaminated the pork. Pork roasts, on the other hand, must hold the same

Exhibit 8f

Alternative Minimum Internal Temperatures for Cooking Beef and Pork Roasts

Temperature	Hold for (in minutes)
130°F (54°C)	112
131°F (55°C)	89
133°F (56°C)	56
135°F (57°C)	36
136°F (58°C)	28
138°F (59°C)	18
140°F (60°C)	12
142°F (61°C)	8
144°F (62°C)	5
145°F (63°C)	4

Chart: Adapted from the FDA Food Code

internal temperature for at least four minutes. Depending on the type of roast used, however, pork can be cooked at different internal temperatures. (See *Exhibit 8f.*)

Beef

Steaks must reach and hold an internal temperature of at least 145°F (63°C) for fifteen seconds. Roasts, on the other hand, must hold the same internal temperature for at least four minutes. Depending on the type of roast used, however, beef can be cooked at different internal temperatures. (See *Exhibit 8f.*)

Ground Meats

Ground beef, pork, and other meat or fish must be cooked to an internal temperature of 155°F (68°C) for at least fifteen seconds. They may also be cooked according to the alternative cooking temperatures indicated in *Exhibit 8g.* Most whole-muscle cuts of meat are likely to have microorganisms only on their surface. When meat is ground, microorganisms on the surface are mixed throughout the product. To make sure microorganisms are destroyed, thorough cooking is a must. Check with your local or state regulatory agency for additional requirements.

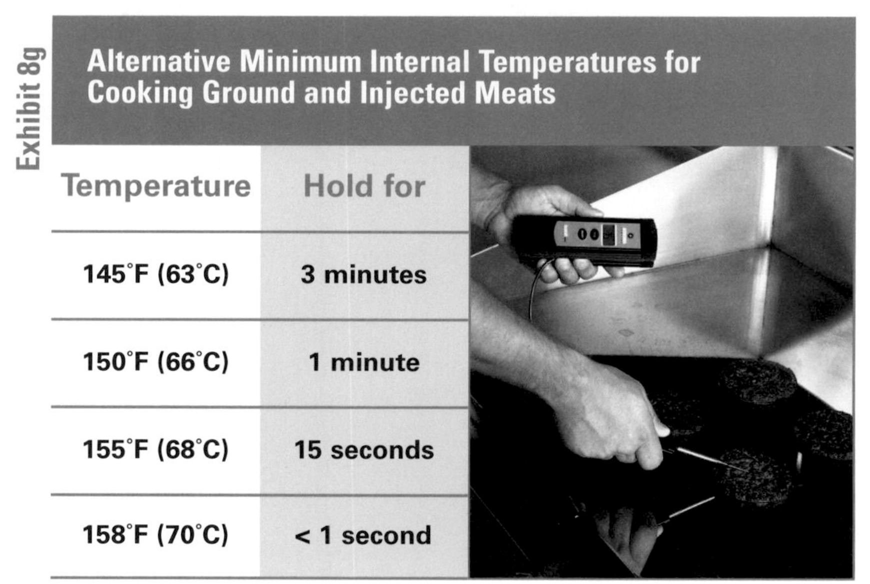

Exhibit 8g

Alternative Minimum Internal Temperatures for Cooking Ground and Injected Meats

Temperature	Hold for
145°F (63°C)	3 minutes
150°F (66°C)	1 minute
155°F (68°C)	15 seconds
158°F (70°C)	< 1 second

Chart: Adapted from the FDA Food Code Photo: Courtesy of Cooper-Atkins Corporation

Injected Meats

When meats are injected, foodborne microorganisms on the surface can be carried into the interior. For this reason, injected meats, such as brined ham or flavor-injected beef roasts, must be cooked to an internal temperature of 155°F (68°C). They may also be cooked according to the alternative cooking temperatures indicated in *Exhibit 8g.*

Game and Ratites

Commercially raised and inspected game animals—such as elk, deer, bison, and rabbit—can be cooked in much the same way as beef. Small cuts, such as steaks, can be cooked to a minimum internal temperature of 145°F (63°C) or higher for fifteen seconds. Ground meat must be cooked to an internal temperature of 155°F (68°C) or higher for fifteen seconds. Stuffed meats should be cooked to 165°F (74°C) or higher for fifteen seconds. Roasts can be cooked in the same way and to the same temperatures as beef and pork roasts.

Although they are birds, ratites (ostrich, emu, and rhea) are not cooked the same way as poultry. They will have a metallic taste if cooked to internal temperatures higher than 165°F (74°C). Ratites are fully cooked when their internal temperature reaches 155°F (68°C) for fifteen seconds. If portions or cutlets of these

Exhibit 8h

Checking the Temperature of Fish

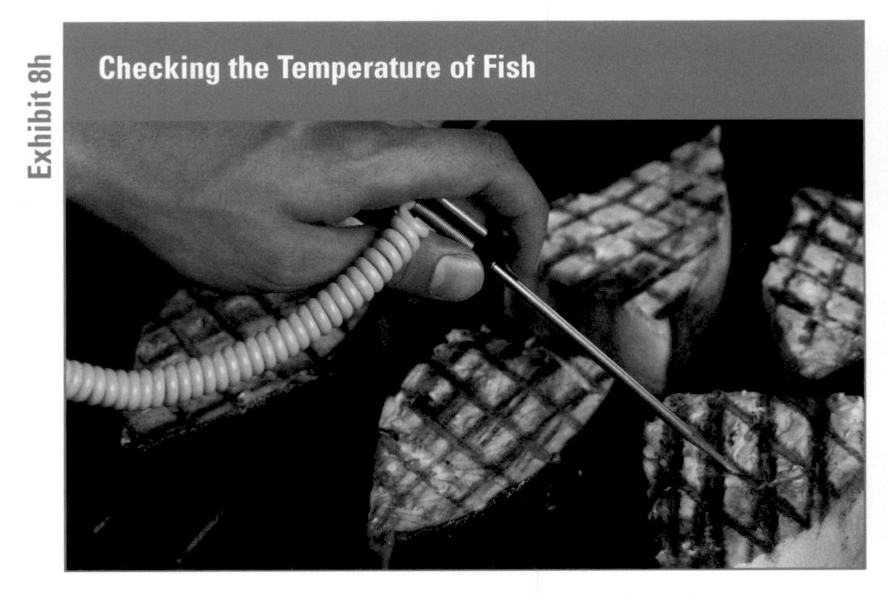

Cook fish to an internal temperature of 145°F (63°C) or higher for fifteen seconds.

birds are stuffed, however, cook them to an internal temperature of 165°F (74°C) for fifteen seconds.

Fish

Cook fish to an internal temperature of 145°F (63°C) or higher for fifteen seconds. (See *Exhibit 8h.*) Stuffed fish should be cooked to 165°F (74°C) or higher for fifteen seconds. This is also true for stuffing containing fish. If fish has been ground, chopped, or minced, it should be cooked to an internal temperature of 155°F (68°C) or higher for fifteen seconds.

Key Point

Shell eggs cooked to order should be cooked to 145°F (63°C) or higher for fifteen seconds.

Eggs and Egg Mixtures

In general, shell eggs cooked for immediate service should be cooked to 145°F (63°C) or higher for fifteen seconds. When eggs are cooked this way, the white is set and the yolk begins to thicken. Properly cooked scrambled eggs are firm, with no visible liquid. To hold eggs for later service, cook them to 155°F (68°C) or higher for fifteen seconds, then hold them at 135°F (57°C).

When cooking eggs, remove from storage only as many eggs as you need for immediate use. Never stack egg trays (flats) near the grill or stove.

Vegetables

Although most vegetables can be eaten raw, many are cooked before they are served. When cooking fruit or vegetables for hot-holding, cook them to 135°F (57°C) or higher. Cooked vegetables must never be left out or held at room temperature. Hold cooked vegetable dishes at 135°F (57°C) or higher, since they are potentially hazardous food.

Tea

Key Point

Never hold brewed tea at room temperature for more than one day.

Dry tea leaves contain low levels of bacteria, yeast, and mold (like most plant-derived food). Using improper brewing temperatures to prepare tea and storing it at room temperature for long periods of time can cause these microorganisms to grow to high levels. Improperly cleaned and sanitized equipment can also promote growth of microorganisms. When handling tea, follow these recommendations:

- Brew only as much tea as you reasonably expect to sell within a few hours.
- Never hold brewed tea at room temperature for more than one day. Discard any unused tea at the end of the day.
- To protect tea flavor and avoid microbial contamination and growth, clean and sanitize tea brewing, storage, and dispensing equipment at least once a day. Equipment should be disassembled, washed, rinsed, and sanitized. Urn spigots should be replaced at the end of each day with freshly cleaned and sanitized ones.
- For any brewing method, use a thermometer to make sure brewing water in your equipment meets the specified temperature.
  - 195°F (91°C) for automatic iced tea and automatic coffee machine equipment. Tea leaves should remain in contact with the water for a minimum of one minute.
  - 175°F (80°C) minimum when using the traditional steeping method. Tea leaves must be exposed to the water for approximately five minutes by this method.

Microwave Cooking

Microwave ovens tend to cook food more unevenly than other methods of cooking. For this reason, there are special rules for using microwave ovens to cook meat, poultry, and fish.

- Cover food to prevent the surface from drying out.
- Rotate or stir food halfway through the cooking process to distribute heat more evenly.
- Let food stand for at least two minutes after cooking to let product temperature equalize.

Eggs, poultry, fish, and meat cooked in a microwave oven must be heated to 165°F (74°C) or higher. Check the internal temperature of the food in several places to make sure it is cooked through.

Key Point

Eggs, poultry, fish, and meat cooked in a microwave oven must be heated to 165°F (74°C) or higher.

Commercially Processed, Ready-to-Eat Food

Commercially processed, ready-to-eat food that will be hot-held for service must be cooked to at least 135°F (57°C). This includes items such as cheese sticks, deep-fried vegetables, chicken wings, etc.

COOLING FOOD

You have already seen how important it is to keep food out of the temperature danger zone. When cooked food will not be served immediately, it is essential to hold it properly or to cool it as quickly as possible.

The FDA Food Code recommends **two-stage cooling**. Cooked food must be cooled from 135°F (57°C) to 70°F (21°C) within two hours and from 70°F (21°C) to 41°F (5°C) or lower in an additional four hours, for a total cooling time of six hours. (Some jurisdictions use one-stage cooling, by which food must be cooled to 41°F [5°C] or lower in less than four hours.)

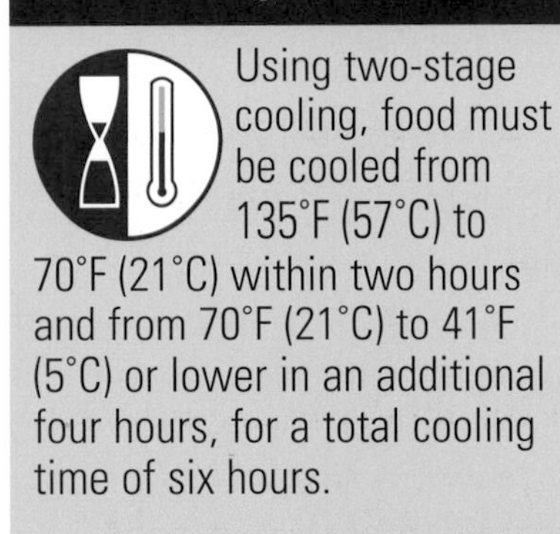

Time & Temperature

Using two-stage cooling, food must be cooled from 135°F (57°C) to 70°F (21°C) within two hours and from 70°F (21°C) to 41°F (5°C) or lower in an additional four hours, for a total cooling time of six hours.

Keep in mind this is a two-stage process (two hours plus four hours). While you have learned that microorganisms grow in the temperature danger zone, the temperature range between 125°F (52°C) and 70°F (21°C) is ideal for the growth of pathogenic microorganisms. Food must pass through this temperature range

quickly during cooling in order to minimize the growth of these microorganisms. Because only two hours are allowed to cool food from 135°F (57°C) to 70°F (21°C), the two-stage cooling process passes potentially hazardous food through this temperature range quickly—and safely.

When using two-stage cooling, if the food has not reached 70°F (21°C) within two hours, the food must be discarded or properly reheated. Reheat the food to 165°F (74°C) for fifteen seconds within two hours and then cool it properly.

Methods for Cooling Food

While common sense may suggest that the quickest way to cool food is to put it in the refrigerator, it is not. Refrigerators are designed to keep cold food cold. They usually do not have the capacity to cool hot food quickly.

In general, the thickness or density of food is the biggest factor in how quickly it cools. The denser the food product, the more slowly it cools. For example, refried beans take longer to cool than vegetable broth since the beans are more dense.

The container in which food is stored also affects how fast it will cool. Stainless steel transfers heat from food faster than plastic. Shallow pans allow the heat from food to disperse faster than deep ones.

There are a number of methods you can use for cooling food quickly and safely. (See *Exhibit 8i.*)

- **Reduce the quantity or size of the food you are cooling.** Cut large food items into smaller pieces, or divide large containers of food into smaller containers or shallow pans.
- **Use ice-water baths to bring food temperature down quickly.** After dividing food into smaller quantities, put the pots or pans into a sink or large pot filled with ice water. (See *Exhibit 8j* on the next page.)
- **Use blast chillers to cool food before placing it into refrigeration.** Since blast chillers are now more affordable, you might consider one for your establishment.
- **If properly equipped, steam-jacketed kettles can be used to cool food by simply running cold water through the jacket.**

Exhibit 8j

Cooling Food with an Ice-Water Bath

Ice-water baths can be used to cool food quickly.

- **Stir food as it cools.** This allows the product to cool faster and more evenly. Some manufacturers make plastic paddles you can fill with water and freeze. Stirring food products with these **cold paddles** cools food quickly. (See *Exhibit 8k*.) Make sure all utensils used to stir food are cleaned and sanitized between each use.
- **Add ice or cool water as an ingredient.** This method works for recipes that require water as an ingredient, such as soup or stew. The recipe can initially be prepared with less water than is required. Cold water or ice can then be added after cooking to cool the product and to provide the remaining water required by the recipe.

Once food has been cooled to at least 70°F (21°C), the refrigerator can handle it more easily. Food will continue to chill faster if you do the following:

- **Keep food in shallow, stainless-steel pans.** You may want to put dense foods, like chili or stew, into pans that are two inches deep. Thinner liquids can be stored in three-inch pans.
- **Always place pans on top shelves in the cooler.** Leave them uncovered while they continue to cool if they are protected from overhead contaminants. Otherwise, cover them loosely with foil or plastic wrap.

Exhibit 8k

Using a Cold Paddle to Cool Food

Stirring products with cold paddles chills food quickly.

- **Position pans so air can circulate around them.** This will help transfer heat from the pan.

All cooked food should be stored in containers labeled with the date the food was prepared and stored.

REHEATING POTENTIALLY HAZARDOUS FOOD

When previously cooked, potentially hazardous food is reheated for hot-holding, take it through the temperature danger zone again as quickly as possible. Food must be reheated to an internal temperature of 165°F (74°C) for fifteen seconds, within two hours. If food has not reached 165°F (74°C) for fifteen seconds within two hours, it must be discarded. When using the microwave oven to reheat previously cooked food, follow the same rules used for microwave cooking. Cover the product. Rotate or stir it midway through cooking. Allow it to stand for two minutes. Check its internal temperature in several places to see if it has reached at least 165°F (74°C).

Food that is reheated for immediate service to a customer, such as a roast beef sandwich, may be served at any temperature as long as the food was cooked properly.

SUMMARY

To protect food during preparation, you must handle it safely. The keys are time and temperature control and the prevention of cross-contamination.

Thaw frozen food in the refrigerator, under cool running water, in a microwave oven, or as part of the cooking process. Never thaw food at room temperature. Have employees prepare food in small batches, use chilled utensils and bowls, and record product temperatures and preparation times.

Cooking can reduce the number of microorganisms in food to safe levels. To ensure that microorganisms are destroyed, food must be cooked to required minimum internal temperatures for a specific amount of time. These temperatures vary from product to product. Cooking does not kill the spores or toxins some microorganisms produce. That is why it is so important to inspect product once it arrives and handle it safely during preparation.

Once food is cooked, it should be served as quickly as possible. If it is going to be stored and served later, it must be cooled rapidly. Cooked food must be cooled from 135°F (57°C) to 70°F (21°C) within two hours and from 70°F (21°C) to 41°F (5°C) or lower in an additional four hours (for a total cooling time of six hours), unless otherwise required by your local health code. Placing large containers of hot food into the refrigerator can put all other stored food in danger. Methods for cooling large quantities of cooked food quickly include dividing it into smaller portions, putting it in shallow stainless steel pans, using an ice-water bath or blast chiller, and stirring it often with cold paddles. When the food is cold enough, store it properly in the refrigerator.

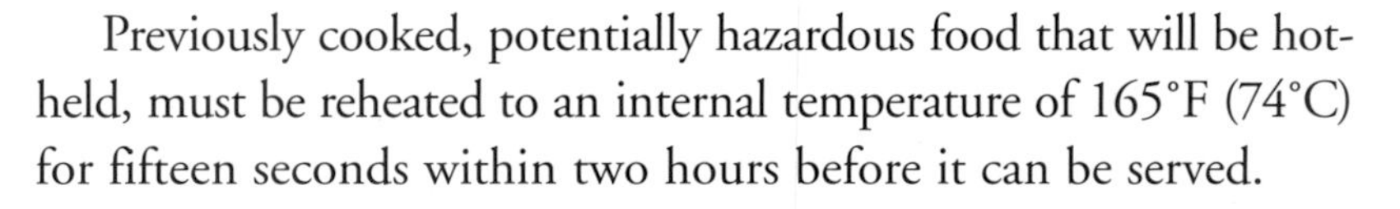

Previously cooked, potentially hazardous food that will be hot-held, must be reheated to an internal temperature of 165°F (74°C) for fifteen seconds within two hours before it can be served.

Apply Your Knowledge

1. What did John do wrong?

For answers, please turn to the Answer Key.

A Case in Point 1

On Friday, John went to work at The Fish House knowing he had a lot to do. After changing clothes and punching in, he took a case of frozen raw shrimp out of the freezer. To thaw it quickly, he put the frozen shrimp into the prep sink and turned on the hot water. While waiting for the shrimp to thaw, John took several fresh, whole fish out of the walk-in refrigerator. He brought them back to the prep area and began to clean and filet them. When he finished, he put the fillets in a pan and returned them to the walk-in refrigerator. He rinsed off the boning knife and cutting board in the sink, and wiped off the worktable with a dish towel.

Next, John transferred the shrimp from the sink to the worktable using a large collander. On the cutting board, he peeled, deveined, and butterflied the shrimp using the boning knife. He put the prepared shrimp in a covered container in the refrigerator, then started preparing fresh produce.

Apply Your Knowledge

❶ What did Angie do wrong?

For answers, please turn to the Answer Key.

A Case in Point 2

By 7:30 P.M., all the residents at Sunnydale Nursing Home had eaten dinner. As she began cleaning up, Angie realized she had a lot of chicken breasts left over. Betty, the new assistant manager, had forgotten to inform Angie that several residents were going to a local festival and would miss dinner.

"No problem," Angie thought. "We can use the leftover chicken to make chicken salad."

Angie left the chicken breasts in a pan on the prep table while she started putting other food away and cleaning up the kitchen. At 9:45 P.M., when everything else was clean, she put her hand over the pan of chicken breasts and decided they were cool enough to handle. She covered the pan with plastic wrap, and put it in the refrigerator.

Three days later, she came in to work on the early shift. Angie decided to make chicken salad from the leftover chicken breasts. After she hung up her coat and put on her apron, Angie took all the ingredients she needed for chicken salad out of the refrigerator and put them on a worktable. Then she started breakfast.

First, she cracked three dozen eggs into a large bowl, added some milk, and set the bowl near the stove. Then she took bacon out of the refrigerator and put it on the worktable next to the chicken salad ingredients. She peeled off strips of bacon onto a sheet pan and put the pan into the oven. After wiping her hands on her apron, she went back to the stove to whisk the eggs and pour them onto the griddle. When they were almost done, Angie scooped the scrambled eggs into a hotel pan and put it in the steam table.

As soon as breakfast was cooked, Angie went back to the prep table to wash and cut up celery and cut up the chicken for chicken salad.

Apply Your Knowledge

Use these questions to review the concepts presented in this chapter.

Discussion Questions

1. What are the required minimum internal cooking temperatures for poultry, fish, pork, and ground beef?
2. What are four proper methods for thawing food?
3. What methods can be used to cool cooked food?
4. What are the steps for properly cooking food in a microwave oven?

For answers, please turn to the Answer Key.

Apply Your Knowledge

Use these questions to test your knowledge of the concepts presented in this chapter.

Multiple-Choice Study Questions

1. **Beef stew must be cooled from 135°F (57°C) to 70°F (21°C) within ____ hours and from 70°F (21°C) to 41°F (5°C) or lower in an additional ____ hours.**
   A. 4, 2
   B. 2, 4
   C. 3, 2
   D. 2, 3

2. **Which of the following is *not* a safe method for thawing frozen food?**
   A. Thawing it by submerging it under running water at 70°F (21°C) or lower
   B. Thawing it in the microwave and cooking it immediately afterward
   C. Thawing it at room temperature
   D. Thawing it in the refrigerator overnight

3. **Stuffed pork chops must be cooked to a minimum internal temperature of**
   A. 135°F (57°C) for fifteen seconds.
   B. 145°F (63°C) for fifteen seconds.
   C. 155°F (68°C) for fifteen seconds.
   D. 165°F (74°C) for fifteen seconds.

4. **All of the following practices can help prevent time and temperature abuse *except***
   A. thawing food in a refrigerator at 41°F (5°C).
   B. chilling all ingredients used to make tuna salad.
   C. leaving food in the refrigerator until all ingredients are ready to be mixed.
   D. thawing steaks in a microwave and promptly refrigerating them for later use.

5. **Meat, poultry, and fish cooked in a microwave must be heated to at least**
   A. 140°F (60°C).
   B. 145°F (63°C).
   C. 155°F (68°C).
   D. 165°F (74°C).

Continued on next page...

Apply Your Knowledge

Multiple-Choice Study Questions *continued*

6. **All of the following practices can help prevent cross-contamination during food preparation *except***
   A. preparing food in small batches.
   B. throwing out unused batter or breading after each shift.
   C. preparing raw meat at a different time than fresh produce.
   D. cleaning and sanitizing pooled egg containers between batches.

7. **What is the proper way to cool a large stockpot of clam chowder?**
   A. Allow the stockpot to cool at room temperature.
   B. Put the hot stockpot into the walk-in refrigerator to cool.
   C. Divide the clam chowder into smaller containers and place them in an ice-water bath.
   D. Put the hot stockpot into the walk-in freezer to cool.

8. **When reheating potentially hazardous food for hot-holding, reheat the food to**
   A. 135°F (57°C) for fifteen seconds within two hours.
   B. 145°F (63°C) for fifteen seconds within two hours.
   C. 155°F (68°C) for fifteen seconds within two hours.
   D. 165°F (74°C) for fifteen seconds within two hours.

Continued on next page...

Apply Your Knowledge — Multiple-Choice Study Questions *continued*

9. Which of the following food has been safely cooked?
   A. A hamburger cooked to an internal temperature of 135˚F (57˚C) for fifteen seconds
   B. A pork chop cooked to an internal temperature of 145˚F (63˚C) for fifteen seconds
   C. A whole turkey cooked to an internal temperature of 155˚F (68˚C) for fifteen seconds
   D. A fish fillet cooked to an internal temperature of 135˚F (57˚C) for fifteen seconds

10. Shell eggs that will be cooked and held for later service must be cooked to an internal temperature of
    A. 140˚F (60˚C) for fifteen seconds.
    B. 145˚F (63˚C) for fifteen seconds.
    C. 155˚F (68˚C) for fifteen seconds.
    D. 165˚F (74˚C) for fifteen seconds.

For answers, please turn to the Answer Key.

ADDITIONAL RESOURCES

Books and Periodicals

A new take on tea: Lipton's patented brewed iced tea system raises iced tea safety and quality. 2003. *Food Safety Illustrated.* 3 (1):18.

Cooking temperature chart. 2003. *Food Safety Illustrated.* 3 (1):3.

Four ways to thaw…safely. 2001. *Food Safety Illustrated.* 1 (4):3.

Golden rules for ground beef. 2001. *Food Safety Illustrated.* 1 (2):14.

How to handle chicken. 2002. *Food Safety Illustrated.* 2 (1):13.

How to "handle" ice safely. 2002. *Food Safety Illustrated.* 2 (3):13.

How to organize your prep area. 2002. *Food Safety Illustrated.* 2 (1):14.

Laconi, D. 1995. *Fundamentals of professional food preparation: A laboratory text workbook.* New York: Wiley.

Proper use of gloves. 2003. *Food Safety Illustrated.* 3 (1):13.

Seafood safety. 2002. *Food Safety Illustrated.* 2 (3):14–15.

Web Sites

American Egg Board

www.aeb.org

As the egg industry's promotional arm, the American Egg Board is the U.S. egg producer's link to the consumer for communicating the value of the egg. Its Web site offers a wealth of information about eggs and egg safety.

American Meat Institute

www.meatami.com

The American Meat Institute is a membership trade association representing the interests of the U.S. meat and poultry industry to the federal government, the media, and the customer. This Web site keeps you up-to-date on what is happening in the meat and poultry processing industry.

Cooper-Atkins Corporation
www.cooperinstrument.com
Cooper-Atkins Corporation is a supplier of temperature, time, and humidity instruments for a variety of global markets. This Web site contains information on how to calibrate thermometers properly and maintain food temperatures during cooking, holding, cooling, etc.

Ecolab, Inc.
www.ecolab.com
An excellent source of information on housekeeping and sanitation supplies for the restaurant and foodservice industry from the world's leading sanitation product supplier.

FDA Center for Food Safety and Applied Nutrition (CFSAN)
http://vm.cfsan.fda.gov
As the center within the FDA responsible for food safety and nutrition, CFSAN promotes and protects public health by researching and implementing guidelines, policies, and standards to ensure that food is safe, nutritious, wholesome, and properly labeled. This Web site provides information on food safety and sanitation, including corresponding guidelines, policies, and standards.

FDA Food Code
http://vm.cfsan.fda.gov/~dms/foodcode.html
As the basis for many local sanitation codes, as well as the basis for information in this textbook, the FDA Food Code, available at this Web address, is a useful resource for information related to food safety for the restaurant and foodservice industry.

FDA Seafood Information and Resources
http://vm.cfsan.fda.gov/seafood1.html
The FDA operates an oversight compliance program for fishery products under which responsibility for product safety, wholesomeness, identity, and economic integrity rests with the processor or importer, who must comply with regulations under the Federal Food, Drug, and Cosmetic (FD&C) Act. This Web site houses information on the seafood program, foodborne pathogens and contaminants associated with seafood, and HACCP compliance.

International Food Information Council (IFIC) Foundation

www.ific.org

IFIC collects and disseminates scientific information on food safety, nutrition, and health. They work with an extensive roster of scientific experts to help translate research into understandable and useful information for opinion leaders and, ultimately, consumers. This Web site provides easy-to-understand information on foodborne illness and health-related stories circulating in the news.

International Fresh-Cut Produce Association (IFPA)

www.fresh-cuts.org

IFPA serves commercial producers, suppliers, and distributors of fresh-cut produce. It provides members with a forum for representation, education, technical support, networking, information, and resources. This site contains information on recent industry happenings and regulations, and provides guidance documents for purchasing, receiving, and storing fresh-cut produce.

KatchAll/San Jamar

www.sanjamar.com

KatchAll is an industry leader in innovative food safety products for the restaurant and foodservice industry. The Web site contains information on all KatchAll products that control time and temperature and help avoid cross-contamination.

National Cattlemen's Beef Association (NCBA)

www.beef.org

NCBA is a consumer-focused, producer-directed organization representing the largest segment of the nation's food and fiber industry. This Web site, as well as its companion site for foodservice, www.beeffoodservice.com, provides excellent information on purchasing, cooking, and serving beef. Both sites also provide important food safety information.

National Chicken Council

www.eatchicken.com

This Web site is brought to you by the National Chicken Council and the U.S. Poultry & Egg Association, representing a wide spectrum of companies and individuals in the chicken business. This fun and interesting site contains chicken safety tips and nutritional and statistical information.

National Frozen Food Association (NFFA)

www.nffa.org

NFFA's mission is to promote the sales and consumption of frozen food through education, training, research, sales planning, and menu development, as well as provide a forum for industry dialogue. Publications and information on frozen food and how to market it to consumers are available at this Web site.

National Pork Producers Council

www.nppc.org

Billing itself as the place for pork lovers, pork producers, and anyone who wants to learn more about pigs and pork, the National Pork Producers Council Web site provides information on pork safety and quality, nutrition and industry information, and publications.

National Restaurant Association

www.restaurant.org

The National Restaurant Association is the leading business association for the restaurant industry. Together with the National Restaurant Association Educational Foundation, the Association's mission is to represent, educate, and promote the rapidly growing restaurant and foodservice industry. This Web site should be your starting place for all issues and concerns related to your restaurant. This Web site has it all, from tips for running your establishment to vital data on your customers' spending habits.

National Turkey Federation

www.eatturkey.com

The National Turkey Federation advocates for all segments of the turkey industry, providing services and conducting activities that increase demand for its members' products by protecting and enhancing their ability to profitably provide wholesome, high-quality, nutritious products. Its Web site contains a special area specifically for the restaurant and foodservice industry, including recipe and menu ideas, cooking demonstrations, and a reference on purchasing, preparing, and promoting turkey.

Pasteurized Eggs, L.P.

www.davidsonseggs.com

This company is the developer of an all-natural process that eliminates the risk of foodborne illness by destroying *Salmonella* spp., which may be present in raw or soft-cooked eggs.

This Web site explains the process of in-shell pasteurization, why in-shell pasteurization is an additional step restaurants and foodservice establishments can take to ensure the safety of food they serve, and where pasteurized eggs are available for purchase in the U.S.

Procter & Gamble

www.pg.com

Procter & Gamble produces cleaning and food and beverage products for the restaurant and foodservice industry. This Web site provides information on its products.

Produce Marketing Association

www.pma.com

The Produce Marketing Association is a not-for-profit trade association serving members who market fresh fruit, vegetables, and floral products worldwide. Its members are involved in the production, distribution, retail, and foodservice sectors of the industry. This Web site provides information on hot topics in the produce industry and promotional ideas.

Tyson Foods, Inc.

www.tyson.com

Tyson Foods is the world's largest poultry producer. This Web site is a great resource about chicken products for the restaurant and foodservice industry. It includes recipes and menu ideas, information on new products, and information about Tyson Foods' efficient distribution and transportation system.

USDA–FSIS

www.fsis.usda.gov

The Food Safety and Inspection Service (FSIS) is the public health agency of the USDA responsible for ensuring that the nation's commercial supply of meat, poultry, and egg products is safe, wholesome, and correctly labeled and packaged. This Web site contains a wealth of information on food safety relating to meat, poultry, and egg.

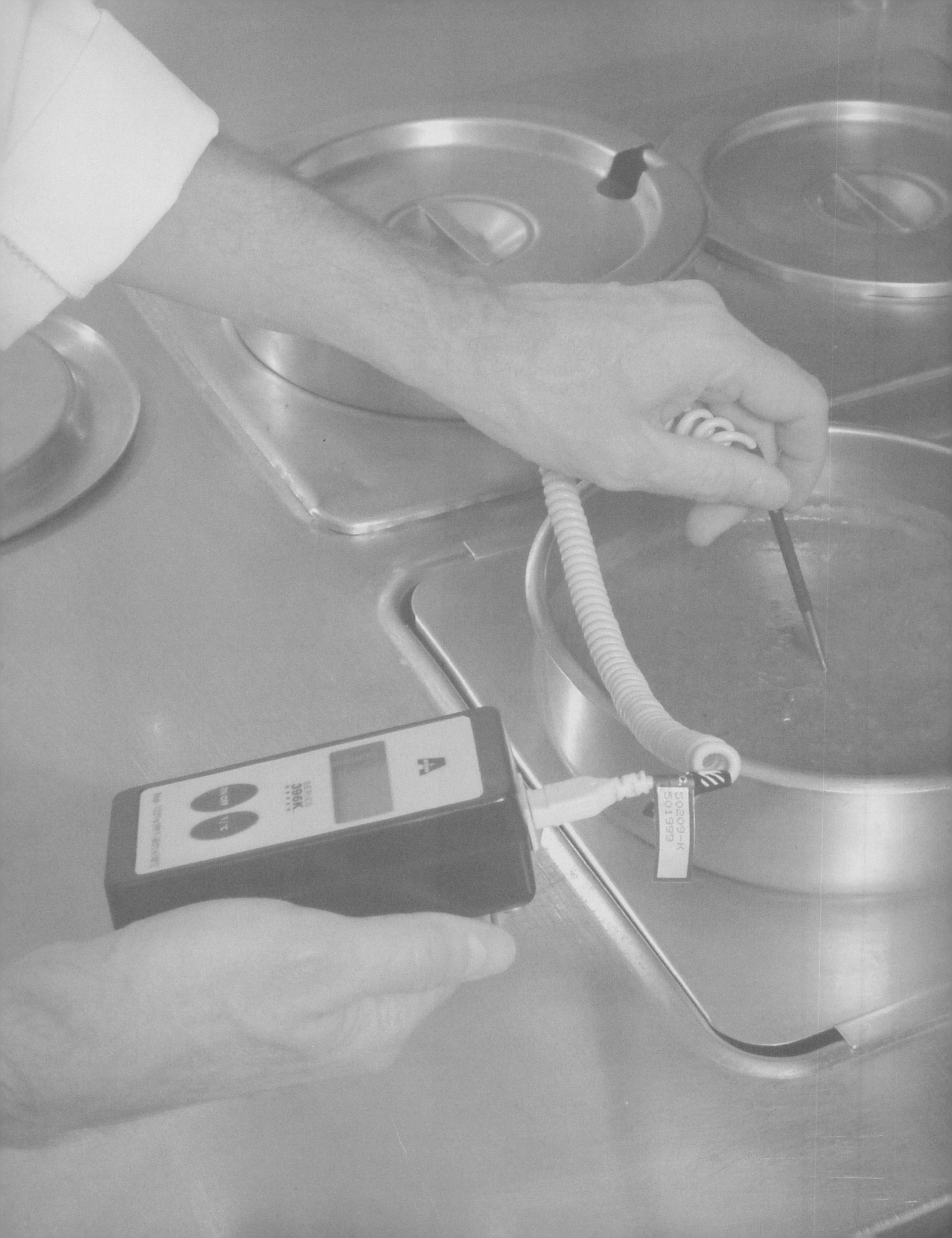
396K
50209-K
501999

9

The Flow of Food: Service

Inside this chapter:

- Holding Food for Service
- Serving Food Safely
- Off-Site Service

After completing this chapter, you should be able to:

- Identify time and temperature requirements for holding hot and cold, potentially hazardous food.
- Identify procedures for preventing time-temperature abuse and cross-contamination when displaying and serving food.
- Identify the requirements for using time rather than temperature as the only method of control when holding ready-to-eat food.
- Implement methods for minimizing bare-hand contact with ready-to-eat food.
- Identify hazards associated with the transportation of food and methods for preventing them.
- Identify hazards associated with the service of food off-site and methods for preventing them.
- Identify hazards associated with vending food and methods for preventing them.
- Prevent customers from contaminating self-service areas.
- Prevent employees from contaminating food.

Key Terms

Hot-holding equipment
Cold-holding equipment
Food bar
Sneeze guard
Off-site service
Single-use item
Mobile unit
Temporary unit
Vending machine

Apply Your Knowledge

Check to see how much you know about the concepts in this chapter. Use the page references provided to explore the topic in each question.

Test Your Food Safety Knowledge

1. **True or False:** Cold, potentially hazardous food must be held at an internal temperature of 41°F (5°C) or lower. *(See page 9-4.)*
2. **True or False:** Hot, potentially hazardous food must be held at an internal temperature of 120°F (49°C) or higher. *(See page 9-3.)*
3. **True or False:** Chicken salad can be held at room temperature if it has a label that specifies it must be discarded after six hours. *(See page 9-4.)*
4. **True or False:** When holding potentially hazardous food for service, the internal temperature must be checked at least every four hours. *(See page 9-3.)*
5. **True or False:** Servers can contaminate food simply by handling the food-contact surface of a plate. *(See page 9-6.)*

For answers, please turn to the Answer Key.

INTRODUCTION

The job of protecting food continues even after it has been prepared and cooked properly, since microorganisms can still contaminate food before it is eaten. The key to serving safe food is to prevent time-temperature abuse and cross-contamination. Hold, display, and serve food at the right temperature, and handle it safely. People do many things without knowing their actions can lead to contamination. Train employees to serve food properly, and make sure food safety rules are followed.

HOLDING FOOD FOR SERVICE

In many establishments, food is cooked to order. Food that is stored, prepared, and cooked properly, and then served immediately, is less likely to cause illness. Even in facilities that cook food to order, many menu items are cooked and held for service. A coffee shop might hold soup in a warming kettle.

Time & Temperature

A steak house might keep prime rib warm on a steam table. Many establishments, such as cafeterias and buffets, hold almost all food they serve.

Kitchen staff might be tempted to hold hot food at a lower temperature that keeps it warm but does not affect quality. However, all employees must remember that microorganisms can grow at temperatures between 41°F (5°C) and 135°F (57°C). To ensure the safety of food that is held hot or cold, specific procedures must be followed.

General Rules For Holding Food

- **Check the internal temperature of food using a thermometer.** The holding equipment's thermostat measures the temperature of the equipment, not the food.
- **Check the temperature of food at least every four hours.** Food that is not at 135°F (57°C) or higher or 41°F (5°C) or lower must be discarded. As an alternative, check the temperature every two hours to leave time for corrective action.
- **Establish a policy to ensure that food being held for service will be discarded after a predetermined amount of time.** For example, a policy may state that a pan of veal can be replenished all day as long as it is discarded at the end of the day.
- **Protect food from contaminants with covers or sneeze guards.** Covers help to maintain temperature and keep out contaminants.
- **Prepare food in small batches so it will be used faster.** Do not prepare food any further in advance than necessary to minimize the potential for time-temperature abuse.

Hot Food

- **Potentially hazardous, hot food must be held at an internal temperature of 135°F (57°C) or higher.** You can also hold it at an even higher temperature of 140°F (60°C) as an additional safeguard.
- **Only use hot-holding equipment that can keep food at the proper temperature.** (See *Exhibit 9a.*)

Exhibit 9a

Hot-Holding Equipment

Only use hot-holding equipment that can keep food at 135°F (57°C) or higher.

Courtesy of Cooper-Atkins Corporation

Exhibit 9b

Cold-holding equipment must keep food at 41°F (5°C) or lower.

- **Never use hot-holding equipment to reheat food if it is not designed to do so.** Reheat food to 165°F (74°C) for fifteen seconds within two hours, then transfer it to holding equipment. Most hot-holding equipment is incapable of passing food through the temperature danger zone quickly enough during the reheating process to prevent the growth of microorganisms.
- **Stir food at regular intervals to distribute heat evenly.**

Cold Food

- **Potentially hazardous, cold food must be held at an internal temperature of 41°F (5°C) or lower.**
- **Only use cold-holding equipment that can keep food at the proper temperature.** (See *Exhibit 9b.)*
- **Do not store food directly on ice.** Whole fruit and vegetables and raw, cut vegetables are the only exceptions. Place food in pans or on plates first. Ice used on a display should be self-draining and drip pans should be cleaned and sanitized after each use.

Holding Food Without Temperature Control

Ready-to-eat, potentially hazardous food can be displayed or held for consumption without temperature control for up to four hours under the following conditions:

- **Prior to removing the food from temperature control it has been held at 41°F (5°C) or lower, or 135°F (57°C) or higher.**
- **The food contains a label that specifies when the item must be discarded.** The label must reflect a time that is four hours after the item was taken out of temperature control. For example, if potato salad served at a picnic was removed from refrigeration at 12:00 P.M., the time indicated on the label must be 4:00 P.M. because the potato salad must be discarded within four hours.
- **The food must be sold, served, or discarded within four hours.**

Before using time as a method of control, check with your regulatory agency for specific requirements in your area.

SERVING FOOD SAFELY

After handling food safely and cooking it properly, you do not want to risk contamination when serving it.

Kitchen Staff

Train your kitchen staff to follow these procedures to serve food safely.

- **Store serving utensils properly**. Serving utensils can be stored in the food, with the handle extended above the rim. (See *Exhibit 9c*.) They can also be placed on a clean, sanitized food-contact surface. Spoons or scoops used to serve food such as ice cream or mashed potatoes can be stored under running water.
- **Use serving utensils with long handles.** Long-handled utensils keep the server's hands away from food.
- **Use clean and sanitized utensils for serving.** Use separate utensils for each food item, and properly clean and sanitize them after each serving task. Utensils should be cleaned and sanitized at least once every four hours during continuous use.
- **Minimize bare-hand contact with food that is cooked or ready-to-eat.** Handle food with tongs, deli sheets, or gloves, for example. Bare-hand contact is allowed in some jurisdictions if the establishment has a verifiable written policy on handwashing procedures. Check with your regulatory agency for requirements in your jurisdiction. (See *Exhibit 9d* on the next page.)

Exhibit 9c

Properly Stored Utensils

If stored in food, utensils should be stored with the handle extended above the rim of the container.

- **Practice good personal hygiene.** Wash hands after using the restroom, or after hands have come in contact with anything that may contaminate food.

Servers

Food servers need to be just as careful as kitchen staff. If they are not careful, they can contaminate food simply by handling the food-contact surfaces of glassware, dishes, and utensils. Servers should use the following guidelines to serve food safely. (See *Exhibit 9e.*)

- **Glassware and dishes should be handled properly.** The food-contact area of plates, bowls, glasses, or cups should not be touched. Dishes should be held by the bottom or the edge. Cups should be held by their handles, and glassware should be held by the middle, bottom, or stem.
- **Glassware and dishes should not be stacked when serving.** The rim or surface of one can be contaminated by the one above it. Stacking china and glassware also can cause it to chip or break.

Exhibit 9d

Handling Cooked and Ready-to-Eat Food

Minimize bare-hand contact with cooked and ready-to-eat food by using gloves or tongs to serve it.

- **Flatware and utensils should be held at the handle.** Store flatware so servers grasp handles, not food-contact surfaces.
- **Minimize bare-hand contact with food that is cooked or ready-to-eat.**
- **Use ice scoops or tongs to get ice.** Servers should never scoop ice with their bare hands or use a glass since it may chip or break. Ice scoops should always be stored in a sanitary location—not in the ice bin.
- **Practice good personal hygiene.** Servers should be neat and clean, and their hair should be pulled back

Exhibit 9e

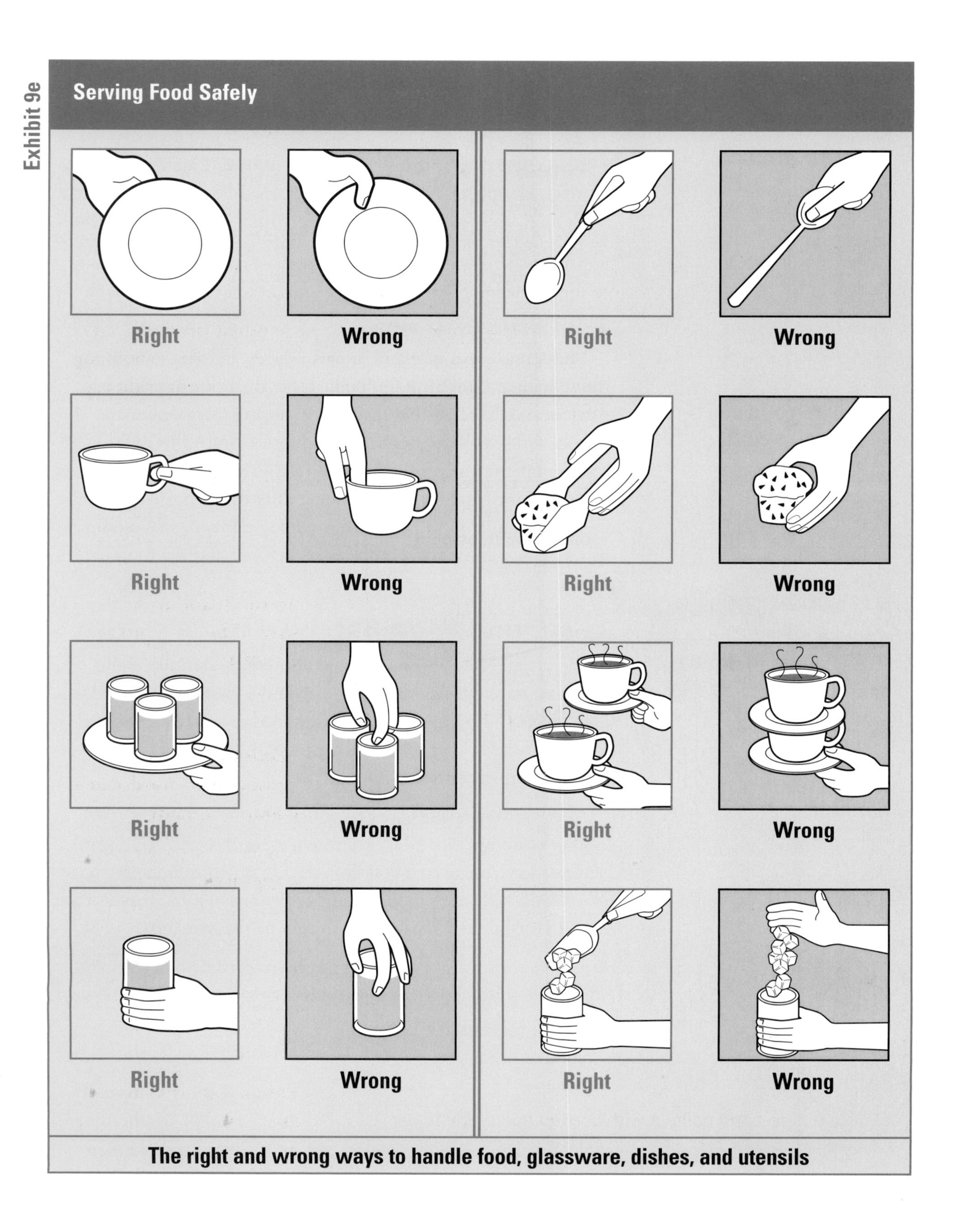

The right and wrong ways to handle food, glassware, dishes, and utensils

and covered. They should avoid touching their hair or face when serving food, and should refrain from habits such as chewing fingernails or licking their fingers.

- **Never use cloths meant for cleaning food spills for any other purpose.** When tables are cleaned between guest seatings, spills should be wiped up with a disposable, dry cloth. The table should then be cleaned with a moist cloth that has been stored in a fresh sanitizer solution.

Division of Labor

Cross-Contamination

Never re-serve plate garnishes, uncovered condiments, or uneaten bread.

To prevent cross-contamination, it is a good idea to schedule staff so they are not assigned to do more than one job during a shift. Serving food, setting tables, and busing dirty dishes are separate tasks with different responsibilities. Since this division of labor is difficult to manage in most establishments, it is important for servers and busers who do double duty to wash their hands often and handle food safely. After wiping tables or busing dirty dishes, servers must wash their hands before handling food or place settings.

Re-serving Food Safely

Servers and kitchen staff should also know the rules about re-serving food. In general, only unopened, prepackaged food can be re-served. Condiment packets, wrapped crackers or wrapped breadsticks, and other sealed food generally have adequate protection from contamination.

Never re-serve plate garnishes, such as fruit or pickles, to another customer. Served, but unused, garnishes must be discarded. Never re-serve uncovered condiments. Do not combine previously served food with fresh food. Opened portions of salsa, mayonnaise, mustard, or butter, for example, should be thrown away.

Uneaten bread or rolls may not be re-served to other customers. Linens used to line bread baskets must be changed after each customer.

Cross-Contamination

Assign a trained staff member to monitor food bars.

Self-Service Areas

Customers choosing food from self-service areas, or **food bars,** often unknowingly serve themselves in ways that can put them and other customers in danger. A customer may eat from her plate or nibble from the food bar while moving through the line. Another might pick up carrot sticks, pickles, and olives with her fingers, or dip a finger into salad dressing to taste it. Another might return unwanted food items, use a soiled plate for a second helping, or put his head under the sneeze guard to reach items in the back of the display.

Buffets and food bars should be monitored closely by employees trained in food safety procedures. Assign a staff member to replenish food-bar items and to hand out fresh plates for return visits. Post signs with polite tips about food-bar etiquette. These practices will go a long way toward keeping self-service areas more sanitary.

Exhibit 9f

Protecting Food on Display

Sneeze guards must be fourteen inches above the food counter, and the shield should extend seven inches beyond the food.

Here are more rules for food bars:

- **Protect food on display with sneeze guards or food shields.** (See *Exhibit 9f.*) These must be fourteen inches above the food counter, and the shield should extend seven inches beyond the food.
- **Identify all food items in the foodbar by labeling containers.** Place names of salad dressings on ladle handles. (See *Exhibit 9g* on the next page.)
- **Maintain proper food temperatures.** Keep hot food hot—135°F (57°C) or higher, and cold food cold—41°F (5°C) or lower.

Exhibit 9g

Identifying Food Items

Label all items on a food bar.

- **Replenish food on a timely basis.** Prepare and replenish small amounts at a time so food is fresher and has less chance of being exposed to contamination. Practice the FIFO method of product rotation.
- **Keep raw meat, fish, and poultry separate from cooked and ready-to-eat food.** Customers can easily spill food when serving themselves. Use separate displays or food bars for raw and cooked food (for example, at a Mongolian barbecue) to reduce the chance of cross-contamination.
- **Do not let customers use soiled plates or silverware for refills.** Encourage all customers to take a clean plate for return trips to the food bar. Customers can use glassware for refills as long as beverage-dispensing equipment does not come in contact with the rim or interior of the glass.

OFF-SITE SERVICE

Establishments use many different ways to deliver food to people. No matter how food is prepared and delivered, operators must follow the same food safety rules as permanent establishments. Food must be protected from contamination and time-temperature abuse, and facilities and equipment used to prepare food must be clean and sanitary. Menu items must lend themselves to safe service. Food must be handled safely. **Off-site services,** including delivery, mobile/temporary kitchens, and vending machines, all present special challenges.

Exhibit 9h

Delivery of Food Off Site

Equipment used to transport food must be designed to maintain safe food temperatures and must be easy to keep clean.

Time & Temperature

When delivering food off-site, check food temperatures regularly, and take corrective action if an item is not at its proper temperature.

Delivery

Many establishments—such as schools, hospitals, caterers, and even restaurants—use a central commissary to prepare food. Food from the central kitchen is then delivered to remote locations for service. The greater the time and distance from the point of preparation to the point of consumption, the greater the risk food will be exposed to contamination or time-temperature abuse. Equipment used to transport food—both containers and vehicles—must be designed to maintain safe food temperatures and be easy to keep clean.

When transporting food from a central kitchen, the following safety procedures should be followed:

- **Use rigid, insulated food containers capable of maintaining food temperatures above 135°F (57°C) or below 41°F (5°C).** Containers should be sectioned so that food does not mix, leak, or spill. Containers must also allow air circulation to keep temperatures even, and should be kept clean and sanitized. (See *Exhibit 9h.*)
- **Clean and sanitize the insides of delivery vehicles regularly.**
- **Make sure employees practice good personal hygiene when distributing food.**
- **Check internal food temperatures regularly.** Take corrective action if food is not at the proper temperature. If containers or delivery vehicles are not maintaining proper food temperatures at the end of each route, reevaluate the length of delivery routes or the efficiency of the equipment being used.
- **Label food with storage, shelf life, and reheating instructions for employees at off-site locations.**
- **Provide food safety guidelines for consumers.** If you or your employees will not be serving the food you deliver, provide customers with information on which items should be eaten immediately, which items may be saved for later, and how to serve all items.

Catering

Caterers provide food for private parties and events, as well as public and corporate functions. They might bring prepared food or cook food on-site in a mobile unit, a temporary unit, or in the customer's own kitchen.

Caterers must follow the same food safety rules as permanent establishments. Food must be protected from contamination and time-temperature abuse. Facilities must be clean and sanitary. Food must be prepared and served safely. Employees must follow good personal hygiene practices.

Caterers must meet the special challenges of off-site foodhandling. They must make sure there is safe drinking water for cooking, warewashing, and handwashing, as well as adequate power for holding and cooking equipment. Caterers must make proper arrangements for garbage disposal.

Outdoor catering for barbecues and cookouts may require special arrangements. When power or running water is not available, caterers may have to change their foodhandling procedures.

- **Use ice chests or insulated containers for all potentially hazardous food.** Raw meat should be wrapped and stored on ice. Deliver milk and dairy products in a refrigerated vehicle or on ice.
- **Serve cold food in containers on ice or in chilled, gel-filled containers.** If that is not desirable, the food may be held without temperature control according to the guidelines specified in this chapter.
- **Keep raw and ready-to-eat products separate during delivery and storage.** For example, store raw chicken separately from ready-to-eat salads.
- **Use single-use items.** Make sure customers get a new set of disposable tableware for refills. Arrange for proper garbage disposal away from food-prep and serving areas.
- **If leftovers are given to customers, provide instructions on how they should be handled.** This may include reheating and storage instructions for the products and shelf-life dates.

Key Point

Mobile kitchens serving potentially hazardous food must follow the same rules as permanent kitchens.

Mobile Units

Mobile units are portable facilities ranging from concession vans to elaborate field kitchens. Those serving only frozen novelties, candy, packaged snacks, and soft drinks have to meet basic sanitation requirements. Mobile kitchens preparing and serving potentially hazardous food, however, must follow the same rules required of permanent foodservice kitchens. Both might be required to apply for a special permit or license from the local regulatory agency.

Like permanent establishments, mobile kitchens must have adequate cooking equipment, mechanical cooling and hot-holding units, and warewashing and handwashing sinks with hot and cold potable water under pressure. Mobile kitchens must provide adequate ventilation, garbage storage and disposal facilities, and pest control. Operators should clean and maintain mobile units.

Temporary Units

Temporary units typically operate in one location for less than fourteen days. Foodservice tents or kiosks set up for food fairs, special celebrations, or sporting events may be temporary units. (See *Exhibit 9i.*) In some areas, the definition also extends to units set up for longer periods of time. Temporary units usually serve

Exhibit 9i

Temporary Unit

It is best to keep the menu simple to limit the amount of on-site food preparation.

Courtesy of the Illinois Restaurant Association

prepackaged food or food requiring limited preparation, such as hot dogs. It is best to keep the menu simple to limit the amount of on-site food preparation. Check with regulatory agencies for operating requirements.

Temporary units should be constructed so dirt and pests are kept out. If floors are made of dirt or gravel, cover them with mats or platforms to control dust and mud. Construct walls and the ceiling with materials that will protect food from weather and windblown dust.

In addition, the same safe-handling rules previously discussed apply to food preparation in temporary units. If food is prepared on-site, the unit must have adequate cooking, cold storage, and hot-holding equipment. If food is prepared off-site, it must be transported and held at proper temperatures below 41°F (5°C) or above 135°F (57°C).

Safe drinking water must be available for cleaning, sanitizing, and handwashing. Since warewashing facilities will most likely be limited, it is best to use disposable, single-use items.

Exhibit 9j

Vending Machines

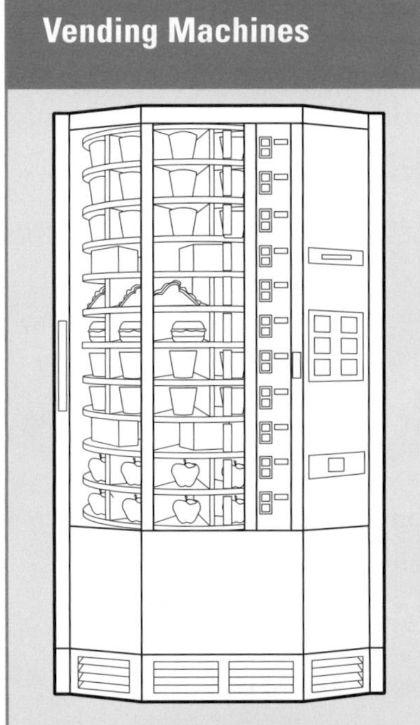

Replace food with expired code dates and discard food if it is not used within seven days of preparation.

Vending Machines

Food prepared and packaged for **vending machines** has to be handled with the same care as any other food served to a customer. Vending operators also have to protect food from contamination and time-temperature abuse during transport, delivery, and service.

Keep food at the right temperature. Machines that contain potentially hazardous food must be able to maintain proper food temperatures of 41°F (5°C) or lower and 135°F (57°C) or higher. They must also have automatic shutoff controls that prevent food from being dispensed if the temperature stays in the danger zone for a specified amount of time.

Check product shelf-life daily. Replace food with expired code dates. If refrigerated food is not used within seven days of preparation, it must be discarded. (See *Exhibit 9j.*) Dispense potentially hazardous food, such as milk, in its original container. Fresh fruit with an edible peel should be washed and wrapped before being put into a machine.

Place machines in appropriate locations, away from garbage containers, sewage drains, and overhead pipes. Make sure the vending area is clean and well lighted. Supply safe drinking water for beverage machines.

Clean and service vending machines regularly. Sanitize food-contact surfaces in machines each time food is replenished. Employees must also wash their hands before and after servicing or refilling machines.

SUMMARY

Safe foodhandling does not stop once food is properly prepared and cooked. You must continue to protect it from time-temperature abuse and contamination until it is eaten. When holding potentially hazardous food for service, keep hot food hot—above 135°F (57°C), and cold food cold—below 41°F (5°C). Check the internal temperature of food being held at least every four hours and discard it if it is not at the proper temperature. Protect food from contaminants with covers and lids, and establish policies to ensure that food being held for service will be discarded after a predetermined amount of time. If food will be held without temperature control, label it with a discard time, and sell, serve, or discard it within four hours. Use clean, sanitized utensils to serve food and minimize bare-hand contact with cooked and ready-to-eat food.

Make sure all employees practice good personal hygiene. Train them to avoid cross-contamination when handling service items and tableware. Teach them about the potential hazards posed by re-serving plate garnishes, breads, or open dishes of condiments.

Customers can unknowingly contaminate food in self-service areas. Post signs to communicate self-service rules, and station employees in these areas to ensure compliance. Protect food in food bars and buffets with sneeze guards, and make sure equipment can hold food at the proper temperature.

Take special precautions when preparing, delivering, or serving food off-site. Catering, mobile kitchens, temporary units, and vending machines pose unique challenges to food safety. Learn and follow all regulations in your jurisdiction. Prepare backup plans for when off-site facilities are not adequate or equipment breaks down.

Apply Your Knowledge

1. What did Jill do wrong?
2. What should have been done?

For answers, please turn to the Answer Key.

A Case In Point 1

Jill, a line cook on the morning shift at Memorial Hospital, was busy helping the kitchen staff put food on display for lunch in the hospital cafeteria. Ann, the kitchen manager who usually supervised lunch in the cafeteria, was at an all-day seminar on food safety. Jill was responsible for making sure meals were trayed and put into food carts for transport to the patients' rooms. The staff also packed two dozen meals each day for a neighborhood group that delivered them to homebound elderly people.

First, Jill looked for insulated food containers for the delivery meals. When she could not find them, she loaded the meals into cardboard boxes she found near the back door, knowing the driver would be there soon to pick them up. To help the cafeteria staff, Jill filled a baine with soup by dipping a two-quart measuring cup into the stockpot and pouring it into the baine. She carried the baine out to the cafeteria, put it into the steam table, and turned it on low.

The lunch hour was hectic. The cafeteria was busy, and the staff had many patient meals to tray and deliver. Halfway through lunch, a cashier came back to the kitchen to tell Jill that the salad bar needed replenishing. Since she was busy, Jill asked a kitchen employee to take pans of prepared ingredients out of the refrigerator and put them on the salad bar. When she looked up a few moments later, she saw the kitchen employee send away two children who were eating carrot sticks from the salad bar.

With lunch almost over, Jill breathed a sigh of relief. She moved down the cafeteria serving line, checking food temperatures. One of the casseroles was about 130°F (54°C). Jill checked the water level in the steam table and turned up the thermostat, then went to clean up the kitchen and finish her shift.

Apply Your Knowledge

1. What errors did Megan make?
2. What should she have done?

For answers, please turn to the Answer Key.

A Case In Point 2

Megan, a new server at The Fish House, reported for work ten minutes early on Thursday. Excited about her new job, she made sure to shower and wash her hair before going to work. When she arrived, she changed into a clean uniform, pulled her hair back tightly into a ponytail, and checked her appearance. She had used makeup sparingly and wore only small hoop earrings and a short chain necklace.

Her shift started off well. One of her customers ordered a menu item that Megan had not tried yet. When the order came up, Megan dipped her finger into the sauce at the edge of the plate for a taste. As the shift progressed, Megan's station got busier. When the hostess came to ask how soon one of Megan's tables could be cleared, Megan decided to do it herself. She took the dirty dishes to a bus station, then wiped down the table with a serving cloth she kept in her apron.

The buser, John, finished another task and came to help her set the table. While he put out silverware and linens, Megan filled water glasses with ice by scooping the glasses into the ice bin in the bus station. Only one slice of bread and one pat of butter were missing from the basket that had been on a previous customer's table, so she put that on a tray with the water glasses and took it to the table.

The table was reset in record time, and Megan soon had more guests. While taking their orders, Megan reached up to scratch a mosquito bite on her neck. Then she went to the kitchen to turn in the order and pick up a dessert order for another table.

Apply Your Knowledge

Use these questions to review the concepts presented in this chapter.

Discussion Questions

1. What can be done to minimize contamination in self-service areas?
2. What hazards are associated with the transportation of food and how can they be prevented?
3. What are the requirements for using time rather than temperature as the only method of control when holding potentially hazardous, ready-to-eat food?
4. What practices should be followed to serve food safely off-site?

For answers, please turn to the Answer Key.

Apply Your Knowledge

Use these questions to test your knowledge of the concepts presented in this chapter.

Multiple-Choice Study Questions

1. **Which of the following is an unsafe serving practice?**
   A. Stacking plates of food before serving them to the customer
   B. Holding flatware by the handles when setting a table
   C. Serving soup with a long-handled ladle
   D. Holding glassware by the stem

2. **Which of the following statements about serving utensils is *not* true?**
   A. They should be cleaned and sanitized at least once every four hours during continuous use.
   B. They can be used to handle more than one food item at a time.
   C. They can be stored in the food with the handle extended above the rim of the container.
   D. They must be cleaned and sanitized after each task.

3. **To keep vended food safe, you should**
   A. leave fresh fruit with edible peels unwrapped.
   B. discard food within fourteen days of preparation.
   C. dispense potentially hazardous food in its original container.
   D. ensure that cold, potentially hazardous food is kept at 50°F (10°C) or lower.

4. **To hold cold food safely, you should**
   A. store it directly on ice.
   B. store it at 41°F (5°C) or lower.
   C. stir it regularly.
   D. leave it uncovered.

Continued on next page...

Apply Your Knowledge

Multiple-Choice Study Questions *continued*

5. **Which of the following is an acceptable serving practice at a self-service bar?**
   A. Holding hot, potentially hazardous food at 120˚F (49˚C)
   B. Storing raw meat next to ready-to-eat food
   C. Allowing customers to use the same plate for a return trip to the self-service bar
   D. Allowing customers to re-use glassware for beverage refills

6. **Hot, potentially hazardous food should be held at an internal temperature of**
   A. 135˚F (57˚C) or higher.
   B. 130˚F (54˚C) or higher.
   C. 120˚F (49˚C) or higher.
   D. 110˚F (43˚C) or higher.

7. **All of the following are conditions for allowing food to be held without temperature control *except***
   A. the food must contain a label specifying when it must be discarded.
   B. gloves must be worn when handling the food.
   C. the local jurisdiction must allow food to be held without temperature control.
   D. the food must be sold, served, or discarded within four hours.

For answers, please turn to the Answer Key.

ADDITIONAL RESOURCES

Books and Periodicals

Avoid buffet blunders. 2001. *Food Safety Illustrated.* 1 (1):13.

Dining room do's and don'ts. 2003. *Food Safety Illustrated.* 3 (2):13.

How to "handle" ice safely. 2002. *Food Safety Illustrated.* 2 (3):13.

Proper use of gloves. 2003. *Food Safety Illustrated.* 3 (1):13.

Web Sites

Bunzl Distribution, Inc.

www.bunzldistribution.com

Bunzl Distribution is a supplier of disposable paper and plastic packaging supplies, as well as specialty items for service and packaging for the restaurant and foodservice industry.

Cooper-Atkins Corporation

www.cooperinstrument.com

Cooper-Atkins Corporation is a supplier of temperature, time, and humidity instruments for a variety of global markets. This Web site contains information on how to calibrate thermometers properly and maintain food temperatures during cooking, holding, cooling, etc.

FDA Center for Food Safety and Applied Nutrition (CFSAN)

www.cfsan.fda.gov

As the center within the FDA responsible for food safety and nutrition, CFSAN promotes and protects public health by researching and implementing guidelines, policies, and standards to ensure that food is safe, nutritious, wholesome, and properly labeled. This Web site provides a wealth of information on food safety and sanitation, including corresponding guidelines, policies, and standards.

FDA Food Code

http://vm.cfsan.fda.gov/~dms/foodcode.html

As the basis for many local sanitation codes, as well as the basis for the information in this textbook, the FDA Food Code, available at this Web address, is a useful resource for information relating to food safety for the restaurant and foodservice industry.

FoodHandler Inc.
www.foodhandler.com
FoodHandler Inc. is a manufacturer of foodservice safety equipment, such as gloves, aprons, and food-storage systems. The site contains information on products, as well as issues related to food safety.

National Automatic Merchandising Association (NAMA)
www.vending.org
NAMA is the national trade association of the merchandising vending and contract-foodservices management business. This Web site provides up-to-date information on all issues concerning this segment of the foodservice industry, as well as provides a resource for conferences and education for those working in this segment.

National Restaurant Association
www.restaurant.org
The National Restaurant Association is the leading business association for the restaurant industry. Together with the National Restaurant Association Educational Foundation, the Association's mission is to represent, educate, and promote the rapidly growing restaurant and foodservice industry. This Web site should be your starting place for all issues and concerns related to your restaurant. This Web site has it all, from tips for running your establishment to vital data on your customers' spending habits.

Apply Your Knowledge

Notes

RANGE GUARD
IN CASE OF FIRE
PULL PIN

Food Safety Systems

Inside this chapter:

- Food Safety Programs
- Active Managerial Control
- HACCP
- When a HACCP Plan is Required
- Crisis Management

After completing this chapter, you should be able to:

- Identify how active managerial control can impact food safety.
- Identify HACCP principles for preventing foodborne illness.
- Implement HACCP principles when applicable.
- Identify when a HACCP plan is required.
- Implement a crisis management program.
- Cooperate with regulatory agencies in the event of a foodborne-illness investigation.

Key Terms

Food safety management system
Active managerial control
HACCP
HACCP plan
Critical control point (CCP)

Apply Your Knowledge

Check to see how much you know about the concepts in this chapter. Use the page references provided to explore the topic in each question.

Test Your Food Safety Knowledge

1. **True or False:** Active managerial control focuses on taking action to control three foodborne-illness risk factors identified by the CDC. *(See page 10-4.)*
2. **True or False:** Purchasing fish from local fishermen would be considered a risk under active managerial control. *(See page 10-4.)*
3. **True or False:** Cooking chicken to a minimum internal temperature of 165°F (74°C) for fifteen seconds would be an appropriate critical limit. *(See page 10-7.)*
4. **True or False:** A critical control point (CCP) is a point in the flow of food where a hazard can be prevented, eliminated, or reduced to safe levels. *(See page 10-7.)*
5. **True or False:** An establishment that cures food must have a HACCP plan. *(See page 10-10.)*

For answers, please turn to the Answer Key.

INTRODUCTION

In Chapters 6 through 9, you learned how to handle food safely as it flows through the establishment from receiving, storage, and preparation through cooking, holding, cooling, and reheating. This accumulated knowledge will help you develop a **food safety management system** to prevent foodborne illness by actively controlling hazards throughout the flow of food. In this chapter, you will be introduced to the fundamentals for developing this type of system.

FOOD SAFETY PROGRAMS

A food safety management system must be built on a solid foundation of programs that supports your efforts to minimize the risk of foodborne illness.

These programs include (see *Exhibit 10a*):

- Personal hygiene (including restriction and exclusion policies for sick employees)
- Facility design
- Supplier selection and specification
- Cleaning and sanitation
- Equipment maintenance
- Manager and employee food safety training

Exhibit 10a

Food Safety Programs

The following programs must be in place for a food safety management system to be effective.

Proper personal hygiene program

Proper facility-design program

Supplier selection and specification programs

Proper cleaning and sanitation programs

Appropriate equipment-maintenance programs

Food safety training programs

Key Point

Active managerial control focuses on establishing policies and procedures to control the CDC's five most common risk factors for foodborne illness.

The development and maintenance of these programs are crucial for addressing the five most common risk factors responsible for foodborne illness as identified by the CDC. The factors are:

- Purchasing food from unsafe sources
- Failing to cook food adequately
- Holding food at improper temperatures
- Using contaminated equipment
- Poor personal hygiene

ACTIVE MANAGERIAL CONTROL

Active managerial control is a proactive, rather than reactive, approach to addressing the CDC's risks. By continuously monitoring and verifying procedures responsible for preventing these risks, you will ensure they are being controlled. (See *Exhibit 10b.*)

Exhibit 10b

Active Managerial Control

Managers must monitor policies and procedures to ensure that they are being followed.

Approach

There are specific steps that should be taken when developing a food safety management system using active managerial control.

Step 1: Establish the necessary food safety programs, and support them through your standard operating procedures.

Step 2: Consider the five risk factors as they apply throughout the flow of food, and identify the potential breakdowns that could impact food safety.

Step 3: If necessary, revise policies and procedures to prevent these breakdowns from occurring.

Step 4: Monitor the policies and procedures to ensure that they are being followed.

Step 5: Verify that the policies and procedures you have established are actually controlling the risk factors. Use feedback from internal sources (records, temperature logs, and self-inspections) and external sources (health inspection reports, customer comments, and quality assurance audits) to adjust your policies and procedures in order to continuously improve the system.

For an example of active managerial control in action, see *Exhibit 10c.*

Exhibit 10c

An Example of Active Managerial Control

A full-service seafood restaurant chain has identified buying seafood from an unsafe source as a risk in their establishments during purchasing. To avoid buying unsafe product, management has decided to do the following:

- Develop criteria for creating an approved list of inspected vendors.
- Create a policy stating that seafood may only be purchased from vendors on this list.
- Periodically monitor invoices and deliveries to ensure they are coming from approved vendors.
- Regularly verify that the criteria set for the vendors are still appropriate for controlling the risk.
- Review the policies and procedures when a breakdown occurs, to determine why, and change them to ensure that it does not happen again.

HACCP

A food safety management system may also include a **Hazard Analysis Critical Control Point (HACCP) system.** A HACCP (pronounced *Hass-ip*) system is based on the idea that if significant biological, chemical, or physical hazards are identified

Key Point

A HACCP system is based on the idea that if significant biological, chemical, or physical hazards are identified at specific points within a product's flow through the operation, they can be prevented, eliminated or reduced to safe levels.

Key Point

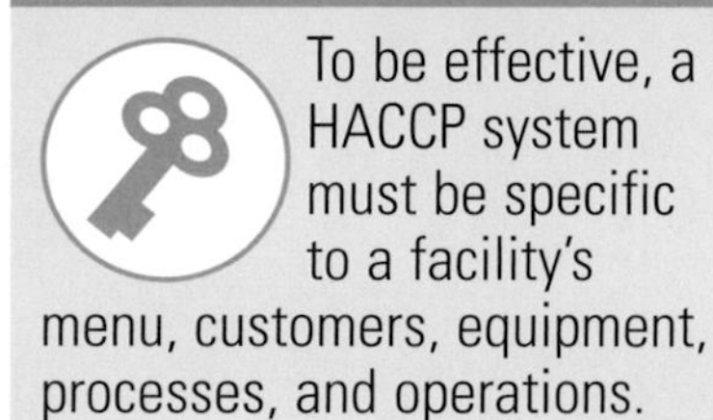

To be effective, a HACCP system must be specific to a facility's menu, customers, equipment, processes, and operations.

at specific points within a product's flow through an operation, they can be prevented, eliminated, or reduced to safe levels.

To be effective, a HACCP system must be based on a written plan that is specific to each facility's menu, customers, equipment, processes, and operations. A **HACCP plan** is based on the seven basic principles outlined by the National Advisory Committee on Microbiological Criteria for Foods.

Approach

The HACCP principles are seven sequential steps that outline how to create a HACCP plan. Since each principle builds on the information gained from the previous principle, when developing your plan you must consider all seven principles in order.

In general terms:

- Principles One and Two help you identify and evaluate your hazards.
- Principles Three, Four, and Five help you establish how you will control those hazards.
- Principles Six and Seven help you maintain both your HACCP plan and system and verify their effectiveness.

The Seven HACCP Principles

Principle One: Conduct a Hazard Analysis

Identify and assess potential hazards in the food you serve by taking a look at how the food is processed, or flows through your establishment. Many types of food are processed similarly. The most common processes include:

- Preparing and serving without cooking
- Preparing and cooking for same-day service
- Preparing, cooking, holding, cooling, reheating, and serving—which is also called complex food preparation

Once common processes have been identified, you can determine where food safety hazards are likely to occur for each one. Hazards include:

- Bacterial, viral, or parasitic contamination
- Contamination by cleaning compounds, sanitizers, and allergens
- General physical contamination

> **Key Point**
>
> Find the points in the process where the identified hazard(s) can be prevented, eliminated, or reduced to safe levels, known as critical control points.

Principle Two: Determine Critical Control Points (CCPs)

Find the points in the process where the identified hazard(s) can be prevented, eliminated, or reduced to safe levels. These are the **critical control points (CCPs).** Depending on the process, there may be more than one CCP.

Principle Three: Establish Critical Limits

For each CCP, establish minimum and maximum limits that must be met to prevent or eliminate the hazard, or to reduce it to a safe level.

Principle Four: Establish Monitoring Procedures

Once critical limits have been established, determine the best way for your operation to check them to make sure they are consistently met. Identify who will monitor them and how often.

Principle Five: Identify Corrective Actions

Identify steps that must be taken when a critical limit is not met. These steps should be determined in advance.

Principle Six: Verify that the System Works

Determine if the plan is working as intended. Plan to evaluate on a regular basis your monitoring charts, records, how you performed your hazard analysis, etc., and determine if your plan adequately prevents, reduces, or eliminates identified hazards.

Principle Seven: Establish Procedures for Record Keeping and Documentation

Maintain your HACCP plan. Keep records obtained while performing monitoring activities, whenever a corrective action is taken, when equipment is validated (checked to make sure it is in good working condition), and when working with suppliers (i.e., shelf-life studies, specifications, challenge studies, etc.). Also keep all documentation created while you were developing the plan.

Now let's look at the development of a HACCP plan at Enrico's, a full-service Italian restaurant, in *Exhibit 10d* on the next page.

Exhibit 10d

Development of a HACCP plan at Enrico's

1 **Principle One—Conduct a Hazard Analysis**	Enrico's management team began by conducting a hazard analysis. Looking at their menu, they noted that several of their dishes—including *Chicken Breast alla Parmigiana* and *Pepper Steak*—used the same process of receiving, storage, preparation, cooking, and same-day service. The team determined that several biological hazards were most likely to affect the food prepared by this process. In *Chicken Breast alla Parmigiana, Salmonella* spp. and *Campylobacter* spp. are the most likely biological hazards, while shiga toxin-producing *E. coli* could affect the *Pepper Steak.*
2 **Principle Two—Determine Critical Control Points**	Enrico's management team identified cooking as a critical control point for this process. While proper food safety practices must be followed throughout the food's flow, proper cooking is the only step that will eliminate or reduce the identified hazards to safe levels. Since the food was prepared for same-day service, it was the only CCP identified.
3 **Principle Three—Establish Critical Limits**	Since cooking was identified as a CCP for the process, the team determined that their critical limit for *Chicken Breast alla Parmigiana* would be cooking the chicken to a minimum internal temperature of 165°F (74°C) for fifteen seconds. For the *Pepper Steak,* the beef must be cooked to a minimum internal temperature of 145°F (63°C) for fifteen seconds. They determined that the critical limits would be met by placing the chicken in a convection oven set to 350°F (177°C) and cooking it for forty-five minutes, and by sautéing the beef to the required temperature.
4 **Principle Four—Establish Monitoring Procedures**	Since *Chicken Breast alla Parmigiana* is cooked to order, Enrico's management team chose to monitor their critical limit by inserting a clean and sanitized thermocouple probe into the thickest part of each breast. Employees were instructed to record the readings in a temperature log. The team chose to monitor the critical limit of the *Pepper Steak* by taking sample temperatures of the beef.

Development of a HACCP plan at Enrico's *continued*

5 **Principle Five—Identify Corrective Actions**	In the event that the chicken breast or the beef has not reached its respective critical limit, Enrico's employees have been instructed to keep cooking the food until it does. This is the corrective action, which is recorded in the temperature log.
6 **Principle Six—Verify That the System Works**	Enrico's management team checked their temperature logs on a weekly basis to verify that their critical limits were being met. They noticed that occasionally the chicken breast was not meeting its critical limit, but that the appropriate corrective action was being taken to ensure that the chicken was properly cooked. The HACCP plan was reevaluated six months after implementation. The reevaluation revealed that the beef consistently met the critical limit set by the management team; however, the chicken routinely failed to meet its set critical limit. Upon reevaluating the cooking process, it was discovered that Enrico's vendor had started supplying a slightly larger chicken breast. This caused the chicken to be undercooked, given the equipment and established cooking parameters. Enrico's cooking process was adjusted to account for the larger chicken breast.
7 **Principle Seven—Establish Procedures for Record Keeping and Documentation**	Enrico's management team determined that time-temperature logs should be kept for three months and that receiving invoices should be kept for sixty days. They used this documentation to support and revise their HACCP plan as needed, such as reflecting the change in the chicken's cooking process.

WHEN A HACCP PLAN IS REQUIRED

The National Restaurant Association and the FDA recommend that all restaurants and foodservice establishments, no matter how large or small, develop and implement a food safety management system. Establishments that perform the following activities, however, must have a HACCP plan in place:

- Smoke or cure food as a method of food preservation
- Use food additives as a method of food preservation
- Package food using a reduced-oxygen packaging method
- Offer live, molluscan shellfish from a display tank
- Custom-process animals for personal use
- Package unpasteurized juice for sale to the consumer without a warning label

CRISIS MANAGEMENT

A food safety system is designed to help you take steps to ensure that the food you serve is safe. Despite your best efforts, however, a foodborne-illness outbreak can occur in your establishment at any time. How you respond when that happens can determine whether or not you end up in the middle of a crisis.

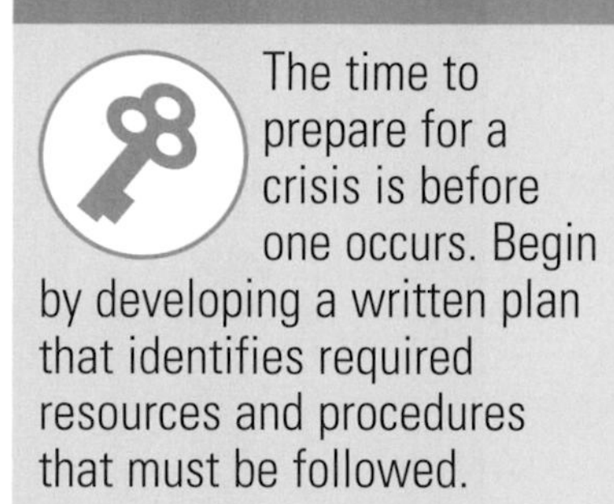

Key Point

The time to prepare for a crisis is before one occurs. Begin by developing a written plan that identifies required resources and procedures that must be followed.

The basis of a successful crisis management program is a written plan that identifies the resources required and procedures that must be followed to handle crises. The time to prepare for a crisis is before one occurs. There is no off-the-shelf disaster plan that works for every establishment. Each plan must meet its operation's individual needs.

Developing a Plan

When developing your plan, start by stating the basic objectives of the plan. These usually include meeting the immediate needs of the operation and keeping the business viable. Then, include the level of detail for the plan. This may consist of a checklist with brief step-by-step procedures to follow or a full-scale plan covering specific tasks, roles, and resources. You should also prepare specific procedures for developing, updating, and distributing the plan.

There are a number of steps you can take to prepare for the possibility of a crisis.

Develop a Crisis Management Team

Develop a team so that every task and role in your plan will be supported in the event of a crisis.

- **Develop a crisis management team.** In large, multi-unit operations, the team may include the president and senior managers from finance, operations, marketing, franchising, human resources, public relations, and training departments. Actual crisis-management teams are usually much smaller and may vary in makeup depending on the establishment and the situation. An independent establishment's team might include the owner, general manager, and chef. (See *Exhibit 10e.*)
- **Identify potential crises.** While the greatest threat to customers may be from foodborne illness, do not forget that other crises can include food security threats, robberies, severe weather, fire, or some other trauma.
- **Develop simple instructions on what to do in each type of crisis.** In a foodborne-illness outbreak, steps include isolating the suspect food, obtaining samples of the suspect food, preventing further sale of the food, excluding suspect employees from handling food, and contacting the local health department.
- **Assemble a contact list of names and numbers, and post it by the phones.** The list should include all crisis-management team members and outside resources, such as police, fire and health departments, testing labs, issues experts, and management or headquarters personnel.

- **Develop a crisis communication plan.** It should include:
  - A list of media responses or a Q&A sheet suggesting what to say in the event of each type of crisis identified. Create sample press releases that can be tailored quickly to each incident.
  - A list of media contacts to call for press conferences or news briefings. Include a media-relations plan with do's and don'ts for dealing with the media.
  - How to communicate with employees. Possibilities include shift meetings, email, a telephone tree, etc.
- **Assign and train a spokesperson to handle media relations.** Appoint a single spokesperson to handle all media queries and communications. Designating a point person usually results in more consistent messages and allows you to control media access to your staff. The spokesperson should be familiar with interview skills so that he or she knows what to expect and how to respond. Crisis situations can be very stressful, and training will enable your spokesperson to handle it better. Make sure all of your staff knows who the spokesperson is, and instruct them to direct questions to that person.
- **Assemble a crisis kit for the establishment.** The kit can be in the form of a three-ring notebook or binder enclosing the plan's materials. Keep the kit in an accessible place, such as the manager's or chef's office.
- **Test the plan by running a simulation.** Hire a public relations or consulting firm with crisis-management experience to enact a crisis and test your team's readiness. In most cases, the firm will design a simulated crisis that will be as close as possible to what could happen in a real-life situation.

Crisis Response

You may be able to avert a crisis by responding quickly when you do receive customer complaints. Take all customer complaints seriously. Express your concern and be sincere, but do not admit responsibility or accept liability. Listen carefully and promise to investigate and respond.

Key Point

Take all customer complaints seriously. Express your concern, listen carefully, and promise to investigate and respond.

With legal guidance, consider developing an incident report to help you through the process. Questions to ask include:

- What did you eat and drink at our establishment and when?
- When did you become ill? What were the symptoms, and how long did you experience them?
- Did you eat anything else before or after eating at our establishment? What and where? Who else ate the same food, and did they become ill?
- Did you seek medical attention? Where and how soon after becoming ill? What diagnosis and treatment did you receive?

Evaluate the complaint. If more than one person has complained, you have the potential for a crisis on your hands. Take steps to control the situation and reassure customers that you are doing everything you can to identify and fix the problem. At this point, call your crisis team together and implement your plan.

- Direct the team to gather information, plan courses of action, and manage events as they unfold.
- Work with, not against, the media. Be as proactive as you can, as early as you can. Make sure the spokesperson is fully informed before arranging a press conference. Contacting the media before they contact you helps you to control what the media reports. Stick to the facts, and be as honest as possible. If you do not have all the facts, say so, and let the media know that you will communicate them as soon as you do know. Keep a cool head and do not be defensive. The easiest way to magnify or prolong a crisis is to deny, lie, or change your story.
- Show concern and be sincere. If health officials have confirmed that your establishment is the source of the illness, accept responsibility. Accepting responsibility is not the same as admitting liability. While customers may have become ill from eating food in your operation, the cause may have been beyond your control and not your fault. If you do not express your concern, and mean it, you will lose credibility with the public, not just customers.
- Communicate the information directly to all of your key audiences. Do not depend on the media to relay all the facts.

Tell your side of the story to employees, customers, stockholders, and the community. Use newsletters, a Web site, flyers, and newspaper or radio advertising.

- Fix the problem and communicate what you have done both to the media and to your customers. Each time you take a step to resolve the problem, let the media know. Hold briefings when you have news, and go into each briefing or press conference with an agenda. Take control. Do not simply respond to questions.

Post Crisis Assessment

Once a crisis is over, it is important to determine the causes and effects of the crisis so that your establishment can implement changes to take advantage of the lessons learned. Some things to evaluate include:

- Obstacles faced in returning to normal operations
- How communication with employees and customers was handled
- Assessment of the damages from the crisis
- Overall assessment of the crisis and your response

A foodborne-illness outbreak has the potential to damage your business beyond repair. Investing time and resources in a crisis management plan can ensure against that. Remember these three key rules of crisis management:

1. Take steps to prevent a crisis from occurring by practicing good food safety habits.
2. Prepare for the possibility of a crisis by developing contingency plans.
3. If a crisis does occur, take control of the situation. Use your plan to manage the crisis thoughtfully, honestly, and as quickly as possible.

SUMMARY

A food safety management system will help you prevent foodborne illness by controlling hazards throughout the flow of food.

Active managerial control focuses on establishing policies and procedures to control five common risk factors responsible for foodborne illness: purchasing food from unsafe sources, failing to cook food adequately, holding food at improper temperatures, using contaminated equipment, and poor personal hygiene. The polices and procedures that an establishment puts in place, or revision of existing ones, will be the result of a careful analysis of potential breakdowns related to these five risk factors as they apply throughout the flow of food. Once procedures have been implemented, they must be continuously monitored, and the system must be verified to ensure that the procedures put in place are controlling the identified risks.

A HACCP (Hazard Analysis Critical Control Point) system focuses on identifying specific points within a product's flow through the operation that are essential to prevent, eliminate, or reduce biological, chemical, or physical hazards to safe levels. To be effective, a HACCP system must be based on a plan specific to a facility's menu, customers, equipment, processes, and operation. The HACCP plan is developed following seven sequential principles—essential steps for building a food safety system.

First, the establishment must identify and assess potential hazards in the food they serve by taking a look at how it is processed. Once common processes have been identified, they can determine where food safety hazards are likely to occur for each one. The establishment must then identify points where they can be prevented, eliminated, or reduced to safe levels. These are the critical control points (CCPs). Next, the establishment must determine minimum and maximum limits that must be met for each CCP to prevent, eliminate, or reduce the hazard. The establishment must determine how they will monitor the CCPs they have identified and what actions will be taken when critical limits have not been met. Finally, the establishment must find ways to verify that the HACCP system is working, and establish procedures for record keeping and documentation.

A food safety system is designed to help you take steps to ensure that the food you serve is safe. Despite your best efforts, however, a foodborne-illness outbreak can occur in your establishment. The time to prepare for a crisis is before one occurs. The key is to start with a written plan that identifies the resources required and procedures that must be followed to handle the crisis. To prepare for a crisis: develop a crisis management team, spell out specific instructions for handling the crisis, develop a crisis communication plan, and assign and train a media spokesperson. When receiving customer complaints, listen carefully, express concern, and be sincere. Do not admit responsibility, but promise to investigate and respond. With legal advice, consider developing an incident report to guide you through the process. Call your crisis management team together and implement your plan.

Apply Your Knowledge

Use these questions to review the concepts presented in this chapter.

Discussion Questions

1. What is active managerial control?
2. What is the difference between active managerial control and HACCP?
3. What are the five foodborne-illness risk factors identified by the CDC?
4. What is a critical limit? Give an example.
5. When is an establishment required to have a HACCP plan?

For answers, please turn to the Answer Key.

Apply Your Knowledge

Use these questions to test your knowledge of the concepts presented in this chapter.

Multiple-Choice Study Questions

1. The temperature of a roast is checked to see if it has met its critical limit of 145°F (63°C). This is an example of which HACCP principle?
   A. Verification
   B. Monitoring
   C. Record keeping
   D. Hazard analysis

2. The temperature of a pot of beef stew is checked during holding. The stew has not met the critical limit of 135°F (57°C) and is discarded according to house policy. This is an example of which HACCP principle?
   A. Monitoring
   B. Corrective action
   C. Hazard analysis
   D. Verification

3. Al's Big Burgers makes some of the juiciest burgers in town. Every hamburger is cooked from fresh ground beef to a minimum internal temperature of 150°F (66°C) for fifteen seconds and then dressed with all the trimmings. Al's establishment is at risk of
   A. using contaminated equipment.
   B. failing to receive food properly.
   C. failing to cook food adequately.
   D. failing to store food properly.

4. Which of the following risks is *not* commonly responsible for foodborne illness?
   A. Failing to cook food adequately
   B. Failing to thaw food properly
   C. Failing to purchase food from safe sources
   D. Failing to hold food at the proper temperature

5. Which of the following is *not* a corrective action?
   A. Continuing to cook a hamburger until it reaches a minimum internal temperature of 155°F (68°C) for fifteen seconds
   B. Discarding cooked chicken that has been held at 120°F (49°C) for five hours
   C. Sanitizing a prep counter before starting a new task
   D. Rejecting a shipment of oysters received at 55°F (13°C) that will be served raw

Continued on next page...

Multiple-Choice Study Questions *continued*

6. Which of the following programs should be in place before you begin developing your food safety system?
   A. Personal hygiene program
   B. Incentive program
   C. Workplace accident prevention program
   D. None of the above

7. The purpose of a food safety management system is to
   A. identify the proper methods for receiving food.
   B. identify and control possible hazards throughout the flow of food.
   C. keep the establishment pest free.
   D. identify faulty equipment within the establishment.

8. A chef sanitized his thermometer probe and checked the temperature of a baine of minestrone soup being held in a hot-holding unit. The temperature was 120°F (49°C), which did not meet the establishment's critical limit of 135°F (57°C). He recorded the temperature in the log and reheated the soup to 165°F (74°C) for fifteen seconds. Which was the corrective action?
   A. Sanitizing the thermometer probe
   B. Taking the temperature of the soup
   C. Reheating the soup
   D. Recording the temperature of the soup in the temperature log

9. A HACCP plan is required when an establishment
   A. serves raw shellfish.
   B. serves undercooked ground beef.
   C. uses mushrooms that have been picked in the wild.
   D. packages unpasteurized juice for sale to consumers without a warning label.

For answers, please turn to the Answer Key.

ADDITIONAL RESOURCES

Books and Periodicals

Bolat, T. 2002. Implementation of the hazard analysis critical control point (HACCP) system in a fast food business. *Food Reviews International.* 18 (4):337.

Food and Drug Administration. 2002. *Managing food safety: A guide for the voluntary use of HACCP principles for operators of food service and retail establishments.* Washington, DC: Food and Drug Administration.

National Advisory Committee on Microbiological Criteria for Foods. 1997. *Hazard analysis and critical control point principles and application guidelines.* Washington, DC: U.S. Department of Agriculture.

Stevenson, K.E. and D.T. Bernard, eds. 1999. *A systematic approach to food safety: A comprehensive manual for developing and implementing a hazard analysis and critical control point plan.* Washington, DC: The Food Processors Institute.

Stevenson, K.E. and D.T. Bernard, eds. 1995. *Establishing hazard analysis critical control point programs: A workshop manual.* Washington, DC: The Food Processors Institute.

Web Sites

FDA Bad Bug Book

http://vm.cfsan.fda.gov/~mow/intro.html

Produced by the FDA's Center for Food Safety and Nutrition (FDA CFSAN), this online handbook provides basic facts regarding pathogenic foodborne microorganisms and natural toxins. It brings together information from the FDA, the CDC, the USDA Food Safety Inspection Service, and the National Institutes of Health.

FDA Center for Food Safety and Applied Nutrition (CFSAN)

www.cfsan.fda.gov

As the center within the FDA responsible for food safety and nutrition, CFSAN promotes and protects public health by researching and implementing guidelines, policies, and standards to ensure that food is safe, nutritious, wholesome, and properly labeled. This Web site provides a wealth of information on food safety and sanitation, including corresponding guidelines, policies, and standards.

FDA Food Code
http://vm.cfsan.fda.gov/~dms/foodcode.html
As the basis for many local sanitation codes, as well as the information in this textbook, the FDA Food Code, available at this Web address, is a useful resource for all food safety-related information for the restaurant and foodservice industry.

FDA Seafood Information and Resources
http://vm.cfsan.fda.gov/seafood1.html
The FDA operates an oversight compliance program for fishery products under which responsibility for the products' safety, wholesomeness, identity, and economic integrity rests with the processor or importer. This Web site houses information on the seafood program, foodborne pathogens and contaminants associated with seafood, and HACCP compliance.

International HACCP Alliance
http://haccpalliance.org
The International HACCP Alliance was formed to assist the meat and poultry industry in preparing for mandatory HACCP. The Alliance promotes public health and safety by facilitating uniform development and implementation of HACCP.

National Advisory Committee for the Microbiological Criteria for Foods (HACCP Principles and Application Guidelines)
http://seafood.ucdavis.edu/Guidelines/nacmcf1.htm
The guidelines found on this Web page are intended to facilitate the development and implementation of effective HACCP plans as appropriate to each segment of the food industry.

USDA–FDA HACCP Training Programs
www.nal.usda.gov/fnic/foodborne/haccp/index.shtml
This database, a joint project of the FDA, FSIS, Cooperative State Research, Education and Extension Service, and the National Agricultural Library, provides up-to-date listings of HACCP training programs and HACCP resource materials.

USDA–FSIS
www.fsis.usda.gov
FSIS is the public health agency of the USDA responsible for ensuring that the nation's commercial supply of meat, poultry, and egg products is safe, wholesome, and correctly labeled and packaged. This Web site contains a wealth of information on food safety relating to meat, poultry, and eggs.

Unit 3

Clean and Sanitary Facilities and Equipment

Traulsen
Chris

11 Sanitary Facilities and Equipment

Inside this chapter:

- Designing a Sanitary Establishment
- Considerations for Other Areas of the Facility
- Sanitation Standards for Equipment
- Installing and Maintaining Kitchen Equipment
- Utilities

After completing this chapter, you should be able to:

- Identify when a plan review is required.
- Identify organizations that certify equipment that meets sanitation standards.
- Identify characteristics of an appropriate food-contact and nonfood-contact surface.
- Identify the requirements for installing stationary and mobile equipment.
- Recognize the importance of maintaining equipment.
- Identify and prevent cross-connection and backflow.
- Identify requirements for handwashing facilities including appropriate locations and numbers.
- Identify the proper response to a waste-water overflow.
- Recognize the importance of properly installing and maintaining grease traps.
- Identify potable water sources and testing requirements.
- Identify lighting-intensity requirements for different areas of the establishment.
- Identify methods for preventing lighting sources from contaminating food.
- Identify methods for preventing ventilation systems from contaminating food and food-contact surfaces.
- Identify requirements for storing indoor and outdoor waste.
- Identify proper methods for cleaning waste receptacles.
- Recognize the need for frequent waste removal to prevent odor and pest problems.
- Identify characteristics of appropriate flooring for food establishments.
- Recognize the importance of complying with ADA requirements for facility design.
- Recognize the importance of keeping physical facilities in proper repair.
- Identify requirements for warewashing facilities.

Key Terms

Americans with Disabilities Act (ADA)
Porosity
Resiliency
Coving
Service sink
NSF International
Underwriters Laboratories (UL)
Blast chiller
Tumble chiller
Cantilever–mounted equipment
Potable water
Booster heater
Cross-connection
Backflow
Flood rim
Vacuum breaker
Air gap
Foot-candle
Pulper

Apply Your Knowledge

Check to see how much you know about the concepts in this chapter. Use the page references provided to explore the topic in each question.

Test Your Food Safety Knowledge

1. **True or False:** A hose attached to a utility-sink faucet and left sitting in a bucket of dirty water could contaminate the water supply. *(See page 11-24.)*
2. **True or False:** There must be a minimum of twenty foot-candles of light (220 lux) in a food-preparation area. *(See page 11-27.)*
3. **True or False:** Handwashing stations are required in warewashing and service areas. *(See page 11-10.)*
4. **True or False:** When mounted on legs, stationary equipment must be at least two inches off the floor. *(See page 11-19.)*
5. **True or False:** Grease on an establishment's ceiling can be a sign of inadequate ventilation. *(See page 11-26.)*

For answers, please turn to the Answer Key.

INTRODUCTION

Many breakdowns in sanitation are caused by facilities and equipment that are simply too difficult to keep clean. Sanitary facilities and equipment are basic parts of a well-designed, food-safety system. In this chapter, you will find a wide range of information on various equipment and facility-related issues that are key to keeping an establishment safe.

DESIGNING A SANITARY ESTABLISHMENT

When designing or remodeling a facility, consider how the building and the equipment in each area will be kept clean and maintained in good repair. Those areas not cleaned properly can allow microorganisms to remain, which can cause serious problems for the food coming in contact with them. Facilities should be arranged so that contact with contaminated sources—such as garbage or dirty tableware, utensils, and equipment—is unlikely to occur.

This chapter focuses on four topics related to the sanitary layout and design of equipment and facilities:

- Arrangement and design of equipment and fixtures to comply with sanitation standards
- Material selection for walls, floors, and ceilings that will make cleaning these surfaces easier
- Design of utilities to prevent contamination and to make cleaning easier
- Proper waste management to avoid contaminating food and attracting pests

The Plan Review

A sanitary foodservice layout and design begins in the planning stage. Prior to starting construction, consult local regulations. Many jurisdictions require approval of layout and design plans by the health department or local regulatory agency prior to new construction or extensive remodeling. These plans should include proposed layout, mechanical plans, type of construction materials to be used, and the types or models of proposed equipment. Specifications for utilities, plumbing, and ventilation will probably be required.

Some local jurisdictions will require the approval of design plans by building and zoning departments. In addition, the **Americans with Disabilities Act (ADA)** requires reasonable accommodation for access to the building by both patrons and employees with disabilities. These guidelines can be found in the ADA Accessibility Guide (ADAAG).

Key Point

Even if local laws do not require it, layout and design plans should be reviewed by the local regulatory agency.

Even if local laws do not require it, layout and design plans should be reviewed by local or state regulatory agencies. In addition to ensuring compliance with sanitation requirements, such reviews can save time and money. This is true for remodeling, as well as for new construction.

To assist establishments, regulatory agencies might provide information they consider necessary for good sanitation. Ask for guides on how to submit plans and specifications.

Once construction is completed, the establishment applies for a permit to operate. Before granting the permit, the regulatory agency might conduct an inspection to make certain all design and installation requirements have been met. After the establishment has passed the inspection and obtains a certificate of operation, it can open for business.

Materials for Interior Construction

Materials used during construction must be selected with several factors in mind. Sound-absorbent surfaces that also resist absorption of grease and moisture and reflect light will probably create an environment acceptable to your regulatory agency.

However, the most important consideration when selecting construction materials is how easy the establishment will be to clean and maintain.

Flooring

Flooring materials in kitchen and service areas should meet requirements for health and safety, strength and durability, and appearance. The floor surfaces should be easy to clean, wear-resistant, slip-resistant, and nonporous. Flooring should be kept in good repair and replaced if damaged or worn.

Key Point

Select nonabsorbent flooring materials for food-prep areas, walk-in refrigerators, warewashing areas, and restrooms.

One of the most important factors to consider when selecting floor covering for an establishment is the porosity of the material. **Porosity** is the extent to which a floor covering can become saturated by liquids. When liquids are absorbed, the flooring can be damaged and microorganisms can grow. It might also cause a potential slip-and-fall situation. The FDA Food Code recommends the use of nonabsorbent flooring in food-preparation areas, walk-in refrigerators, warewashing areas, restrooms, and other areas subject to moisture, flushing, or spray cleaning.

Nonporous, Resilient Flooring

In most areas of the establishment, nonporous, resilient flooring is the best choice. **Resiliency** means a material has the ability to react to a shock without breaking or cracking.

Nonporous, resilient materials are relatively inexpensive and are easy to clean and maintain. They are rated for light, moderate, and heavy traffic, as well as for resistance to grease and alkalis. Some are easily damaged by cigarette burns or sharp objects, but they are also easy to repair or replace. They tend to be slippery when wet. Rubber tile and light- and medium-weight vinyl are good considerations. Vinyl tile is not recommended in dining rooms or public areas because it requires a high level of maintenance, such as waxing and frequent machine buffing. Vinyl tile is a practical choice in employee dressing rooms and break rooms, and foodservice offices. See *Exhibit 11a* for characteristics and recommended uses of nonporous, resilient flooring.

Hard-Surface Flooring

Hard-surface flooring is also commonly used in establishments. It includes quarry tile, ceramic tile, brick, terrazzo, marble, and hardwood. These materials are nonporous, but are not resilient.

Hard-surface floors are very durable, but may crack or chip if heavy objects are dropped on them. In addition, breakable objects dropped on hard-surface flooring will be more likely to shatter than those dropped on resilient flooring. These surfaces do not absorb sound and are somewhat difficult to clean compared to resilient surfaces. Some types, such as marble, are slippery.

Exhibit 11a

Characteristics of Resilient Flooring

Material	Where to Use	Durability	Advantages	Disadvantages
Rubber tile	Kitchens; restrooms	Less durable and less resistant to grease and alkalis	Nonslip; resilient	Can only be used in moderate traffic areas
Vinyl sheet	Offices; kitchens; corridors	Less resistant to grease and alkalis	Very resilient	Can only be used in light or moderate traffic areas
Vinyl tile	Offices; employee restrooms	Wears out quickly with high traffic	Very resilient	Requires waxing and machine buffing

Quarry and ceramic tile are excellent for use in public restrooms or high-soil areas, but unglazed tiles should be selected for these areas due to their slip-resistant qualities. In general, hard-surface flooring is very heavy and more expensive to install and maintain than resilient flooring. See *Exhibit 11c* for characteristics and recommended uses of hard-surface flooring.

Carpeting

Carpeting is popular in dining rooms because it absorbs sound. However, it is not recommended in high-soil areas, such as waitstaff service areas, tray and dish drop-off areas, beverage stations, and major traffic aisles. Carpet can be maintained by simple vacuum cleaning. Areas prone to heavy traffic and moisture will require routine cleaning. Where sanitation, soiling, moisture, and fire safety are concerns, special carpet can be purchased.

Special Flooring Needs

Each area of an establishment has its particular flooring needs. Nonslip surfaces should be used in traffic areas. In fact, nonslip surfaces are best for the entire kitchen, since slips and falls are a potential hazard. Rubber mats are allowed for safety reasons in areas where standing water may occur, such as the dish room. Rubber mats should be picked up and cleaned separately when scrubbing floors.

Coving is required in establishments using resilient or hard-surface flooring materials. **Coving** is a curved, sealed edge placed between the floor and the wall to eliminate sharp corners or gaps that would be impossible to clean. (See *Exhibit 11b.*) The coving tile or strip must adhere tightly to the wall to eliminate hiding places for insects and to prevent moisture from deteriorating the wall.

Exhibit 11b

Coving

Coving is a curved, sealed edge placed between the floor and the wall to eliminate sharp corners or gaps.

Finishes for Interior Walls and Ceilings

Interior finishes are the materials used on the surface of walls, partitions, or ceilings of an establishment. As with flooring, the most important criteria when choosing interior finishes are ease of cleaning and porosity.

Exhibit 11c

Hard-Surface Flooring

Material	Where to Use	Durability	Advantages	Disadvantages
Marble; Terrazzo	Public corridors; dining rooms; public restrooms	Wear-resistant	Nonporous; good appearance	Nonresilient; expensive; requires special care, such as buffing and polishing; heavy and difficult to install
Quarry tile	Kitchen; dishwashing areas; receiving areas; offices; restrooms; dining rooms; service areas	Wear-resistant	Nonporous	Nonresilient; heavy; relatively expensive; slippery when wet unless an abrasive is added
Wood	Offices; dining rooms	Durable in lower traffic areas	Good appearance and sound absorption	Requires frequent polishing and periodic refinishing to maintain surface qualities
Acrylic wood (plastic absorbed into wood)	Offices; dining rooms	Highly abrasion-resistant	Less vulnerable to stains, scratches, chemical damage; high resistance to bacterial growth	Much like resilient flooring

When selecting finishes for walls and ceilings, consider location. One type of material might be suitable in one area, but may be a poor choice for another. Walls and ceilings in food-preparation areas must be light in color to distribute light and to make it easier to spot soil when cleaning. They should be kept in good repair, free of cracks, holes, or peeling paint. The best wall finish in cooking areas is ceramic tile; however, it must be monitored for grout loss and regrouted whenever necessary. Stainless steel is used occasionally because of its durability and resistance to moisture. The most common ceiling materials are acoustic tile, painted drywall, painted plaster, or exposed concrete.

The support structures for walls and ceilings (studs, joists, and rafters) and pipes should not be exposed unless they are finished and sealed for cleaning. Flexible materials such as paper, vinyl, and thin wood veneers are often used for walls and ceilings. Vinyl wall coverings are used in many areas of an establishment because they are attractive, relatively inexpensive, easy to clean, and durable. Vinyl wall coverings are rated for flammability by testing agencies. Plaster or cinder-block walls sealed and painted with oil-resistant and easy-to-wash, glossy paints are appropriate for dry areas of the facility.

CONSIDERATIONS FOR OTHER AREAS OF THE FACILITY

Dry Storage

Dry storerooms should be constructed of easy-to-clean materials that allow good air circulation. (See *Exhibit 11d.*) Shelving, table tops, and bins for dry ingredients should be made of corrosion-resistant metal or food-grade plastic.

Any windows in the storeroom should have frosted glass or shades. Direct sunlight can increase the temperature of the room and affect food quality.

Steam pipes, water lines, and other conduits have no place in a well-designed storeroom. Dripping condensation or leaks in overhead pipes can promote microbial growth in such normally stable items as crackers, flour, and baking powder. Leaking overhead sewer lines can be a source of contamination for any

Exhibit 11d

Acceptable Dry-Storage Facility

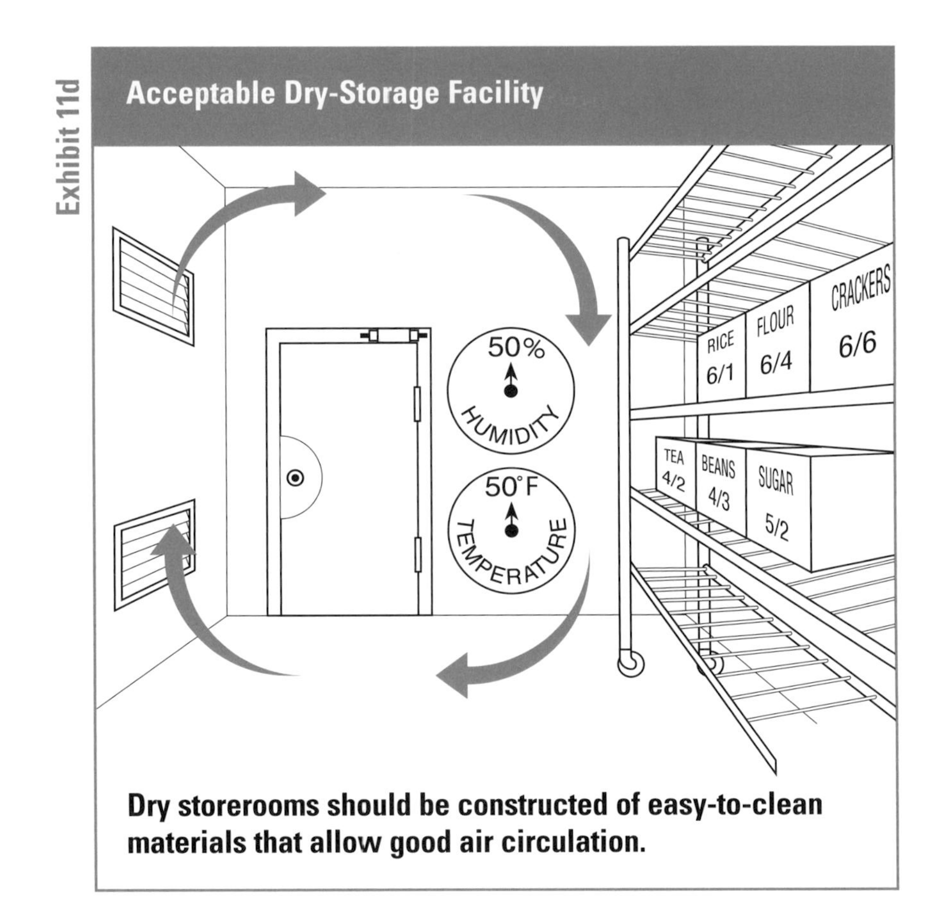

Dry storerooms should be constructed of easy-to-clean materials that allow good air circulation.

food. Hot water heaters or steam pipes can increase the temperature of the storeroom to levels that will allow foodborne pathogens to grow.

Dry food is especially susceptible to attack by insects and rodents. Cracks and crevices in floors or walls should be filled. Doors leading to the exterior of the building should be self-closing. Screens for windows and doors should be sixteen mesh to the inch, without holes or tears.

Restrooms and Handwashing Stations

Local building and health codes usually specify how many sinks, stalls, toilets, and urinals are required in an establishment. It is best if separate restrooms are provided for employees and customers. If this is not possible, the establishment must be designed so patrons do not pass through food-preparation areas to reach the restroom, since they could contaminate food or food-contact surfaces.

Restrooms should be convenient, sanitary, and have a fully equipped handwashing station, as well as self-closing doors. They must be adequately stocked with toilet paper, and trash receptacles must be provided if disposable paper towels are used. Covered waste containers must be provided in women's restrooms for the disposal of sanitary supplies.

Handwashing Stations

Key Point

Handwashing stations must be conveniently located in food-preparation areas, service areas, warewashing areas, and restrooms.

Handwashing stations must be conveniently located so employees will be encouraged to wash their hands often. They are required in food-preparation areas, service areas, warewashing areas, and restrooms. These stations must be operable and must be stocked and maintained. A handwashing station must be equipped with the following items: (See *Exhibit 11e.*)

- **Hot and cold running water.** Hot and cold water should be supplied through a mixing valve or combination faucet at a temperature of at least 100°F (38°C).
- **Soap.** The soap can be liquid, bar, or powder. Liquid soap is generally preferred, and some local codes require it.
- **A means to dry hands.** Most local codes require establishments to supply disposable paper towels in handwashing stations. Continuous-cloth towel systems, if allowed, should be used only if the unit is working properly and the towel rolls are checked and changed regularly. Installing at least one warm-air dryer in a handwashing station will provide an alternate method for drying hands if paper towels run out. The use of common cloth towels is not permitted because they can transmit contaminants from one person's hands to another.
- **A waste container.** Waste containers are required if disposable paper towels are provided.
- **Signage indicating employees are required to wash hands before returning to work.**

Sinks

Each sink in an establishment must be used for its intended purpose only. Handwashing sinks are used for handwashing.

Exhibit 11e

Acceptable Handwashing Station

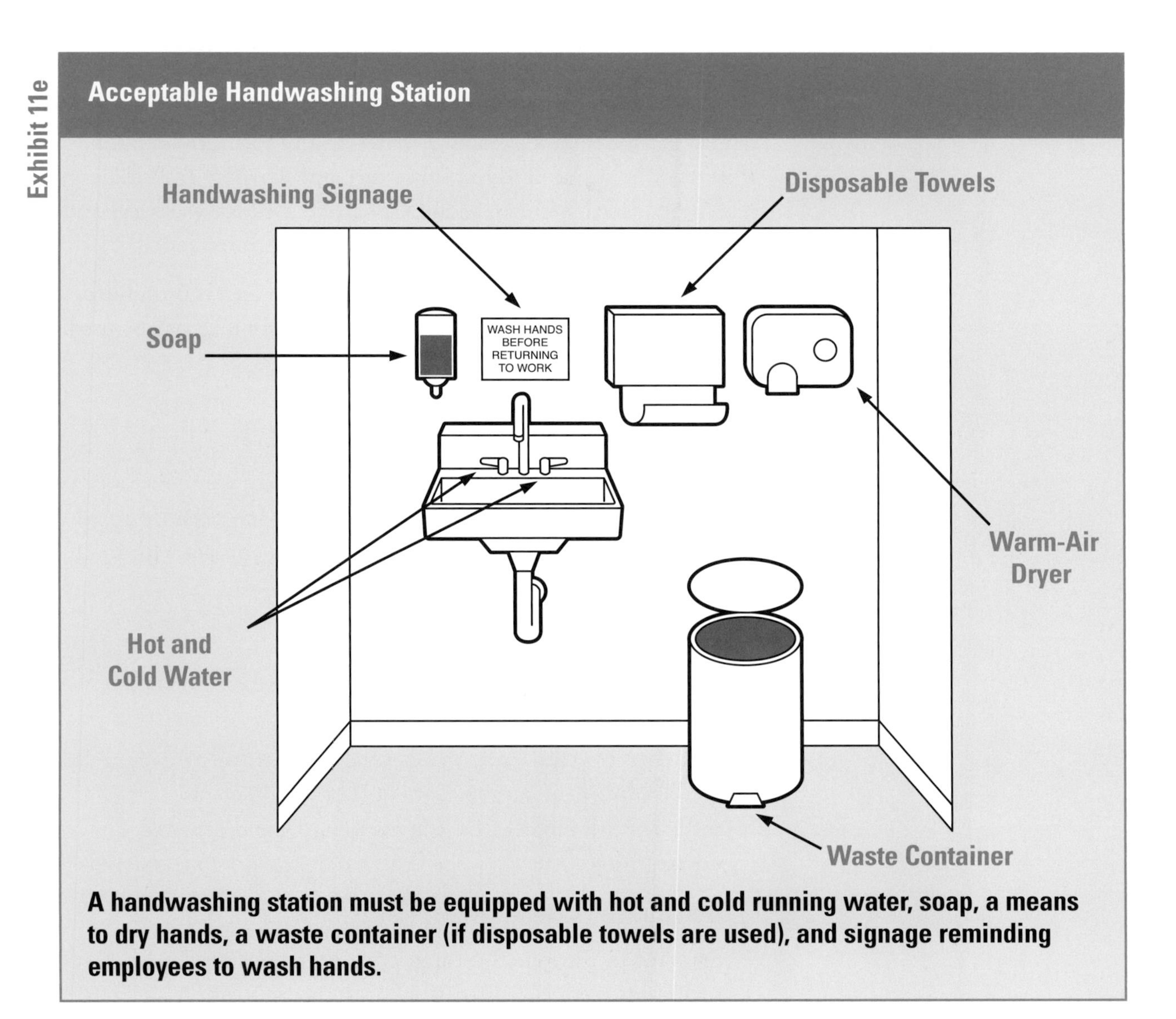

A handwashing station must be equipped with hot and cold running water, soap, a means to dry hands, a waste container (if disposable towels are used), and signage reminding employees to wash hands.

Food-preparation sinks are used for food preparation. **Service sinks** are used for cleaning mops and disposing of waste water, and must be kept separate. At least one service sink or curbed drain area is required in an establishment to dispose of soiled water.

Dressing Rooms and Lockers

Dressing rooms are not required. If available, they must not be used for food preparation, storage, or utensil washing. Lockers should be located in a separate room or a room where contamination of food, equipment, utensils, linens, and single-service items will not occur.

Premises

The parking lot and walkways should be kept free of litter and graded so that standing pools of water do not form. In addition, they must be surfaced to minimize dirt and blowing dust. It is recommended that concrete and asphalt be used for walkways and parking lots. Gravel, while acceptable, is not recommended.

Patron traffic through the food-preparation area is prohibited, although guided tours are allowed. The premises may not be used for living or sleeping quarters.

SANITATION STANDARDS FOR EQUIPMENT

It is important to purchase equipment that has been designed with sanitation in mind. Surfaces that come in contact with food must be

- safe.
- durable.
- corrosion-resistant.
- nonabsorbent.
- sufficient in weight and thickness to withstand repeated warewashing.
- smooth, and easy to clean.
- resistant to pitting, chipping, crazing (spider cracks), scratching, scoring, distortion, and decomposition.

Equipment surfaces that are not designed to come in contact with food, but which are exposed to splash, spillage, or other food soiling, or that require frequent cleaning, must be

- constructed of smooth, nonabsorbent, corrosion-resistant material.
- free of unnecessary ledges, projections, and crevices.
- designed and constructed to allow easy cleaning and maintenance.

Exhibit 11f

NSF and UL EPH Marks

Look for the NSF International mark and UL EPH product marks on sanitary equipment.

The task of choosing equipment designed for sanitation has been simplified by organizations such as **NSF International** and **Underwriters Laboratories (UL).** NSF International develops and publishes standards for sanitary equipment design. They also assess and certify that equipment has met these standards. The presence of the NSF mark on foodservice equipment means it has been evaluated, tested, and certified by NSF International as meeting international, commercial food equipment standards. UL similarly provides sanitation classification listings for equipment found in compliance with NSF International standards. UL also lists products complying with their own published environmental and public health standards. Restaurant and foodservice managers should look for the NSF International mark or the UL EPH product mark on commercial foodservice equipment. *Exhibit 11f* shows examples of the NSF International mark and the UL EPH product marks.

Only commercial foodservice equipment should be used in establishments since household equipment is not built to withstand heavy use.

Although all equipment used in an establishment must meet standards such as those set by NSF International, certain equipment requires particular attention.

Warewashing Machines

Warewashing machines vary widely by size, style, and method of sanitizing. High-temperature machines sanitize with extremely hot water; chemical-sanitizing machines use a chemical solution. Some states require the local regulatory agency's approval before installing a chemical warewashing system.

The size and type of machine you choose depends upon the nature and volume of the items to be cleaned and the required turnaround time for clean tableware and utensils. Because a warewashing machine is a big investment, the manager must carefully match the machine to the establishment's needs.

The following types of warewashing machines are common in restaurant and foodservice establishments:

- **Single-tank, stationary-rack, with doors.** This machine holds a stationary rack of tableware and utensils. Items are washed by detergent and water from below and, sometimes, from above the rack. The wash cycle is followed by a hot-water or chemical-sanitizer final rinse.
- **Conveyor machine.** With this machine, a conveyor moves racks of items through the various cycles of washing, rinsing, and sanitizing. The machine may have a single tank or multiple tanks.
- **Carousel or circular conveyor machine.** This multiple-tank machine moves tableware and utensils on a peg-type conveyor or in racks. Some models have an automatic stop after items go through the final rinse cycle. In other models, items must be removed after the final rinse, or they will continue to travel through the machine.
- **Flight-type**. This is a high-capacity, multiple-tank machine with a peg-type conveyor. It may also have a built-in dryer. It is commonly used in institutions and very large establishments.
- **Batch-type, dump.** This stationary-rack machine combines the wash and rinse cycles in a single tank. Each cycle is timed, and the machine automatically dispenses both the detergent and the sanitizing chemical or hot water. Wash and rinse water are drained after each cycle.
- **Recirculating, door-type, non-dump.** This stationary-rack machine is not completely drained of water between cycles. The wash water is diluted with fresh water and reused from cycle to cycle.

Consider the following general guidelines regarding the installation and use of warewashing machines.

- Water pipes to the warewashing machine should be as short as possible to prevent the loss of heat from hot water entering the machine.
- The machine must be raised at least six inches off the floor to permit easy cleaning underneath.

- Materials used in warewashing machines should be able to withstand wear, including the action of detergents and sanitizers.
- Information should be posted on or near the machine regarding proper water temperature, conveyer speed, water pressure, and chemical concentration.
- The machine's thermometer should be located so it is readable. The thermometer should have a scale in increments no greater than 2°F (1°C).

Clean-in-Place Equipment

Some equipment is designed to be cleaned and sanitized by having a detergent solution, hot-water rinse, and sanitizing solution passed through it. Certain soft-serve ice cream and frozen yogurt dispensers are cleaned and sanitized this way. These machines must be constructed so the cleaning and sanitizing solution remains within the fixed system of tubes and pipes for a predetermined amount of time. All food-contact surfaces must be reached by the solutions, but they must not leak into the rest of the machine.

Clean-in-place equipment must be self-draining. Some means of inspection must be provided to make sure the machine has been completely emptied of cleaning solution and has been thoroughly rinsed with fresh water. Manufacturers' instructions should be followed carefully.

Refrigerators and Freezers

There are several types of foodservice refrigerator and freezer units. The two most common types are walk-in and reach-in refrigerators and freezers. These units should be made of stainless steel or a combination of stainless steel and aluminum. The doors should be constructed to withstand heavy use and should close with a slight nudge. Door gaskets can be fixed in place, or removable for easy cleaning. A drain must be provided and maintained for disposal of condensation and defrost water. A properly plumbed, indirect drain can be used in the walk-in refrigerator. Excess condensation can be minimized by maintaining a flush-fitting floor sweep (gasket) under the door.

Exhibit 11g

Commercial Type Walk-In Refrigerator

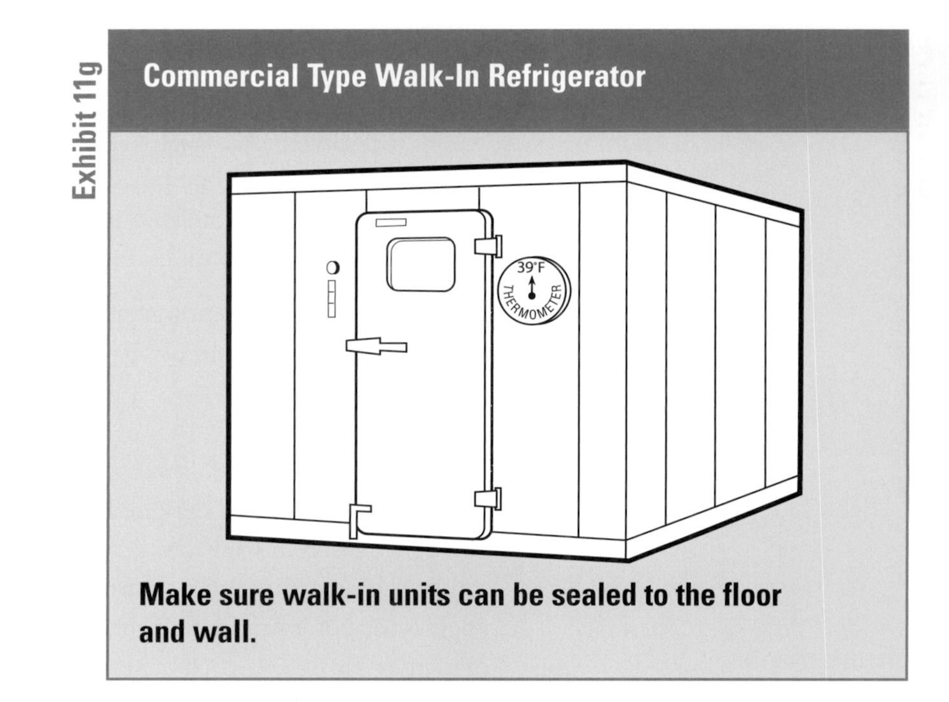

Make sure walk-in units can be sealed to the floor and wall.

Walk-in refrigerators and freezers with windows in the door may reduce unnecessary opening. Forced-air circulating fans are essential to help provide a quick recovery time so refrigerator and freezer temperatures remain at the appropriate level.

When purchasing a refrigerator or freezer unit, make sure it carries the NSF International mark, the UL EPH product mark, or their equivalent. This will ensure that the unit is designed to protect food and simplify cleaning. In addition to the NSF International standards mentioned earlier in this chapter, consider these factors when purchasing a refrigerator or freezer unit:

- **Choose a unit with adequate storage space**. An uncrowded unit can maintain required holding temperatures, is easier to clean, will prevent moisture buildup, and can minimize breakdowns.
- **Make sure walk-in units can be sealed to the floor and wall.** (See *Exhibit 11g*.) They should offer no access to moisture or rodents. Flooring materials must be able to withstand heavy impact.
- **Purchase reach-in refrigerator or freezer units with legs that elevate them six inches off the floor.** Otherwise mount and seal them on a masonry base. Casters making it easy to

Exhibit 11h

Commercial Type Reach-In Freezer

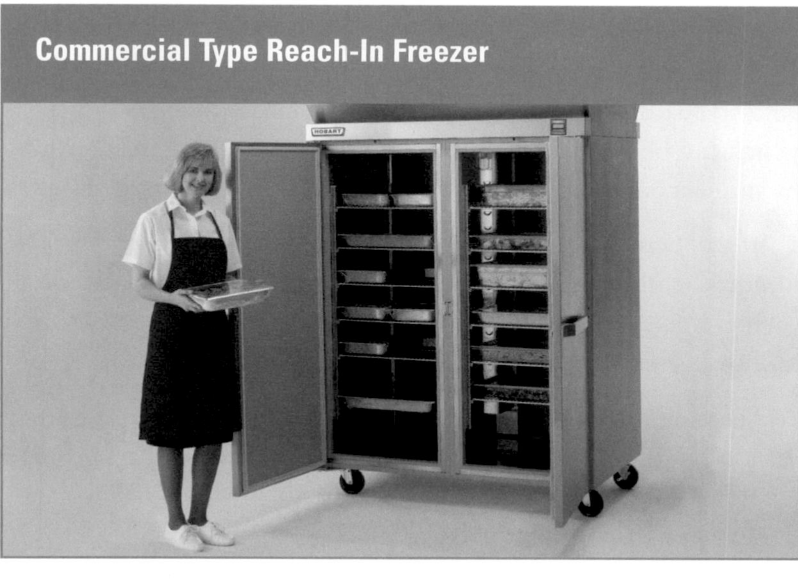

Note the caster wheels, which make it easy to move the unit for cleaning.

Courtesy of Hobart Corporation

move the unit for cleaning are often preferred or required by local regulatory agencies. (See *Exhibit 11h.*)

- **Make sure the unit meets the temperature requirements of the food you store.** Built-in thermometers should be easy to locate and read, and be accurate to within 3°F (2°C).

Blast Chillers and Tumble Chillers

Blast chillers are designed to move food through the temperature danger zone quickly. (See *Exhibit 11i.*) Many blast chillers are able to cool food from 135°F to 37°F (57°C to 3°C) within ninety minutes. Most units allow the operator to set target chill temperatures and monitor the temperature of food throughout the chill cycle. Once chilled to safe temperatures, the food then can be stored in conventional refrigerators or freezers.

Exhibit 11i

Blast Chiller

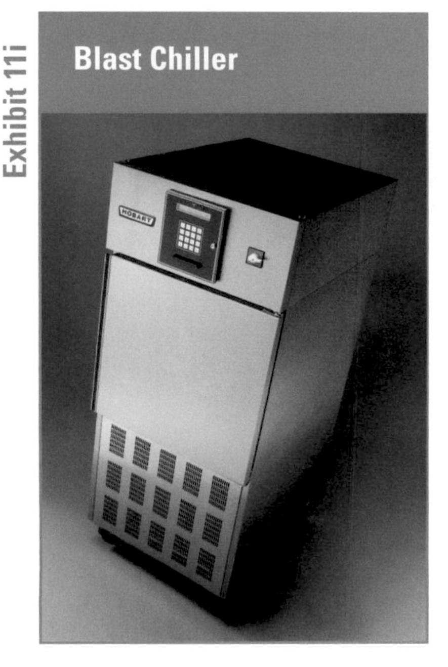

Many blast chillers can cool food from 135°F to 37°F (57°C to 3°C) within ninety minutes.

Courtesy of Hobart Corporation

Tumble chillers are also designed to cool food quickly. Prepackaged hot food is placed into a drum, which rotates inside a reservoir of chilled water. The tumbling action increases the effectiveness of the chilled water in cooling the food.

Cook-Chill Equipment

Some operations prepare food using a cook-chill system. By this method, food is partially cooked, rapidly chilled, and then held in refrigerated storage. When needed, the food simply is reheated. A cook-chill unit is an integrated piece of equipment capable of cooking, cooling, and reheating food.

Cross-Contamination

Synthetic cutting boards are generally preferred because they can be cleaned and sanitized in a warewashing machine.

Cutting Boards

Many jurisdictions allow the use of either wooden or synthetic cutting boards; however, some experts prefer synthetic boards, which can be cleaned and sanitized in a warewashing machine or by immersion in a three-compartment sink.

If wooden cutting boards and baker's tables are allowed by local codes, they must be made from a nonabsorbent hardwood, such as maple or oak. They must also be free of seams and cracks, nontoxic, and must not transfer any odor or taste to food.

Separate cutting boards should be used for raw and ready-to-eat food in order to prevent cross-contamination. Cutting boards must be washed, rinsed, and sanitized between uses. Due to the high risk of cross-contamination, procedures for cleaning and sanitizing cutting boards should be included in your standard operating procedures.

Exhibit 11j

Simplified Foodservice Floor Plan

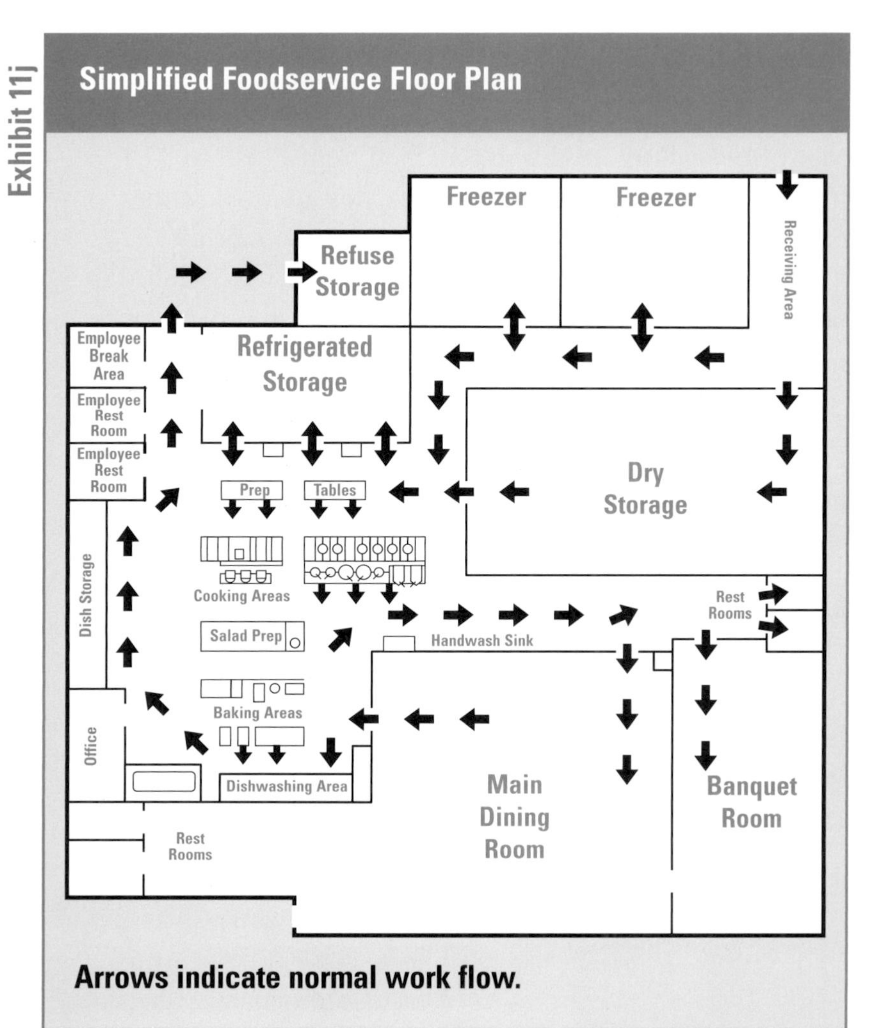

Arrows indicate normal work flow.

INSTALLING AND MAINTAINING KITCHEN EQUIPMENT

Well-designed kitchens make it easier to keep food safe. Generally, an efficient kitchen design is a more sanitary kitchen design.

Layout

A well-designed kitchen will address the following factors:

- **Work flow.** A work flow must be established that will minimize the amount of time food spends in the temperature danger zone. It must also minimize the number of times food is handled. (See *Exhibit 11j.*) For example, storage areas should be located near the receiving area to prevent delays in storing food. Prep tables should be located near refrigerators and freezers for the same reason.

- **Contamination.** A good layout will minimize the risk of cross-contamination. Dirty equipment should not be placed where it will touch clean equipment or food. For example, it is not a good practice to place the soiled-utensil table next to the salad-preparation sink.
- **Equipment accessibility.** A well-planned layout will ensure that equipment is accessible for cleaning. Hard-to-reach areas are less likely to be cleaned.

Equipment

Key Point

Install equipment so it is easy to clean both the equipment and surrounding areas.

When installing equipment, keep in mind that it should be easy for employees to clean both the equipment and surrounding floors, walls, and tabletops. Portable equipment makes it much simpler to do this.

When installing stationary equipment, it must be mounted on legs at least six inches off the floor or it must be sealed to a masonry base (allowing a toe space of four inches). (See *Exhibit 11k.*) How far the equipment is installed from the wall or from other equipment depends upon its size and the amount of surface to be cleaned. Follow specifications supplied by the manufacturer.

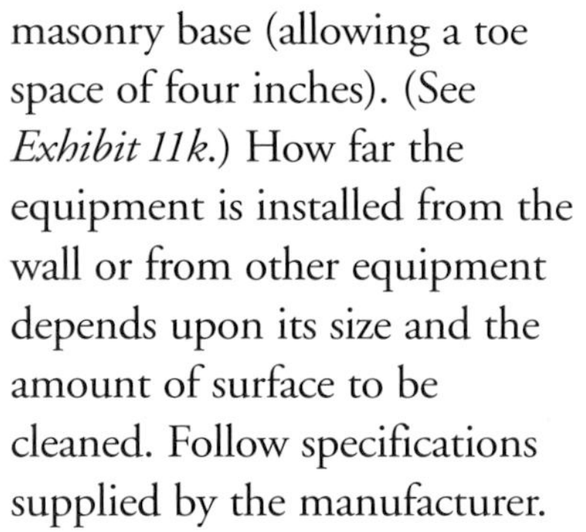

Exhibit 11k

Installing Stationary Equipment

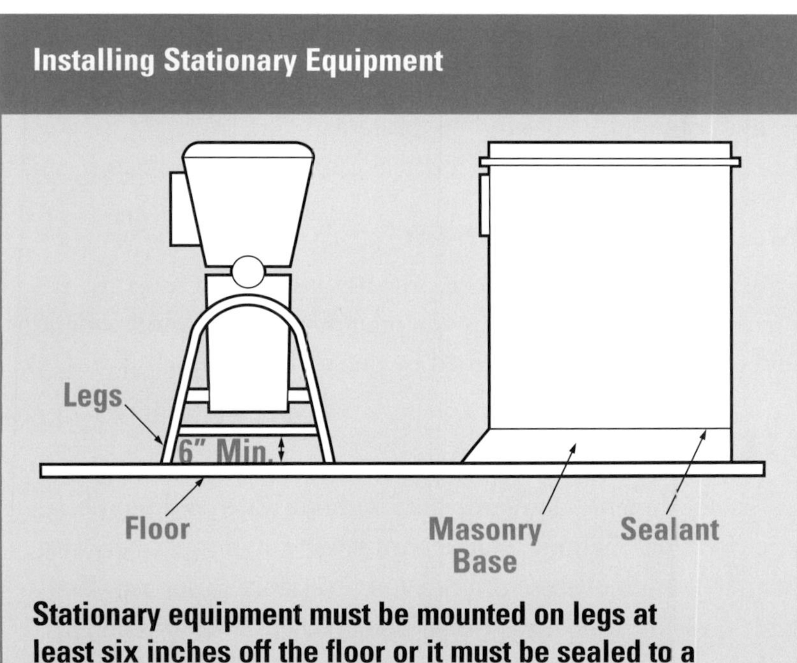

Stationary equipment must be mounted on legs at least six inches off the floor or it must be sealed to a masonry base.

Stationary tabletop equipment should be mounted on legs, providing a minimum clearance of four inches between the base of the equipment and the tabletop. Alternatively, the equipment should be tiltable, or it should be sealed to the countertop with a nontoxic, food-grade sealant.

When equipment is sealed to the floor, wall, or counter, any crack or seam greater than 1/32″ (1mm) must be filled with a nontoxic sealant to prevent food buildup or pests. Sealant should not be used to cover wide gaps resulting from faulty construction or repairs. These gaps should be properly repaired before equipment is installed.

Cantilever-Mounted Equipment

- Attached to the wall with a bracket, **cantilever-mounted equipment** allows for easier cleaning of surfaces behind and underneath it. (See *Exhibit 11l.*)

Exhibit 11l

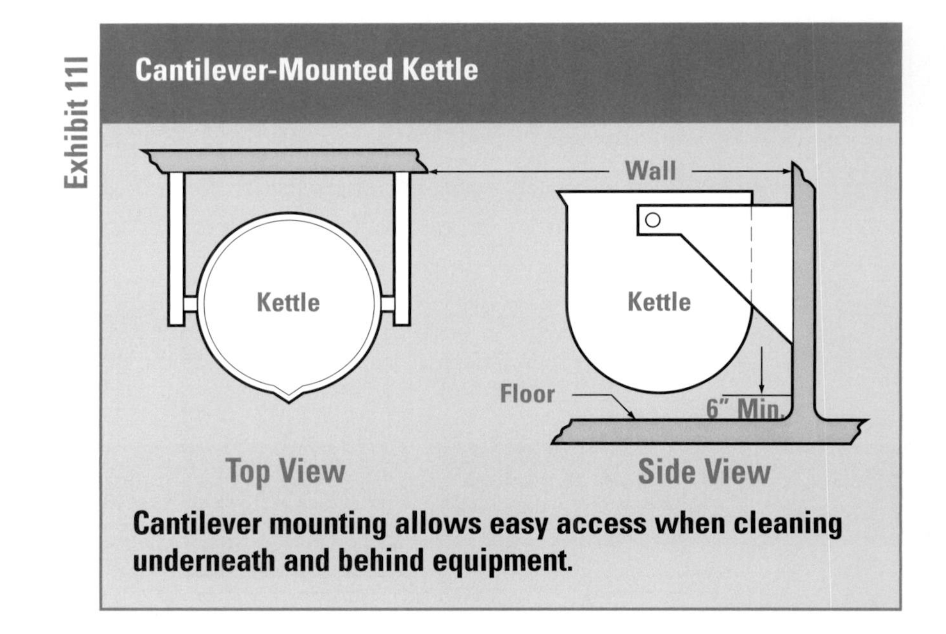

Cantilever mounting allows easy access when cleaning underneath and behind equipment.

Maintaining Equipment

Once equipment has been properly installed, it must receive regular maintenance. Follow the manufacturer's recommendations and make sure it is maintained by qualified personnel.

UTILITIES

Establishments cannot operate without water and plumbing, electricity, gas, lighting, ventilation, sewage, and waste handling. Sanitary design of these utilities and services is important. Two basic goals must be met in their design: there must be enough utilities to meet the cleaning needs of the establishment, and the utilities themselves must not contribute to contamination.

Water Supply

Unsafe water can carry foodborne pathogens. Therefore, safe water is vital in every establishment. It is used as a beverage and in ice, as an ingredient in food, and for handwashing, cleaning, laundry, and showers, as well as for flushing in restrooms.

Water that is safe to drink is called **potable water.** Sources of potable water include approved public water mains, private water sources that are regularly maintained and tested, and bottled drinking water. Other sources include closed, portable water containers filled with potable water, on-premises water storage tanks, and properly maintained water-transport vehicles.

If your establishment uses a private water supply, such as a well, rather than an approved public source, you should check with your local regulatory agency for information on inspections, testing, and other requirements. Generally, nonpublic water systems should be tested at least annually and the report kept on file in the establishment.

The use of nonpotable water is extremely limited. If nonpotable water is allowed by local codes, the uses are generally limited to air conditioning, cooling equipment (nonfood), fire protection, and irrigation.

Water Emergencies

Occasionally an emergency occurs that causes the water supply to become unusable. Examples include natural disasters like floods, problems with a city water supply, or local problems with broken pipes. If there is a problem with the potable water supply, the establishment might have to be closed or an alternate source of water found. Anytime a water emergency occurs, water systems and equipment (e.g., ice machines) should be flushed and disinfected.

Establishments might not want to close during a water emergency due to the loss of revenue. Some establishments might need to stay open during the emergency, such as hospitals or nursing homes. In these cases, the regulatory agency might allow the establishment to continue to operate if certain precautions are followed.

The following are issues an establishment must consider in order to continue serving food safely during a water emergency. Contact your local regulatory agency if you have any questions about the safety of a particular practice.

Water Used as a Beverage or Ingredient

If potable water must be obtained from an alternate source for use as a beverage or ingredient, there are several options, including buying bottled water and boiling water (the local regulatory authority should be contacted to determine the proper length of time for boiling). If you have been notified of an upcoming water emergency (such as water being shut off for repairs), you can store potable water in advance using closed, food-grade water containers.

Ice

Most ice machines make new ice and drop it onto previously made ice. Ice made from contaminated water will, therefore, contaminate any previously made ice already in the bin. If the establishment knows in advance that the water supply might become unsafe, previously made ice can be stored before the emergency occurs. In an unexpected water emergency, ice can be purchased. If purchased, ice must be contained in single-use, food-grade plastic bags or wet-strength paper bags filled and sealed at the point of manufacture.

Cleaning

If a water emergency occurs, nonessential cleaning in the establishment should be minimized. When performing essential tasks, such as cleaning and sanitizing pots and pans, the water used should first be boiled. Establishments might consider using single-use items to eliminate the need for using a warewasher, which requires large amounts of water.

Handwashing

Warm, potable water is required for handwashing. Boiled water can be placed in large plastic containers that dispense water through a spout. Replenish the supply frequently to keep it warm.

Restrooms, Showers, and Laundries

Restrooms, showers, and laundries require large amounts of water. Managing them would be difficult without a potable water system. Portable toilets can be used, but it might not be possible to supply water to showers and laundry facilities. The local regulatory agency should be contacted to determine possible alternatives. Some codes allow the use of nonpotable water for flushing toilets and operating washing machines.

Hot Water

Providing a continuous supply of hot water can be a problem for many establishments serving the public. They must have enough hot water to meet peak demand. Water heaters should be evaluated regularly to make sure they can meet this demand. Consider how quickly the heater produces hot water, the size of the holding tanks, and the location of the heater in relation to sinks or warewashing machines.

Since most general-purpose water heaters will not heat water to temperatures required for hot-water sanitizing, a **booster heater** may be needed to maintain a water temperature of 180°F (82°C) for heat-sanitizing tableware and utensils. Many warewashing machines now come with booster heaters.

Plumbing

In almost every community in the U.S., plumbing design is regulated by law, and with good reason. Improper plumbing design can have serious consequences. Improperly installed or poorly maintained plumbing that allows the mixing of potable and nonpotable water has been implicated in outbreaks of typhoid fever, dysentery, hepatitis A, Norovirus, and other gastrointestinal illnesses. Improperly installed water pipes can also lead to contamination from metals, such as toxic-metal poisoning from beverage dispensers, or contamination from chemicals used in the system, such as detergents, sanitizers, or drain cleaners.

Key Point

Only licensed plumbers should install and maintain plumbing systems in an establishment.

Local plumbing codes vary widely regarding the types of connections permitted and the kind of protection required. If in doubt about the plumbing regulations governing your establishment, contact the local regulatory agency. Only licensed plumbers should install and maintain plumbing systems in an establishment.

Common Cross-Connection

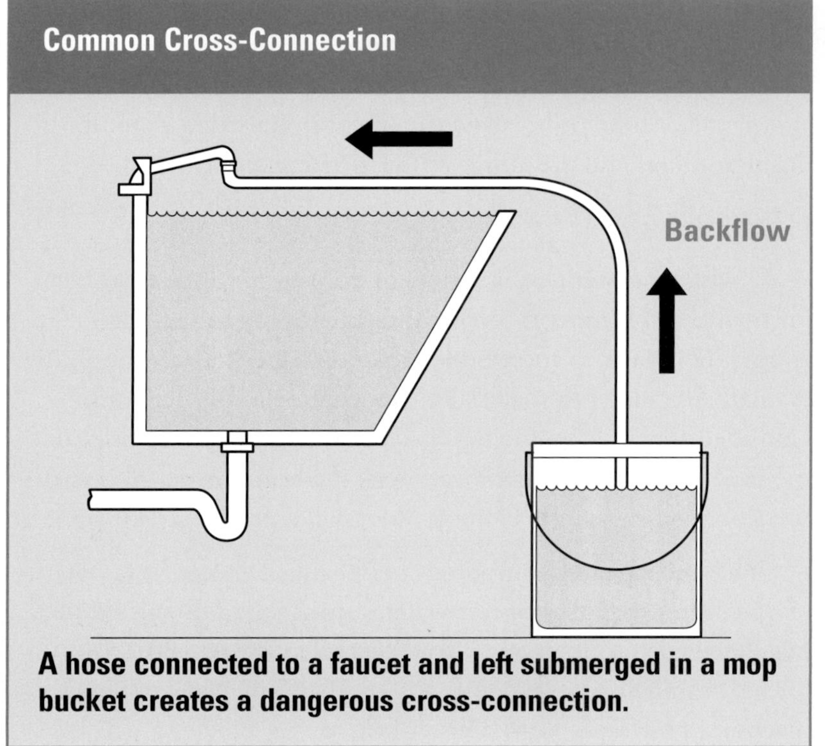

A hose connected to a faucet and left submerged in a mop bucket creates a dangerous cross-connection.

Cross-Connections

The greatest challenge to water safety comes from cross-connections. A **cross-connection** is a physical link through which contaminants from drains, sewers, or other waste-water sources can enter a potable water supply. Cross-connection is dangerous because it allows the possibility of **backflow,** or the unwanted reverse flow of contaminants through a cross-connection into a potable water system. Backflow can occur whenever the pressure in the potable water supply drops below the pressure of the contaminated supply. A running faucet located below the **flood rim** of a sink, or a running hose in a mop bucket, are examples of a cross-connection.

Exhibit 11n

Vacuum Breaker

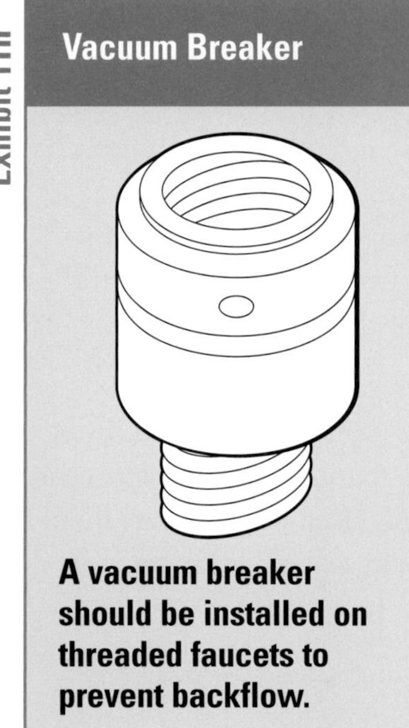

A vacuum breaker should be installed on threaded faucets to prevent backflow.

Exhibit 11m illustrates a situation in which an employee has attached a hose to the faucet of a utility sink to add hot water to a partially filled mop bucket. The nozzle of the hose is left submerged in the bucket of dirty water, accidentally creating a cross-connection. Because of heavy water usage somewhere else in the facility, the water pressure could drop low enough that contaminated water from the mop bucket would be drawn back through the hose and into the potable water supply.

To prevent cross-connections like this, do not attach a hose to a faucet unless a backflow-prevention device—such as a **vacuum breaker**—is attached. (See *Exhibit 11n.*) Threaded faucets and connections between two piping systems must have a vacuum breaker or other approved backflow-prevention device.

The only completely reliable method for preventing backflow is creating an **air gap.** An air gap is an air space used to separate a water supply outlet from any potentially contaminated source. A properly designed and installed sink typically has two air gaps to

Air Gaps to Prevent Backflow in a Sink

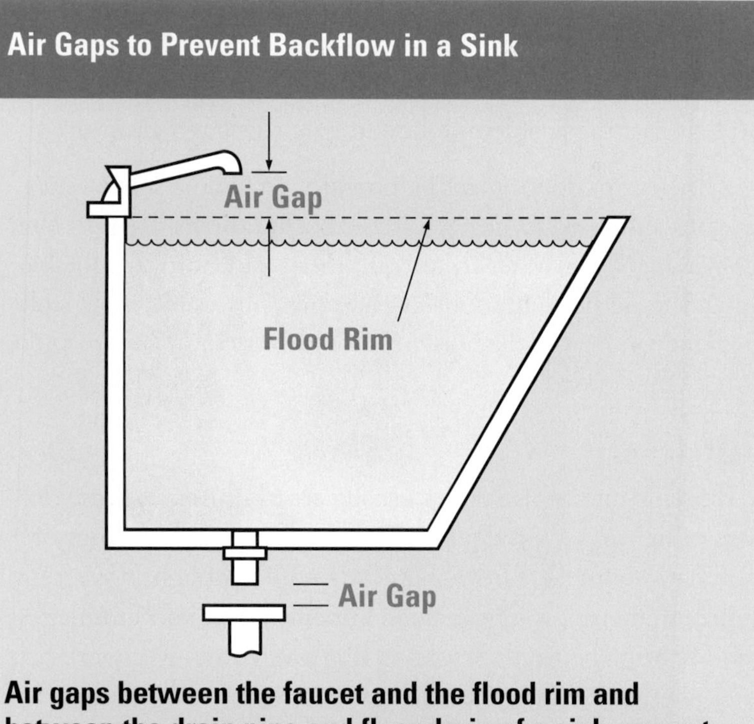

Air gaps between the faucet and the flood rim and between the drain pipe and floor drain of a sink prevent backflow.

prevent backflow. The air space between the faucet and the flood rim of the sink is one. Another is located between the drain pipe of the sink and the floor drain of the establishment. (See *Exhibit 11o.*) Typically, the size of the air gap should be twice the diameter of the water supply outlet. For example, if a faucet has an opening with a diameter of one inch, the air gap between the faucet and the flood rim of the sink must be at least two inches.

Grease Condensation and Leaking Pipes

Grease condensation in pipes is another common problem in plumbing systems. Grease traps are often installed to prevent buildup from creating a drain blockage. If used, grease traps must be easily accessible, installed by a licensed plumber, and cleaned periodically according to manufacturers' recommendations. If the traps are not cleaned, or it is not done properly, a backup of waste water could lead to odor and contamination.

Overhead, waste-water pipes or fire-safety sprinkler systems can leak and become a source of contamination. Even overhead lines carrying potable water can be a problem, since water can condense on the pipes and drip onto food. All piping should be serviced immediately when leaks occur.

Key Point

A backup of raw sewage on the floor is cause for immediate closure of the establishment, correction of the problem, and thorough cleaning.

Sewage

Sewage and waste water are reservoirs of pathogens, soil, and chemicals. It is absolutely essential to prevent any possible contamination of food or food-contact surfaces by waste water.

If there is a backup of waste water, prompt action must be taken. The action taken depends on the type of backup. A backup of raw sewage on the floor is cause for immediate closure of the establishment, correction of the problem, and thorough cleaning. Other backups might not be as serious, but require an immediate correction of the problem, followed by a thorough clean-up.

Sufficient drainage must be provided to handle waste water. Any area subjected to heavy water exposure should have its own floor drain. Waste water from equipment and from the potable supply should be channeled into an open, accessible waste sink or floor drain. The drainage system should be designed to keep floors from being flooded.

Lighting

Building and health codes usually set minimum acceptable levels of lighting, typically based on the **foot-candle,** a unit of illumination one foot from a uniform source of light. Other units of measurement for light include lumens, luxes, and luminaires. Good lighting generally results in improved employee work habits, easier and more effective cleaning, and a safer work environment. Lighting intensity requirements are different for various areas of the establishment. (See *Exhibit 11p.*)

Overhead or ceiling lights above workstations should be positioned so employees do not cast shadows on the work surface. Using fluorescent lights helps minimize such shadows. Use shatter-resistant light bulbs and protective covers made of metal mesh or plastic to prevent broken glass from contaminating food or food-contact surfaces. Shields should also be provided for heat lamps.

Ventilation

Proper ventilation helps maintain an establishment's indoor air quality by removing steam, smoke, grease, and heat from the establishment. Adequate ventilation is particularly important in the food-preparation area because it reduces the level of odors, gases, dirt, mold, humidity, grease, and fumes present in the air, all of which can contribute to contamination. Excess humidity can cause condensation on walls and ceilings, which may drip onto food. An accumulation of grease can cause fires. If ventilation is adequate, there will be little or no buildup of grease and condensation on walls and ceilings.

Exhibit 11p

Minimum Lighting Intensity Requirements for Different Areas of the Establishment

Minimum Lighting Intensity	Area
50 foot-candles (540 lux)	▶ Food-preparation areas
20 foot-candles (220 lux)	▶ Handwashing or warewashing areas ▶ Buffets and salad bars ▶ Displays for produce or packaged food ▶ Utensil-storage areas ▶ Wait stations ▶ Restrooms ▶ Inside some pieces of equipment (e.g., reach-in refrigerators)
10 foot-candles (110 lux)	▶ Inside walk-in refrigerator and freezer units ▶ Dry-storage areas ▶ Dining rooms (for cleaning)

Mechanical ventilation must be used in areas for cooking, frying, and grilling. Ventilation must be designed so that hoods, fans, guards, and ductwork do not drip onto food or equipment. Hood filters or grease extractors must be tight-fitting and easy to remove, and should be cleaned on a regular basis. Thorough cleaning of the hood and ductwork should also be done periodically by a professional company.

Since so much air is moved through exhaust hoods, clean air must be taken in to replace it. This replacement air is called make-up air, which must be replaced without creating drafts. All outside air intakes must be screened to keep pests out.

In many areas of the U.S., clean-air ordinances restrict the use of exhaust fans. Exhaust air containing food odors, smoke, and grease might have to be purified by filters or other devices. It is the establishment's responsibility to see that the ventilation system meets local regulations.

Solid Waste Management

Waste management is an important issue in establishments. There are many things foodservice managers can do to improve their waste management practices. The Environmental Protection Agency (EPA) has recommended three approaches to managing waste.

1. **Reduce the amount of waste produced.** Eliminate unnecessary packaging.
2. **Re-use when possible.** Re-used containers must be cleaned and sanitized. Never re-use chemical containers as food containers.
3. **Recycle materials.** Store recyclables so they cannot contaminate food or equipment or attract pests.

When these practices are followed, the amount of waste can be greatly reduced.

Garbage Disposal

Garbage is wet waste matter, usually containing food, that cannot be recycled. It attracts pests and has the potential to contaminate food, equipment, and utensils.

Garbage containers must be leak-proof, waterproof, pest-proof, easy to clean, and durable. They can be made of galvanized metal or an approved plastic. They must have tight-fitting lids and must be kept covered when not in use. Plastic bags and wet-strength paper bags may be used to line these containers.

Garbage should be removed from food-preparation areas as soon as possible. Frequent disposal prevents odor and pest problems. Garbage-storage areas, inside or outside, should be large enough to contain all garbage and must be located away from food-preparation and storage areas. When removing garbage, employees should not carry it above or across a food-preparation area.

Exhibit 11q

Outdoor Trash Receptacles

Receptacles and compactor systems should be located on or above a smooth surface of nonabsorbent material, such as concrete or machine-laid asphalt.

All garbage containers should be cleaned frequently and thoroughly. Both the inside and outside of containers must be cleaned. A cleaning area equipped with hot and cold water and a floor drain is recommended for indoor cleaning. It must be located so food being prepared or in storage will not be contaminated when garbage containers are being cleaned.

Food waste can also be disposed of through the use of in-drain garbage disposals, which reduce the amount of waste that goes into garbage containers. However, local regulations often limit their use in establishments. These disposals create huge amounts of food and water waste, which might overload local waste-water systems.

Pulpers, or grinders, are another garbage-disposal alternative. Pulpers grind food and some other types of waste (such as paper) into small parts that are flushed with water. The water is then removed so the processed solid waste weighs less and is more compact for easier disposal.

Outdoor trash receptacles should be kept covered (with their drain plugs in place) at all times, except during cleaning. Receptacles and compactor systems should be located on or above a smooth surface of nonabsorbent material such as concrete or machine-laid asphalt. The area must be kept clean. (See *Exhibit 11q*.)

SUMMARY

An establishment that is difficult to clean will not be cleaned well. Sanitation efforts will be more effective if the establishment is designed and equipped with ease of cleaning in mind. In most communities, plans for new construction or extensive remodeling are subject to review and approval by local regulatory agencies.

Sanitation can be built into the facility through the proper design and construction of floors, walls, and ceilings. The selection of materials should be based on ease of cleaning and durability, as well as appearance.

Separate restrooms should be provided for employees and patrons. If this is not possible, restrooms should be positioned so that patrons do not pass through food-preparation areas to reach them. Restrooms should be cleaned regularly and have a fully equipped handwashing station and self-closing doors. They should also be stocked adequately with toilet paper, and have trash receptacles.

Handwashing stations must be conveniently located, operable, and fully stocked and maintained. They are required in food-preparation areas, service areas, warewashing areas, and restrooms. They must be equipped with hot and cold running water, soap, a means to dry hands, a waste container, and signage reminding employees to wash hands.

It is important to purchase equipment that has been designed with sanitation in mind. Food-contact surfaces must be corrosion-resistant, nonabsorbent, smooth, and must resist pitting and scratching. Equipment should be installed so that both the equipment and the area surrounding it can be cleaned easily. Stationary equipment must be mounted on legs at least six inches off the floor, or it must be sealed to a masonry base. Stationary tabletop equipment should be mounted on legs with a clearance of four inches between the equipment and the tabletop, or it should be sealed to the tabletop. Any gap between a piece of equipment and the floor, wall, or tabletop greater than ⅟₃₂″ (1mm) should be filled with a nontoxic sealant to prevent food buildup and pests.

Utilities must be designed to meet the establishment's cleaning needs, and they must not contribute to contamination.

Potable water—water that is safe to drink—is vital in an establishment. Sources include public water mains, private water sources regularly maintained and tested, and bottled drinking water. In a water emergency, an establishment might be allowed to remain open if certain precautions are followed. These could include boiling or purchasing water, storing potable water and ice in advance, and boiling water for handwashing and essential tasks. The local regulatory agency should be contacted any time there is a question about the safety of a particular practice.

Improperly installed or maintained plumbing can have serious consequences. Only licensed plumbers should install and maintain plumbing systems. The greatest challenge to water safety comes from cross-connections—a physical link through which contaminants from drains, sewers, and other waste-water sources can flow back into the potable-water supply. Vacuum breakers and air gaps can be used to prevent backflow. It is essential to prevent waste water from contaminating food and food-contact surfaces. A backup of raw sewage is cause for immediate closure of the establishment, correction of the problem, and thorough cleaning.

Good lighting in an establishment generally results in improved employee work habits, easier and more effective cleaning, and a safer work environment. Follow the lighting intensity requirements for each area of the establishment. Use shatter-resistant bulbs and protective covers to prevent broken glass from contaminating food or food-contact surfaces. Proper ventilation improves the indoor air quality of the establishment by removing smoke, grease, steam, and heat. If ventilation is adequate, there will be little or no buildup of grease and condensation on walls and ceilings. Ventilation must be designed so hoods, fans, guards, and ductwork do not drip onto food or equipment. Hood filters and grease extractors must be cleaned regularly.

Garbage containers must be leak-proof, waterproof, pest-proof, easy to clean, and durable. They must have tight-fitting lids and must be kept covered when not in use. All garbage containers should be cleaned frequently and thoroughly both inside and out. Garbage should be removed from food-preparation areas as soon as possible, and must not be carried above or across a food-preparation area.

Apply Your Knowledge

1. Why did the people become ill?
2. What should Carlos do to correct the problem?

For answers, please turn to the Answer Key.

A Case in Point

Several people became ill shortly after drinking beverages at the bar in a local restaurant. They all complained that their iced drinks had an odd taste. At the time of the incident, the glasswasher in the bar had been out of service.

When interviewed, Carlos, the manager, explained that the glasswasher was functional, but that the unit could not be used because the large volume of water discharged after each wash load was worsening a recent drain-blockage problem. Carlos also mentioned there had been intermittent backups in the plumbing during the previous week, and a large pool of water was found under the glasswasher. Drain cleaners had been used repeatedly with no change in the blockage.

The icemaker shared piping with the glasswasher in the bar and the grease trap on the sink in the restaurant. Carlos revealed he had installed the plumbing himself.

When ice cubes were removed from the icemaker, congealed grease and food debris were found on them. Grease and debris also covered the bottom of the ice bin.

Apply Your Knowledge

Use these questions to review the concepts presented in this chapter.

Discussion Questions

1. What is one of the most important considerations when choosing flooring for food-preparation areas?
2. What action must be taken in the event of a backup of raw sewage in an establishment?
3. What can be done to prevent backflow in an establishment?
4. What are some potable-water sources for an establishment? What are the testing requirements for non-public water systems?
5. What are the requirements of a handwashing station? In what areas of an establishment are handwashing stations required?
6. What are the requirements for installing stationary equipment?

For answers, please turn to the Answer Key.

Apply Your Knowledge

Use these questions to test your knowledge of the concepts presented in this chapter.

Multiple-Choice Study Questions

1. **Generally, establishments that use a private water source, such as a well, must have it tested at least**
   A. once a year.
   B. every two years.
   C. every three years.
   D. every five years.

2. **An establishment should respond to a backup of raw sewage by**
   A. closing.
   B. correcting the problem that caused the backup.
   C. thoroughly cleaning affected areas of the establishment.
   D. All of the above

3. **Which of the following will *not* prevent backflow?**
   A. An air gap between the sink drain pipe and the floor drain
   B. The air space between the faucet and the flood rim of a sink
   C. A vacuum breaker
   D. A cross-connection

4. **When designing the layout of a foodservice establishment, what is the most important consideration for keeping food safe?**
   A. Where the establishment will be located
   B. How the facility and equipment will be maintained
   C. The number of employees that will be required to run the establishment
   D. The number of customers that will patronize the establishment each night

Continued on next page...

Apply Your Knowledge — Multiple-Choice Study Questions *continued*

5. When mounting tabletop equipment on legs, the clearance between the base of the equipment and the tabletop must be at least
   A. one inch.
   B. two inches.
   C. four inches.
   D. six inches.

6. Equipment food-contact surfaces must meet all of the following conditions *except*
   A. they must be corrosion-resistant.
   B. they must be absorbent.
   C. they must be smooth.
   D. they must be resistant to pitting.

7. Which of the following statements about warewashing areas is true?
   A. Water pipes to the warewashing machine should be as long as possible.
   B. The machine should be raised two inches from the floor to permit easy cleaning.
   C. Information should be posted on the machine regarding proper water temperature.
   D. The machine's thermometer should have a scale in increments no greater than 5˚F (–15˚C).

8. What is the purpose of a grease trap?
   A. To prevent backflow
   B. To replace an air gap
   C. To store used fryer shortening
   D. To prevent grease buildup from blocking a drain

Continued on next page...

Apply Your Knowledge

Multiple-Choice Study Questions *continued*

9. **Which of the following practices will prevent overhead lights from contaminating food?**
   A. Using shatter-resistant bulbs
   B. Using flourescent bulbs
   C. Using extended-life bulbs
   D. Using glass covers over bulbs

10. **All of the following are true about garbage containers *except***
    A. they must be leak-proof.
    B. they must be made of galvanized metal or an approved plastic.
    C. they must have tight-fitting lids.
    D. they must fit within the warewashing machine or three-compartment sink.

11. **The lighting intensity in a dry storage area should be at least**
    A. 50 foot-candles (540 lux).
    B. 20 foot-candles (220 lux).
    C. 10 foot-candles (110 lux).
    D. 5 foot-candles (54 lux).

12. **Which of the following organizations certifies equipment that meets sanitation standards?**
    A. NSF International
    B. FDA
    C. EPA
    D. OSHA

Continued on next page...

Multiple-Choice Study Questions *continued*

13. Nonabsorbent flooring should be used in all of the following areas *except*
    A. restrooms.
    B. food-preparation areas.
    C. warewashing areas.
    D. dining rooms.

14. Which of the following is *not* true about garbage?
    A. It is wet waste matter.
    B. It usually does not attract pests or rodents.
    C. It contains food that cannot be recycled.
    D. It has the potential to contaminate food, equipment, and utensils.

For answers, please turn to the Answer Key.

ADDITIONAL RESOURCES

Books and Periodicals

Designed for food safety. 2003. *Food Safety Illustrated.* 3 (1):8–12.

Longree, K and G. Armbruster. 1996. *Quantity Food Sanitation.* New York: Wiley.

Look, Ma, no hands: Touch-free devices gain popularity among the handwashing crowd. 2002. *Food Safety illustrated.* 2 (1):18.

Marriott, N. G. 1994. *Principles of Food Sanitation.* New York: Chapman & Hall.

The well-stocked handsink. 2002. *Food Safety Illustrated.* 2 (1):5.

Troller, J. A. 1993. *Sanitation in Food Processing.* San Diego, CA: Academic Press.

Web Sites

American National Standards Institute (ANSI)
www.ansi.org
ANSI is a private, nonprofit organization administering and coordinating the U.S. voluntary standardization and conformity assessment system. Use this Web site to order ANSI/NSF standards outlining the materials to be used in the construction of commercial food equipment.

Americans with Disabilities Act (ADA) Home Page
www.usdoj.gov/crt/ada/adahom1.htm
ADA publications and information, including requirements, design standards for new construction and alterations, status reports, featured settlements and consent agreements, technical assistance materials, and new or proposed regulations, are all found on this Web site.

FDA Food Code
http://vm.cfsan.fda.gov/~dms/foodcode.html
As the basis for many local sanitation codes, as well as the basis for information in this textbook, the FDA Food Code, available at this Web address, is a useful resource for information relating to food for the restaurant and foodservice industry.

FDA and CFP Food Establishment Plan Review Guide 2000
http://vm.cfsan.fda.gov/~dms/prev-toc.html
The food establishment plan review document found at this site was developed for the purpose of assisting both regulatory and industry personnel in achieving greater uniformity in the plan review process.

National Restaurant Association
www.restaurant.org
The National Restaurant Association is the leading business association for the restaurant industry. Together with the National Restaurant Association Educational Foundation, the Association's mission is to represent, educate, and promote the rapidly growing restaurant and foodservice industry. This Web site should be your starting place for all issues and concerns related to your restaurant. This Web site has it all, from tips for running your establishment to vital data on your customers' spending habits.

NSF International
www.nsf.org
Use this site to find information on food standards and resources. NSF works with the food-equipment industry, regulators, and foodservice organizations to develop public health safety standards used by the restaurant and foodservice industry.

Occupational Safety and Health Administration (OSHA)
www.osha.gov
OSHA's mission is to prevent work-related injuries, illnesses, and deaths. This Web site provides information related to workplace-safety conferences, technical links, and regulation and compliance.

Underwriters Laboratories, Inc. (UL)
www.ul.com/eph
UL is an independent, not-for-profit product-safety testing and certification organization offering certification programs to demonstrate compliance with public health codes to manufacturers of foodservice equipment. UL classifies products in accordance with food-related NSF International standards. This Web site contains a foodservice-equipment and drinking water product directory, information on the indoor air quality program, ISO 14001, and other information related to the foodservice industry.

POT AND PAN CLEANING AND SANITIZING PROCEDURES
PROCEDIMIENTOS DE LIMPIEZA Y SANEAMIENTO DE OLLAS Y SARTENES
PRE-SCRAPE & WASH
CÓMO TALLAR PREVIA-
MENTE Y LAVAR
RINSE & SANITIZE
CÓMO ENJUAGAR
Y SANEAR
ECOLAB
SPARTA

Cleaning and Sanitizing

Inside this chapter:

- Cleaning and Sanitizing
- Machine Warewashing
- Manual Warewashing
- Cleaning and Sanitizing Equipment
- Cleaning the Kitchen
- Cleaning the Premises
- Tools for Cleaning
- Storing Utensils, Tableware, and Equipment
- Using Hazardous Materials
- Implementing a Cleaning Program

After completing this chapter, you should be able to:

- Explain the difference between cleaning and sanitizing.
- Identify approved sanitizers.
- Identify factors affecting the efficiency of sanitizers (i.e., time, temperature, concentration, water hardness, and pH).
- Use the appropriate test kit for each sanitizer.
- Follow the requirements for frequency of cleaning and sanitizing food-contact surfaces.
- Properly clean and sanitize items in a three-compartment sink.
- Properly clean and sanitize food-contact surfaces.
- Properly clean nonfood-contact surfaces.
- Identify proper machine-warewashing techniques.
- Identify storage requirements for poisonous or toxic materials.
- Dispose of poisonous or toxic materials according to legal requirements.
- Follow the legal requirements for the use of poisonous or toxic material in a food establishment.
- Properly store tools, equipment, and utensils that have been sanitized.

Key Terms

Cleaning
Sanitizing
Cleaning agent
Detergent
Solvent cleaners
Acid cleaners
Abrasive cleaners
Heat sanitizing
Chemical sanitizing
Sanitizer
Chlorine
Iodine
Quaternary ammonium compounds (quats)
Hazard Communication Standard (HCS)
Master cleaning schedule

Apply Your Knowledge

Check to see how much you know about the concepts in this chapter. Use the page references provided to explore the topic in each question.

Test Your Food Safety Knowledge

1. **True or False:** Chemicals can be stored in food-preparation areas if they are properly labeled. *(See page 12-18.)*
2. **True or False:** The temperature of the final sanitizing rinse in a high-temperature warewashing machine should be 140°F (60°C). *(See page 12-8.)*
3. **True or False:** Cleaning reduces the number of microorganisms on a surface to safe levels. *(See page 12-3.)*
4. **True or False:** Utensils cleaned and sanitized in a three-compartment sink should be dried with a clean towel. *(See page 12-11.)*
5. **True or False:** Tableware and utensils that have been cleaned and sanitized should be stored at least two inches off of the floor. *(See page 12-18.)*

For answers, please turn to the Answer Key.

INTRODUCTION

In Chapter 10, you learned that a good food safety system depends on food safety programs. A cleaning and sanitation program is one of the most important of these.

If you do not maintain a high standard of cleanliness and sanitation, food can easily become contaminated. No matter how carefully you prepare and cook food, without a clean and sanitary environment, bacteria and viruses—such as those that cause salmonellosis and hepatitis A—can spread quickly to both cooked and uncooked food. Cleaning and sanitizing must be done carefully and correctly. If not used properly, cleaning and sanitizing chemicals can be just as harmful to customers and employees as the illnesses they help prevent.

Cleaning & Sanitizing

Surfaces must *first* be cleaned and rinsed *before* being sanitized.

CLEANING AND SANITIZING

It is important to understand the difference between **cleaning** and **sanitizing**. Cleaning is the process of removing food and other types of soil from a surface, such as a countertop or plate. Sanitizing is the process of reducing the number of microorganisms on that surface to safe levels. To be effective, cleaning and sanitizing must be a two-step process. Surfaces must *first* be cleaned and rinsed *before* being sanitized.

Everything in your operation must be kept clean; however, any surface that comes in contact with food must be cleaned and sanitized. All food-contact surfaces must be washed, rinsed, and sanitized

- after each use.
- any time you begin working with another type of food.
- any time you are interrupted during a task and the tools or items you have been working with may have been contaminated.
- at four-hour intervals, if the items are in constant use.

Cleaning

Several factors affect the cleaning process. The table in *Exhibit 12a* on the next page lists these factors and provides a brief explanation of each.

Cleaning Agents

Cleaning agents are chemical compounds that remove food, soil, rust, stains, minerals, or other deposits. They must be stable, noncorrosive, and safe for employee use. Since cleaning agents have different cleaning properties, ask your supplier to help you select those that will best meet your needs.

Cleaning agents work best when used as directed. They can be ineffective and even dangerous if misused. Follow manufacturers' instructions carefully. Employees should never combine compounds or attempt to make up their own cleaning agents. Combining ammonia and chlorine bleach, for example, produces chlorine gas, which can be fatal. Also, do not substitute one type of detergent for another unless the intended use is stated clearly on the label. Detergents used for warewashing machines, for example, can cause severe burns to the skin if used for manual warewashing.

Cleaning agents are divided into four categories: **detergents, solvent cleaners, acid cleaners,** and **abrasive cleaners.** Some categories may overlap. For example, most abrasive cleaners and some acid cleaners contain detergents. Some detergents also may contain solvents.

Exhibit 12a

Factors Affecting the Cleaning Process

Factor	Effect on Cleaning Process
Type of Soil	Certain types of soil require special cleaning methods.
Condition of Soil	The condition of the soil or stain affects how easily it can be removed. Dried or baked-on stains will be more difficult to remove than soft, fresh stains.
Water Hardness	Cleaning is more difficult in hard water because minerals react with the detergent, decreasing its effectiveness. Hard water can cause scale or lime deposits to build up on equipment, requiring the use of lime-removal cleaners.
Water Temperature	In general, the higher the water temperature, the better a detergent will dissolve and the more effective it will be in loosening dirt.
Cleaning Agent and Surface Being Cleaned	Different surfaces require different cleaning agents. Some cleaners work well in one situation, but might not work well or might even damage equipment when used in another.
Agitation or Pressure	Scouring or scrubbing a surface helps remove the outer layer of soil, allowing a cleaning agent to penetrate deeper.
Length of Treatment	The longer soil on a surface is exposed to a cleaning agent, the easier it is to remove.

Detergents

There are different types of detergents for different types of cleaning jobs. All detergents contain surfactants (surface-acting agents) that reduce surface tension between the soil and the surface, so the detergent can quickly penetrate and soften the soil. General-purpose detergents are mildly alkaline and are used to clean fresh soil from floors, walls, ceilings, prep surfaces, and most equipment and utensils. Heavy-duty detergents are highly alkaline and are used to remove wax, aged or dried soil, and baked-on grease. Warewashing detergents also are highly alkaline.

Solvent Cleaners

Solvent cleaners, often called degreasers, are alkaline detergents containing a grease-dissolving agent. These cleaners work well in areas where grease has been burned on, such as grill backsplashes, oven doors, and range hoods. Solvents are usually effective only at full strength, so they are costly to use on large areas.

Acid Cleaners

Acid cleaners are used on mineral deposits and other soils alkaline cleaners cannot remove. These cleaners are often used to remove scale in warewashing machines and steam tables, as well as rust stains and tarnish on copper and brass. The type and strength of the acid varies with the cleaner's purpose. Follow the instructions carefully and use acid cleaners with caution.

Abrasive Cleaners

Abrasive cleaners contain a scouring agent like silica that helps scrub off hard-to-remove soil. These cleaners are often used on floors or to remove baked-on food in pots and pans. Use abrasives with caution since they can scratch surfaces.

Sanitizing

There are two methods used to sanitize surfaces: **heat sanitizing** and **chemical sanitizing.** Which you use depends on the application.

Heat Sanitizing

The higher the heat, the shorter the time required to kill microorganisms. The most common way to heat-sanitize tableware, utensils, or equipment is to immerse or spray the items with hot water. Use a thermometer to check water temperature when heat sanitizing by immersion. To check water temperature in a high-temperature warewashing machine, attach a temperature-sensitive label or tape, or a high-temperature probe to items being run through the machine.

Chemical Sanitizing

Chemical **sanitizers** are widely used in establishments because they are effective, reasonably priced, and easy to use. They are regulated by state and federal environmental protection agencies (EPAs). The three most common types are **chlorine, iodine,** and **quaternary ammonium compounds (quats).** The advantages and disadvantages of each type are listed in *Exhibit 12b* on the next page.

Cleaning & Sanitizing

Refer to your local or state regulatory agency for recommendations on selecting a sanitizer.

Refer to your local or state regulatory agency for recommendations on selecting a sanitizer. For a list of approved sanitizers, check the *Code of Federal Regulations (CFR),* 21CFR178.1010—"Sanitizing Solutions."

Chemical sanitizing is done in two ways: either by immersing a clean object in a specific concentration of sanitizing solution for a required period of time or by rinsing, swabbing, or spraying the object with a specific concentration of sanitizing solution.

In some instances, detergent sanitizer blends may be used to sanitize surfaces, but items still must be cleaned and rinsed first. Scented or oxygen bleaches are not acceptable as sanitizers for

Exhibit 12b

Advantages and Disadvantages of Different Sanitizers

Types	Advantages	Disadvantages
Chlorine	▶ Most commonly used sanitizer ▶ Kills a wide range of vegetative microorganisms ▶ Least expensive of the three ▶ Effective in hard water	▶ Inactivated by presence of soil ▶ Corrosive to some metals, such as stainless steel and aluminum, when used improperly ▶ Can be irritating to skin ▶ Does not remain active after it has dried
Iodine	▶ Remains active for a short period of time after it has dried ▶ Not as quickly inactivated by soil as chlorine ▶ Nonirritating to skin	▶ Less effective in reducing microorganisms than chlorine ▶ Somewhat corrosive to surfaces ▶ Most expensive of the three ▶ Slightly affected by the presence of soil
Quaternary ammonium compounds	▶ Not as quickly inactivated by soil as chlorine ▶ Remains active for a short period of time after it has dried ▶ Noncorrosive to surfaces ▶ Nonirritating to skin	▶ Easily affected by presence of detergent residue ▶ Less effective against certain types of microorganisms ▶ Hard water reduces effectiveness

food-contact surfaces. Household bleaches are acceptable only if the labels indicate they are EPA-registered.

Factors Influencing the Effectiveness of Sanitizers

Different factors influence the effectiveness of chemical sanitizers. The most critical include contact time, temperature, and concentration. (See *Exhibit 12d* on the next page.)

Cleaning & Sanitizing

The effectiveness of a chemical-sanitizing solution depends upon its contact time, temperature, and concentration.

Contact Time

In order for a sanitizing solution to kill microorganisms, it must make contact with the object for a specific amount of time. Since minimum times may differ for each sanitizer, check with your supplier.

Temperature

Generally, sanitizers work best at temperatures between 55°F and 120°F (13°C and 49°C). Some may not be effective at temperatures lower than 55°F (13°C), while others may corrode metals or evaporate at temperatures higher than 120°F (49°C).

Concentration

Chemical sanitizers are mixed with water until the proper concentration—ratio of sanitizer to water—is reached. Concentration is measured using a sanitizer test kit and is expressed in parts per million (ppm). The test kit should be designed for the sanitizer you are using and is usually available from the manufacturer or your supplier. (See *Exhibit 12c.*)

Exhibit 12c

Sanitizer Test Kit

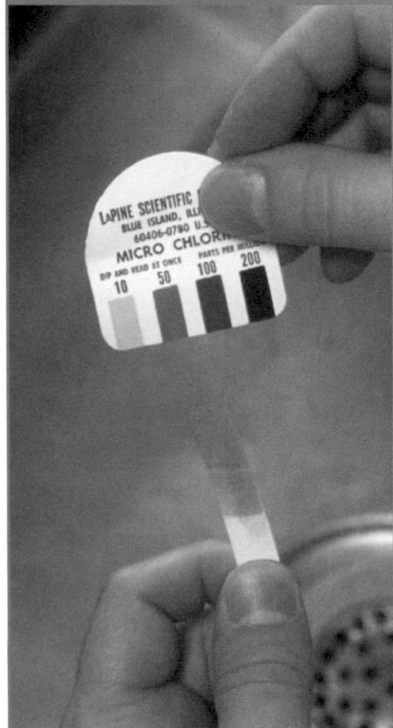

Use a test kit to check the concentration of a sanitizing solution.

Concentrations below either those required in your jurisdiction or recommended by the manufacturer could fail to sanitize objects. Concentrations higher than recommended can be unsafe, leave an odor or bad taste on objects, and might corrode metals.

The concentration of a sanitizing solution must be checked frequently. This is important since a sanitizer is depleted during use. It can also become bound up by hard water, food particles, or detergent not adequately rinsed from a surface.

A sanitizing solution must be changed when it is visibly dirty or when the concentration of the sanitizer has dropped below the level required.

Exhibit 12d

General Guidelines for Chemical Sanitizers

	Chlorine					Iodine	Quats
Temperature	120°F (49°C)	120°F (49°C)	100°F (38°C)	75°F (24°C)	55°F (13°C)	75°F (24°C)	75°F (24°C)
Concentration	25 ppm	25 ppm	50 ppm	50 ppm	100 ppm	12.5–25 ppm	Up to 200 ppm, or as per manufacturer's recommendations
pH	10	8	10	8	8–10	≤5.0	as per manufacturer's recommendations
Contact Time	10 sec	10 sec	7 sec	7 sec	10 sec	30 sec	30 sec

MACHINE WAREWASHING

Most tableware, utensils, and even pots and pans can be cleaned and sanitized in warewashing machines. Warewashing machines sanitize by using either hot water or a chemical-sanitizing solution.

High-Temperature Machines

High-temperature machines rely on hot water to clean and sanitize. Water temperature is critical and may vary by model. The temperature of the final sanitizing rinse must be at least 180°F (82°C). For stationary-rack, single-temperature machines, the temperature of the final sanitizing rinse must be at least 165°F (74°C). Water that is too hot might vaporize before sanitizing items, or might bake food onto tableware and utensils, making it even harder to get them clean. You may need a booster heater to provide enough hot water to sanitize a high volume of tableware. Make sure your warewasher has a built-in thermometer to measure the temperature of water at the manifold, where it sprays into the tank.

Chemical-Sanitizing Machines

Warewashing machines that use chemical sanitizing often wash at much lower temperatures, but not lower than 120°F (49°C).

Rinse-water temperature in these machines should be between 75°F and 120°F (24°C and 49°C) for the sanitizer to be effective. Items washed and rinsed at these lower temperatures may take longer to air-dry, so you may need more room at the clean end of the machine and more tableware during peak periods.

The effectiveness of your warewashing program will depend on a number of factors:

- A well-planned layout (see *Exhibit 12e*) in the warewashing area, with a scraping and soaking area and adequate space for both soiled and clean items
- A sufficient water supply, especially hot water
- A separate area for cleaning pots and pans
- Devices that indicate water pressure and temperature of the wash and rinse cycles

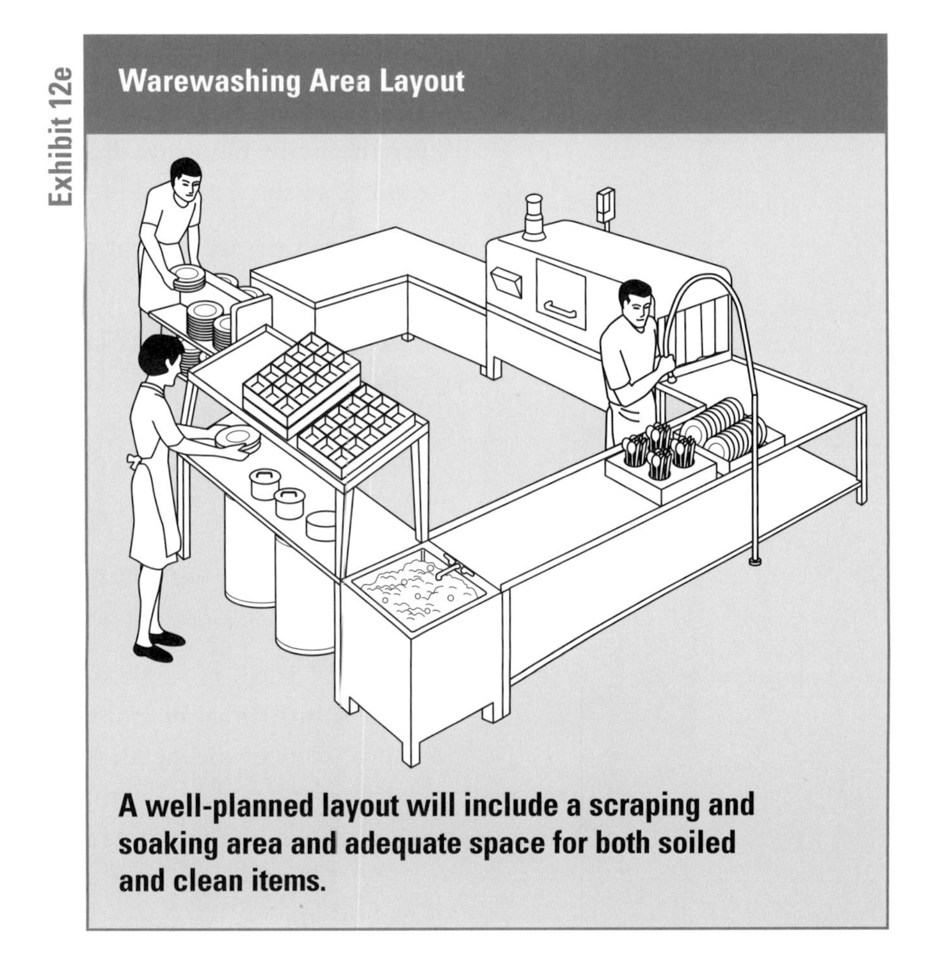

Exhibit 12e

Warewashing Area Layout

A well-planned layout will include a scraping and soaking area and adequate space for both soiled and clean items.

- Protected storage areas for clean tableware and utensils
- Employees who are trained to operate and maintain the equipment and use the proper chemicals

Cleaning & Sanitizing

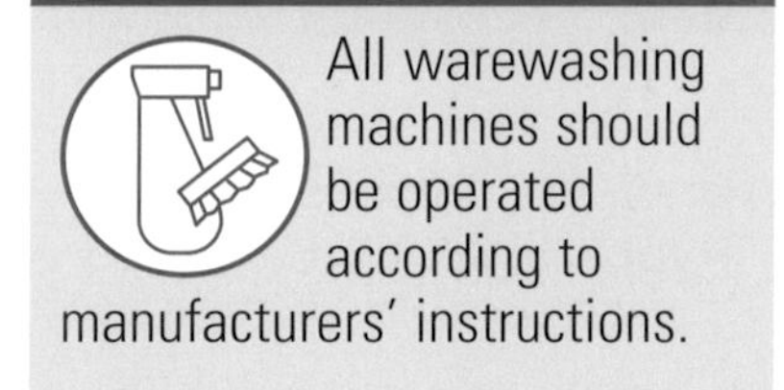

All warewashing machines should be operated according to manufacturers' instructions. These instructions will typically be located on the machine. No matter what type of machine you use, there are some general procedures to follow to clean and sanitize tableware, utensils, and related items.

- **Check the machine for cleanliness at least once a day, cleaning it as often as needed.** Fill tanks with clean water. Clear detergent trays and spray nozzles of food and foreign objects. Use an acid cleaner on the machine whenever necessary to remove mineral deposits caused by hard water.
- **Make sure detergent and sanitizer dispensers are properly filled.**
- **Scrape, rinse, or soak items before washing.** Presoak items containing dried-on food.
- **Load warewasher racks correctly and use racks designed for the items being washed.** Make sure all surfaces are exposed to the spray action. Never overload racks.
- **Check temperatures and pressure.** Follow manufacturers' recommendations.
- **Check each rack as it comes out of the machine for soiled items.** Run dirty items through again until they are clean. Most items will need only one pass if the water temperature is correct and proper procedures are followed.
- **Air-dry all items.** Towels can recontaminate items.
- **Keep your warewashing machine in good repair.**

MANUAL WAREWASHING

Establishments that do not have a warewashing machine may use a three-compartment sink to wash items (some local regulatory agencies allow the use of two-compartment sinks; others require four-compartment sinks). These sinks may also be used to wash larger items. A properly set up warewashing station includes:

- An area for rinsing away food or for scraping food into garbage containers
- Drain boards to hold both soiled and clean items
- A thermometer to measure water temperature
- A clock with a second hand, allowing employees to time how long items have been immersed in the sanitizing sink

Cleaning and Sanitizing in a Three-Compartment Sink

Before cleaning and sanitizing items in a three-compartment sink, each sink and all work surfaces must be cleaned and sanitized. Follow the steps listed below when cleaning and sanitizing tableware, utensils, and equipment. (See *Exhibit 12f* on the next page.)

1. **Rinse, scrape, or soak all items before washing.**
2. **Wash items in the first sink in a detergent solution at least 110°F (43°C).** Use a brush, cloth, or nylon scrub pad to loosen the remaining soil. Replace the detergent solution when the suds are gone or the water is dirty.
3. **Immerse or spray-rinse items in the second sink, using water at least 110°F (43°C).** Remove all traces of food and detergent. If using the immersion method, replace the rinse water when it becomes cloudy or dirty.
4. **Immerse items in the third sink in hot water or a chemical-sanitizing solution.** If hot-water immersion is used, the water must be at least 171°F (77°C)—some health codes require a temperature of 180°F (82°C)—and the items must be immersed for thirty seconds. If chemical sanitizing is used, the sanitizer must be mixed at the proper concentration and the water temperature must be correct. Check the concentration of the sanitizing solution at regular intervals with a test kit.
5. **Air-dry all items.**

Wood surfaces—such as cutting boards, handles, and bakers' tables—need special care. After each use, scour them in a

Exhibit 12f

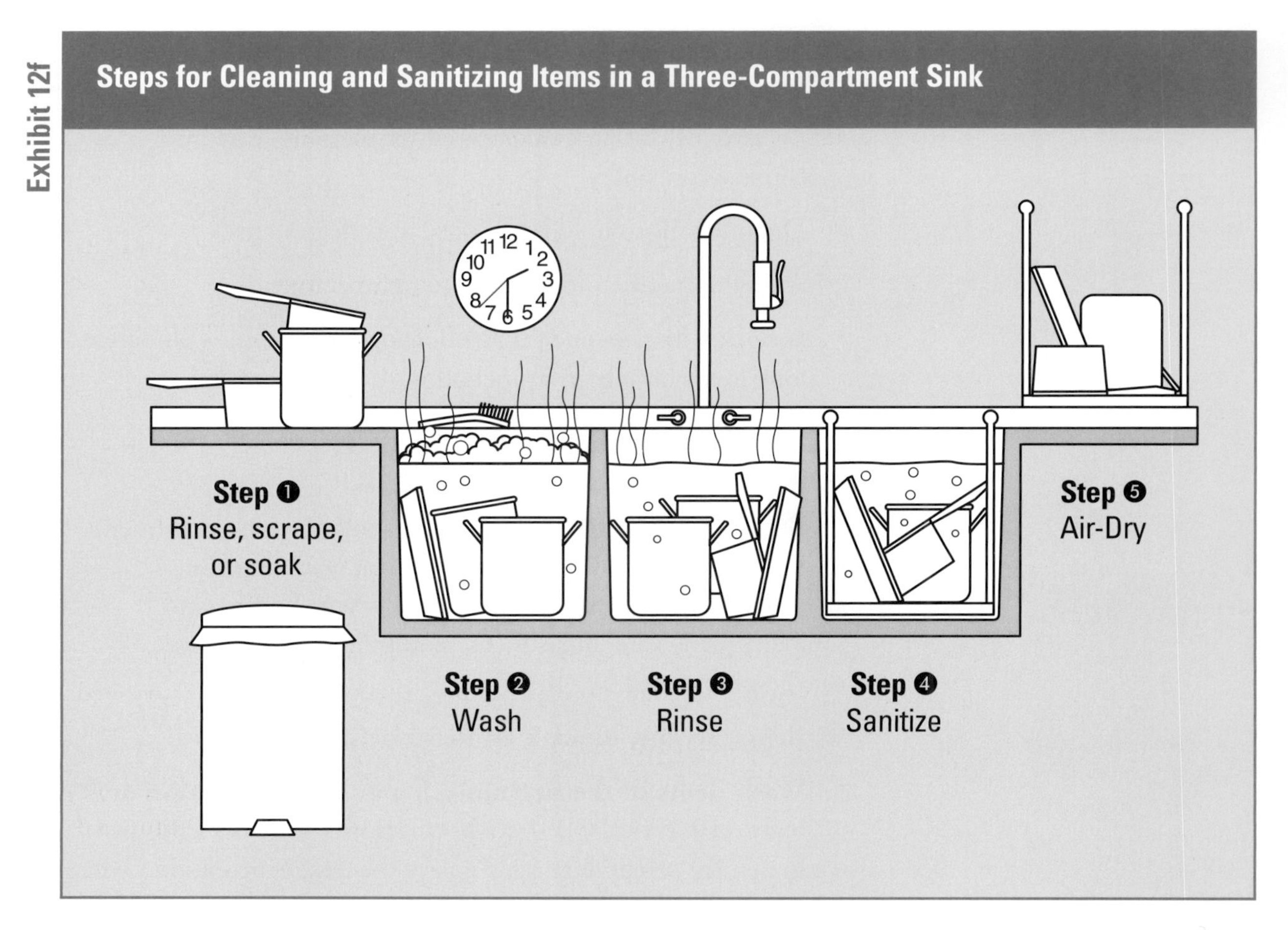

detergent solution with a stiff-bristle nylon brush, rinse them in clean water, and sanitize them. Do not soak wood surfaces in detergent or sanitizing solutions.

CLEANING AND SANITIZING EQUIPMENT

Because equipment must be kept clean and food-contact surfaces must be cleaned and sanitized, employees should be taught how to clean each type of equipment properly.

Clean-in-Place Equipment

Some pieces of equipment, such as soft-serve yogurt machines, are designed to have cleaning and sanitizing solutions pumped through them. Since many of them hold and dispense potentially hazardous food, they must be cleaned and sanitized everyday unless otherwise indicated by the manufacturer.

Cleaning & Sanitizing

Clean-in-place equipment used to hold and dispense potentially hazardous food must be cleaned and sanitized every day unless otherwise indicated by the manufacturer.

Stationary Equipment

Equipment manufacturers will usually provide cleaning instructions. In general, follow these steps:

- Turn off and unplug equipment before cleaning.
- Remove food and soil from under and around the equipment.
- Remove detachable parts and manually wash, rinse, and sanitize them, or run them through a warewasher if permitted. Allow them to air-dry.
- Wash and rinse fixed food-contact surfaces, then wipe or spray them with a chemical-sanitizing solution.
- Keep cloths used for food-contact and nonfood-contact surfaces in separate, properly marked containers of sanitizing solution.
- Air-dry all parts, then reassemble according to directions. Tighten all parts and guards. Test equipment at recommended settings, then turn it off.
- Resanitize food-contact surfaces handled when putting the unit back together by wiping with a cloth that has been submerged in sanitizing solution.

Cleaning & Sanitizing

Cloths used for food-contact and nonfood-contact surfaces should be stored between uses in separate, properly marked containers of sanitizing solution.

In some cases, you may be able to spray-clean fixed equipment. Check with the manufacturer. If allowed, spray each part with solution in the right concentration and let it sit for the recommended amount of time.

Refrigerated Units

Clean up spills in refrigerators and freezers immediately. These units should be thoroughly cleaned and sanitized regularly to remove soil, mold, and odors. When cleaning and sanitizing refrigeration units, follow these suggestions:

- Clean before storing deliveries so less food has to be moved.
- Move food to another unit before starting to clean.
- Clean shelves regularly. Thoroughly clean walls, floors, door edges, and gaskets.

CLEANING THE KITCHEN

Kitchen floors, walls, shelves, equipment exteriors, ceilings, light fixtures, drains, and restrooms are nonfood-contact surfaces, but they still require regular cleaning to prevent accumulation of dust, dirt, food residue, and other debris. Sanitizing these surfaces is not required, but is a good practice. How often these surfaces are cleaned will depend on several factors, including the type of food served, surfaces to be cleaned, and rate of ventilation in the kitchen. Spills should be cleaned up immediately. Floors and walls around food-preparation and cooking areas should be cleaned at least daily—before or after each shift is preferable. Other nonfood-contact surfaces may need to be cleaned less often.

Floors

Floors are a safety hazard if they are not cleaned regularly or rinsed properly. They can also be a source of cross-contamination since soil and spills can be tracked through the entire operation. To clean floors, follow these steps:

- Mark the area being cleaned with signs or safety cones to prevent slips and falls.
- Sweep the floor.
- Use a deck or scrub brush and full-strength detergent on extra-soiled areas to remove grease and dirt.
- Mop or pressure-spray the area, working from the walls toward the floor drain. Soak the mop in a bucket of detergent solution and wring it out. Clean a ten-foot by ten-foot area with both sides of the mop, using a figure-eight motion. Soak and wring out the mop, then clean the same area again.
- Remove excess water with a damp mop or squeegee, working away from the walls and toward the floor drain.
- Rinse thoroughly with clean water, using the same mopping procedure.

Walls and Shelves

Clean tile and stainless-steel surfaces by spraying or sponging with a detergent solution. Use a nylon scrub brush to clean dried-on soil or grease. Rinse with clean water. When spray-cleaning, take care not to damage walls. Protect food, equipment, and nearby supplies. Use a sponge or wet cloth to clean other wall surfaces, such as painted drywall.

Ceilings and Light Fixtures

Ceilings do not need to be cleaned as often as floors or walls. Check ceilings and light fixtures daily to ensure that cobwebs, dust, dirt, or condensation will not fall and contaminate food or food-contact surfaces below. Wipe and rinse ceilings and light fixtures with a sponge or cloth.

CLEANING THE PREMISES

Although the kitchen requires the most attention, all areas of your operation need to be kept clean. Areas such as restrooms, busing or serving stations, and tables and booths need to be kept clean and sanitary as well. Pay attention to the exterior of the facility too. Dirty receiving docks and garbage areas can attract pests. A dirty parking lot or exterior is unsightly and can hurt business.

Tables

Tables, booths, and counters are nonfood-contact surfaces, since food is usually served on dishes. However, tables should always be kept clean. If you use table linens or butcher's paper, change them before seating new customers. If flatware is placed directly on the table surface, tables should also be sanitized.

- Use a dry wiping cloth to clean crumbs and dry-food spills from tables. Replace the cloth whenever it becomes soiled or sticky, so that it does not transfer soil and microorganisms to other surfaces.
- Use a moist cloth to clean up other types of food spills. Keep moist cloths in a bucket of chemical-sanitizing solution. Replace both the sanitizing solution and wiping cloths as necessary. Check the concentration of the sanitizing solution often.

Serving Stations

Serving stations might be little more than a small space to store tableware and linens. In many operations, however, they are also used to dispense water, coffee, and other beverages, to prepare and serve bread and condiment baskets, and to serve desserts. Serving stations often contain a sink, a coffee brewer, beverage dispensers, ice makers or bins, and even small coolers to hold butter, cream, condiments, and more. To keep serving stations clean, follow these procedures:

- Clean up spills immediately. Wipe or sweep up dry food with a dry cloth or broom. Clean wet spills with a damp cloth kept in sanitizing solution.
- Wash, rinse, and sanitize sinks and countertops at least daily or after each shift. If work areas are used to prepare potentially hazardous food, such as cheesecake, clean at least every four hours.
- Clean equipment daily or as often as recommended by the manufacturer. Items such as ice bins, beverage-dispensing nozzles and lines, and coffee grinders should be cleaned as often as needed to prevent dirt or mold from accumulating.
- Wash, rinse, and sanitize bus tubs manually or in the warewashing machine at least daily or after every shift.

Public Restrooms

Cleaning & Sanitizing

Many customers associate the cleanliness of the restroom with the establishment's commitment to food safety.

Unsanitary restrooms pose a danger to your customers and your business. Never underestimate cleaning needs in this area. Restrooms can quickly become visibly dirty and may harbor unpleasant odors and disease-causing microorganisms. Clean restrooms are important to customers. Many customers associate the cleanliness of the restroom with the establishment's commitment to food safety. When maintaining restrooms, consider the following suggestions:

- Check the condition of public and employee restrooms regularly.
- Restock soap, toilet paper, and towels before they run out.

- Clean sinks, mirrors, walls, floors, counters, dispensers, toilets, urinals, and waste receptacles at least daily. Sanitize toilets and urinals at least once daily. Clean up accidents and spills as often as necessary.
- Remove trash at least once daily, or as often as necessary.

Exterior Premises

The exterior of the premises should be kept clean. Windows, walls, and fixtures should be cleaned on a regular basis, and should be included on the master cleaning schedule. Check the grounds at least daily, pick up trash, and sweep walkways. Clean garbage areas as often as necessary to prevent odors or trash from attracting pests.

TOOLS FOR CLEANING

Cleaning & Sanitizing

Use a separate set of cleaning tools for the restroom.

Cleaning is easier when you have the right tools at hand. For example, worn-out tools will not give you the pressure or friction needed for cleaning, and tools that are the wrong size will be ineffective.

Keep tools used for cleaning separate from those used for sanitizing, and tools used for food-contact surfaces separate from those used to clean nonfood-contact surfaces. Using color-coded tools is one way to accomplish this. Always use a separate set of tools for the restroom.

Brushes

Brushes apply more effective pressure than wiping cloths, and the bristles loosen soil more easily. Worn brushes will not clean effectively and can be a source of contamination.

Brushes come in different shapes and sizes for each task. Lacquered wood or plastic brushes with synthetic bristles are preferred. They do not absorb moisture, are nonabrasive, and last longer. Use the right brush for the job.

Scouring Pads

Steel wool and other abrasives are sometimes used to clean heavily soiled pots and pans, equipment, or floors. However, metal scouring pads can break apart and leave residue on surfaces, which can later contaminate food. Nylon scouring pads can provide an alternative.

Mops and Brooms

Keep both light-and heavy-duty mops and brooms on hand. Mop heads can be all-cotton or synthetic blends. It makes sense to have a bucket and wringer for both the front and back of the facility. Both vertical and push-type brooms will come in handy.

STORING UTENSILS, TABLEWARE, AND EQUIPMENT

Once tableware, utensils, and equipment are clean and sanitary, store them so they stay that way. It is equally important to ensure that cleaning tools and supplies are stored properly.

Tableware and Equipment

- Store tableware and utensils at least six inches off the floor. Keep them covered or otherwise protected from dirt and condensation.
- Clean and sanitize drawers and shelves before clean items are stored.
- Clean and sanitize trays and carts used to carry clean tableware and utensils. Do this daily or as often as necessary.
- Store glasses and cups upside down. Store flatware and utensils with handles up so employees can pick them up without touching food-contact surfaces.
- Keep the food-contact surfaces of clean-in-place equipment covered until ready for use.

Key Point

Store flatware and utensils with handles up so employees can pick them up without touching food-contact surfaces.

Cleaning Tools and Supplies

Cleaning tools and supplies should be cleaned and sanitized before being put away. Tools and chemicals should be stored in a locked area away from food and food-preparation areas. The area should be well lighted so employees can identify chemicals easily. It should also be equipped with hooks for hanging mops, brooms, and other cleaning tools, a utility sink for filling buckets and cleaning tools, and a floor drain. (See *Exhibit 12g.*) Never use handwashing sinks, food-preparation sinks, or warewashing sinks to clean mops, brushes, or tools.

Exhibit 12g

Storage Area for Cleaning Tools and Supplies

Tools and chemicals should be stored in a locked area away from food and food-preparation areas.

When storing tools and supplies, consider the following suggestions:

- Air-dry wiping cloths overnight.
- Hang mops, brooms, and brushes on hooks to air-dry. Do not leave brooms or brushes standing on their bristles.
- Clean, rinse, and sanitize buckets. Let them air-dry, and store them with other tools.

USING HAZARDOUS MATERIALS

Chemicals are both useful and necessary to keep an establishment clean, sanitary, and pest free. Used properly, they pose little threat to an employee's safety. Used improperly, they can become a health hazard that can cause injury. To reduce this risk, you should only purchase chemicals that are approved for use in a restaurant or foodservice establishment.

Key Point

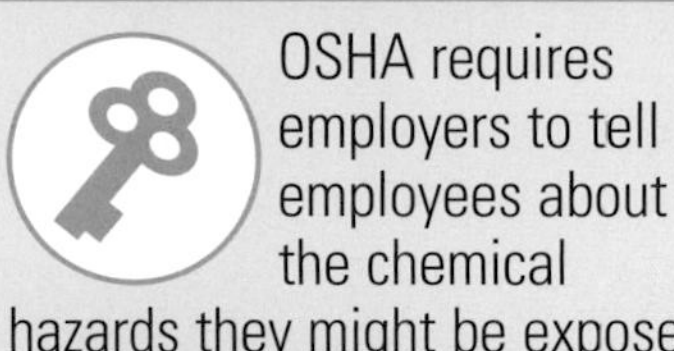

OSHA requires employers to tell employees about the chemical hazards they might be exposed to and to train employees to use the chemicals they work with safely.

Because of the potential dangers of chemicals used in the workplace, the Occupational Safety and Health Administration (OSHA) requires employers to comply with their **Hazard Communication Standard (HCS).** This standard, also known as Right-to-Know or HAZCOM, requires employers to tell their employees about chemical hazards to which they might be exposed at the establishment. It also requires employers to train employees on how to use the chemicals they work with safely. Employers must comply with OSHA's HCS by developing a hazard communication program for their establishment.

A hazard-communication program must include the following components:

- An inventory of hazardous chemicals used at the establishment
- Chemical labeling procedures
- Material Safety Data Sheets (MSDS)
- Employee training
- A written plan addressing the HCS

Inventory of Hazardous Chemicals

A hazardous chemical is any chemical that poses a physical or health hazard to humans. Chemicals known to have acute or chronic health effects, or are explosive, flammable, or unstable, are considered hazardous. Virtually any chemical with toxic properties should be included in an establishment's HAZCOM program.

Take an inventory of the hazardous chemicals stored in your establishment. List the name of the chemical and where it is stored. Update the list when chemicals are added or no longer used.

Labeling Procedures

OSHA requires chemical manufacturers to clearly label the outside of containers with the chemical name, manufacturer's name and address, and possible hazards. When receiving chemicals, only accept containers with proper labels, and make sure they remain readable and attached to the container.

If a chemical is transferred from a manufacturer's container to another container, the new container's label must contain the following information:

- Chemical name
- Manufacturer's name and address
- Potential hazards of the chemical

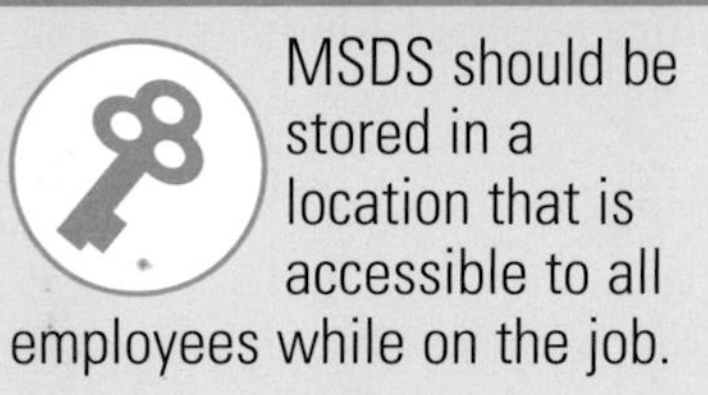

Material Safety Data Sheets (MSDS)

OSHA requires chemical suppliers and manufacturers to provide Material Safety Data Sheets (see *Exhibit 12h*) for each hazardous chemical at your establishment. These sheets are sent periodically with shipments or can be requested by the establishment. MSDS are part of employees' right to know about

Exhibit 12h

Sample Material Safety Data Sheets

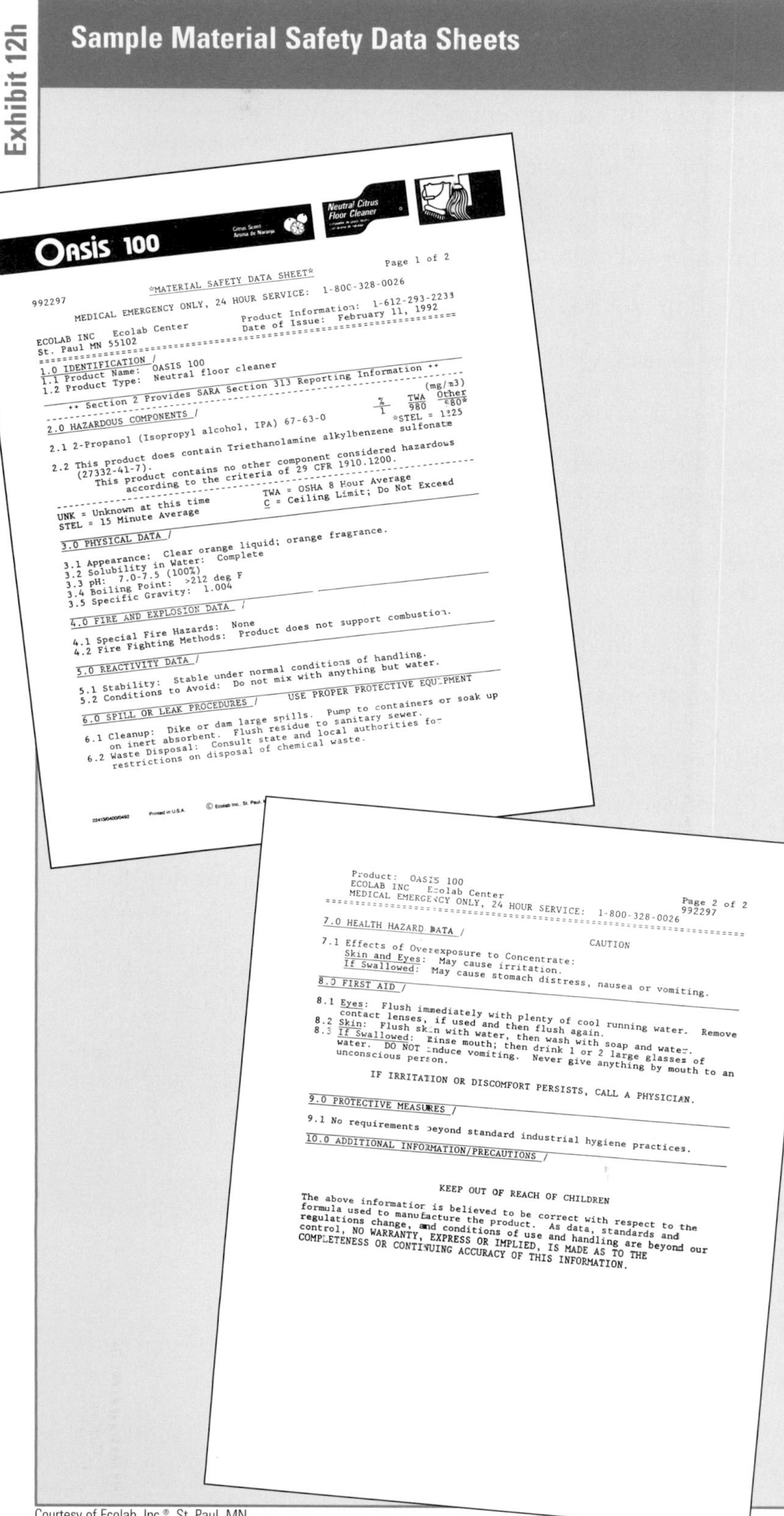

Page 1 of 2

MATERIAL SAFETY DATA SHEET

992297

MEDICAL EMERGENCY ONLY, 24 HOUR SERVICE: 1-800-328-0026

Product Information: 1-612-293-2233

ECOLAB INC Ecolab Center
St. Paul MN 55102

Date of Issue: February 11, 1992

1.0 IDENTIFICATION /
1.1 Product Name: OASIS 100
1.2 Product Type: Neutral floor cleaner

++ Section 2 Provides SARA Section 313 Reporting Information ++

2.0 HAZARDOUS COMPONENTS /

	%	TWA	Other
2.1 2-Propanol (Isopropyl alcohol, IPA) 67-63-0	1	980	980*

(mg/m3)
*STEL = 1225

2.2 This product does contain Triethanolamine alkylbenzene sulfonate (27332-41-7).
This product contains no other component considered hazardous according to the criteria of 29 CFR 1910.1200.

UNK = Unknown at this time
STEL = 15 Minute Average
TWA = OSHA 8 Hour Average
C = Ceiling Limit; Do Not Exceed

3.0 PHYSICAL DATA /
3.1 Appearance: Clear orange liquid; orange fragrance.
3.2 Solubility in Water: Complete
3.3 pH: 7.0-7.5 (100%)
3.4 Boiling Point: >212 deg F
3.5 Specific Gravity: 1.004

4.0 FIRE AND EXPLOSION DATA /
4.1 Special Fire Hazards: None
4.2 Fire Fighting Methods: Product does not support combustion.

5.0 REACTIVITY DATA /
5.1 Stability: Stable under normal conditions of handling.
5.2 Conditions to Avoid: Do not mix with anything but water.

6.0 SPILL OR LEAK PROCEDURES / USE PROPER PROTECTIVE EQUIPMENT
6.1 Cleanup: Dike or dam large spills. Pump to containers or soak up on inert absorbent. Flush residue to sanitary sewer.
6.2 Waste Disposal: Consult state and local authorities for restrictions on disposal of chemical waste.

22413/0400/0492 Printed in U.S.A. © Ecolab Inc., St. Paul,

Product: OASIS 100
ECOLAB INC Ecolab Center
MEDICAL EMERGENCY ONLY, 24 HOUR SERVICE: 1-800-328-0026
Page 2 of 2
992297

7.0 HEALTH HAZARD DATA /

CAUTION

7.1 Effects of Overexposure to Concentrate:
Skin and Eyes: May cause irritation.
If Swallowed: May cause stomach distress, nausea or vomiting.

8.0 FIRST AID /
8.1 Eyes: Flush immediately with plenty of cool running water. Remove contact lenses, if used and then flush again.
8.2 Skin: Flush skin with water, then wash with soap and water.
8.3 If Swallowed: Rinse mouth; then drink 1 or 2 large glasses of water. DO NOT induce vomiting. Never give anything by mouth to an unconscious person.

IF IRRITATION OR DISCOMFORT PERSISTS, CALL A PHYSICIAN.

9.0 PROTECTIVE MEASURES /
9.1 No requirements beyond standard industrial hygiene practices.

10.0 ADDITIONAL INFORMATION/PRECAUTIONS /

KEEP OUT OF REACH OF CHILDREN

The above information is believed to be correct with respect to the formula used to manufacture the product. As data, standards and regulations change, and conditions of use and handling are beyond our control, NO WARRANTY, EXPRESS OR IMPLIED, IS MADE AS TO THE COMPLETENESS OR CONTINUING ACCURACY OF THIS INFORMATION.

Courtesy of Ecolab, Inc.®, St. Paul, MN

the hazardous chemicals they work with, and therefore, must be stored in a location accessible to all employees while on the job. MSDS contain the following information about the chemical:

- Information about safe use and handling
- Physical, health, fire, and reactivity hazards
- Precautions
- Appropriate personal protective equipment (PPE) to wear when using the chemical
- First-aid information and steps to take in an emergency
- Manufacturer's name, address, and phone number
- Preparation date of MSDS
- Hazardous ingredients and identity information

Training

OSHA requires that every employee who might be exposed to hazardous chemicals during normal or regular working conditions be informed of the hazards and trained to use chemicals properly. Employees should receive this training annually, and new employees must receive it when first assigned to a new department or area.

The following topics should be covered during training:

- Existence and requirements of the HCS
- How the HCS is implemented in the workplace
- Operations and processes in which hazardous chemicals are used
- Inventory of chemicals in your establishment
- Location of MSDS
- How to read MSDS and product labels
- Physical and health hazards of all chemicals used
- Specific procedures adopted to provide protection, such as work practices and the use of engineering controls
- Use of PPE, and steps to prevent or reduce exposure to chemicals
- Safety and emergency procedures
- Information on the normal use of chemicals

Written Plan

Key Point

OSHA requires employers to develop a written plan that describes how they will meet the requirements of the Hazard Communication Standard (HCS) in their establishments.

OSHA requires employers to develop a written plan describing how they will meet the requirements of the HCS in their establishment. The following items should be included in your written plan:

- List of hazardous chemicals stored on the premises, and their amounts
- Purchasing specifications for chemicals
- Procedures for receiving and storing chemicals
- Labeling requirements in your establishment
- Procedures for accessing MSDS
- List of PPE
- Employee training procedures
- Reporting and record-keeping procedures
- How the employer will inform employees of the hazards of nonroutine tasks

Disposing of Hazardous Materials

Many chemicals used in the establishment pose a hazard to people and the environment if not disposed of properly. When disposing of chemicals, follow the instructions on the label and any local regulations that may apply.

IMPLEMENTING A CLEANING PROGRAM

A clean and sanitary establishment is a prerequisite for an effective food safety management program. It also takes commitment from management and the involvement of employees. The following are some basic steps for designing and implementing a cleaning program.

Cleaning & Sanitizing

An effective cleaning program takes commitment from management and the involvement of employees.

Identify Cleaning Needs

- **Identify all surfaces, tools, and equipment in the facility that need cleaning.** Walk through every area of the facility.
- **Look at the way cleaning is done currently.** Get input from employees. Ask them how and why they clean a certain way. Find out which procedures can be improved.
- **Estimate the time and skills needed for each task.** Some jobs may be done more efficiently by two or more people. Others might require an outside contractor. Determine how often things need to be cleaned.

Create a Master Cleaning Schedule

Take information gathered while identifying your cleaning needs and develop a **master cleaning schedule.** (See *Exhibit 12i* on the next page.) Organize the schedule by area and list the items that need to be cleaned and how often. Provide a brief description of how to do the job. Assign responsibility by job title. The schedule should include the following:

Cleaning & Sanitizing

When developing a master cleaning schedule for equipment and the facility, consider what should be cleaned, when, how, and who should do it.

- **What should be cleaned.** Arrange the schedule in a logical way so nothing is left out. List all cleaning jobs in one area, or list jobs in the order they should be performed. Keep the schedule flexible enough so you can make changes if needed.

Exhibit 12i

Sample Cleaning Schedule for a Food-Preparation Area

Item	What	When	Use	Who
Floors	▶ Wipe up spills	▶ Immediately	▶ Cloth mop and bucket, broom and dustpan	Busers
	▶ Damp mop	▶ Once per shift, between rushes	▶ Mop, bucket, safety signs	
	▶ Scrub	▶ Daily, at closing	▶ Brushes, squeegee, bucket, detergent, safety signs	
	▶ Strip, reseal	▶ Every 6 months	▶ Check written procedure	
Walls and Ceilings	▶ Wipe up splashes	▶ As soon as possible	▶ Clean cloth, detergent	Dishwashing staff
	▶ Wash walls	▶ Food-prep and cooking areas: daily ▶ All other areas: first of month		
Worktables	▶ Clean and sanitize tops	▶ Between uses and at the end of day	▶ See cleaning procedure for each table	Prep cooks
	▶ Empty, clean, and sanitize drawers	▶ Weekly	▶ See cleaning procedure for table	

- **Who should clean it.** Assign each task to a specific individual. In general, employees should clean their own areas. Rotate other cleaning tasks to distribute them fairly.
- **When it should be cleaned.** Employees should clean as they go, and clean and sanitize at the end of their shifts. Schedule major cleaning when food will not be contaminated or service interrupted—usually after closing. Schedule work shifts to allow enough time. Employees rushing to clean before their shifts end may cut corners.

Exhibit 12j

Sample Cleaning Procedure: Cleaning a Food Slicer

Step	Process	Notes
1	Turn off slicer and unplug	
2	Set blade control to zero	
3	Remove meat carriage	
4	Remove the back blade guard	
5	Remove the top blade guard	
6	Scrub meat carriage and guards in pot-and-pan sink	Use detergent solution and a scrub brush
7	Rinse parts	Immerse in clean water at 171°F (77°C) for 30 seconds. Use an S-hook to remove parts from water
8	Allow parts to air-dry on a clean and sanitary surface.	

- **How it should be cleaned.** Provide clearly written procedures for cleaning. Lead employees through the process step by step. Always follow manufacturers' instructions when cleaning equipment. Specify cleaning tools and chemicals by name. Post cleaning instructions near the item to be cleaned. (See *Exhibit 12j.*)

Choose Cleaning Materials

When selecting cleaning tools for your establishment, consider the following:

- **Select tools and cleaning agents according to the needs identified on the master cleaning schedule.** Talk to suppliers for suggestions on which tools and supplies are appropriate for your operation. Make sure your supplies match the needs listed on the master schedule.
- **Replace worn-out tools.** Equipment that is worn or soiled may not clean or sanitize surfaces properly.
- **Provide employees with the right protective gear.** Make sure there is an adequate supply of rubber gloves, aprons, goggles, and other supplies.

Training Employees

Training is critical to the success of the cleaning program. Employees must understand what you want them to be able to do, how well, and under what conditions. To ensure the success of the program, follow these guidelines:

- **Schedule a kick-off meeting to introduce the program to employees.** Explain the reason behind it. Stress how important cleanliness is to food safety. If people understand why they are supposed to do something, they are more likely to do it.
- **Schedule enough time for proper training.** Work with small groups or conduct training by area. Show employees how to clean equipment and surfaces in each area.
- **Provide plenty of motivation.** Reward employees for any job well done. Create small incentives for individuals or teams, such as "Clean Team of the Month" awards. Tie performance to specific measurements or goals, such as achieving high marks during health code inspections.

Monitor the Program

Once you have implemented the cleaning program, you must monitor it to make sure it is working. This includes:

- **Supervising daily cleaning routines**
- **Monitoring the daily completion of all cleaning tasks against the master cleaning schedule**
- **Reviewing the master schedule every time there is a change in menu, procedures, or equipment.** Make sure the cleaning program addresses any changes.
- **Requesting employee input on the program during staff meetings.** Ask employees if they need additional equipment, supplies, manpower, time, or training to get cleaning jobs done. Find out if they have suggestions for improving the program.
- **Conducting spot inspections**

Cleaning & Sanitizing

Once a cleaning program has been established, it must be monitored. Supervise cleaning routines, conduct spot inspections, and review the master schedule often.

SUMMARY

All the work that goes into a food safety management system can be undermined if you do not keep your utensils, equipment, and facility clean and sanitary. Cleaning is the process of removing food and other types of soil from a surface. Sanitizing is the process of reducing the number of harmful microorganisms on a clean surface to safe levels. You must clean and rinse a surface before it can be sanitized effectively. Surfaces can be sanitized with hot water or with a chemical-sanitizing solution.

All surfaces should be cleaned on a regular basis. Food-contact surfaces must be cleaned and then sanitized after every use, any time a task is interrupted, and at four-hour intervals if items are in constant use.

Warewashing machines can be used to clean, rinse, and sanitize most tableware and utensils. Follow manufacturers' instructions and make sure your machine is clean and in good working condition. Check the temperature and pressure of wash and rinse cycles daily.

Items that are too large to be placed into a warewashing machine can be cleaned and sanitized manually. This may be done in a three-compartment sink or, if the items are stationary, by cleaning and then spraying them with a sanitizing solution. Items cleaned in a three-compartment sink should be presoaked or scraped clean, washed in a detergent solution, rinsed in clean water, and sanitized in either hot water or in a chemical sanitizing solution for a predetermined amount of time. All items should then be air-dried.

Clean all nonfood-contact surfaces regularly. Areas such as public restrooms, floors, shelves, and floor drains should be cleaned daily (or more often as needed), and sanitized when appropriate. Ceilings, walls, and fixtures, as well as exterior areas such as docks, garbage containers, driveways, and parking lots, can be cleaned less frequently.

Keep all cleaning tools clean and sanitary. Cleaning tools and supplies should be stored in a well-lighted, locked room separate from food-storage and food-preparation areas. Make sure chemicals are clearly labeled. Keep MSDS for each chemical in a location accessible to all employees while on the job.

Develop and implement a cleaning program. Identify cleaning needs by walking through the operation and talking to employees. Create a master cleaning schedule listing all cleaning tasks, as well as when and how they are to be cleaned. Assign responsibility for each task by job title. Enlist employee support by including their input in the program's design and rewarding good performance. Explain to employees the important relationship between cleaning and sanitizing and food safety.

Monitor the cleaning program to keep it effective. Supervise cleaning procedures. Check completion of each job against the master schedule. Adjust cleaning and sanitizing procedures when there is a change in menu, equipment, or procedures.

Apply Your Knowledge

1. What do you think went wrong?
2. How can Tim prevent this from happening again?
3. How should any cleaning policy changes be introduced?

For answers, please turn to the Answer Key.

A Case in Point 1

It was only 9:05 A.M., but already, Tim, the day shift manager, could tell it was going to be a bad day. While the executives from American Widget munched unenthusiastically on complimentary doughnuts, they threw annoyed looks at Tim and the buser, who were cleaning the banquet room American Widget had reserved for 9:00 A.M. Tim had opened the room at 8:50 A.M. and found, to his dismay, the remains of the annual banquet of the Pine Valley Martial Arts Club still strewn about the room.

Later, Tim sat down with his cleaning schedule and tried to figure out what had gone wrong. The banquet had been scheduled to end at 11:30 P.M. the previous night. The buser was scheduled to clean the room at midnight. But a note from Norman, the night shift manager, told Tim the banquet had been a wild one and the last guest had left long after the 1:00 A.M. closing time. The buser had punched out at 12:30 A.M., as he always did. Tim sighed.

Apply Your Knowledge

1. What alternative methods can be used to clean and sanitize the tableware since the machine is not functioning properly?

For answers, please turn to the Answer Key.

A Case in Point 2

The kitchen employees in the university cafeteria had scraped and rinsed every piece of tableware, placed them in racks, and fed them into the high-temperature warewashing machine. When the wash and rinse cycles were complete, one of the employees noticed that the items were spotted. The thermometer registering the final rinse temperature for the sanitizing cycle had a reading of 140°F (60°C), rather than the required 180°F (82°C) indicated on the manufacturer's label. The employee went to inform the manager, who called for service.

Apply Your Knowledge

Use these questions to review the concepts presented in this chapter.

Discussion Questions

1. When should food-contact surfaces be cleaned and sanitized?
2. What is the difference between cleaning and sanitizing?
3. What are the steps that should be taken (in order) when cleaning and sanitizing items in a three-compartment sink?
4. How should clean and sanitized tableware, utensils, and equipment be stored?
5. What factors affect the efficiency of a sanitizer?
6. What are the requirements for storing cleaning materials?

For answers, please turn to the Answer Key.

Apply Your Knowledge

Use these questions to test your knowledge of the concepts presented in this chapter.

Multiple-Choice Study Questions

1. **Which factor influences the effectiveness of a chemical sanitizer?**
   A. Amount of time the sanitizer is in contact with the item
   B. Temperature of the sanitizing solution
   C. Concentration of the sanitizer in the solution
   D. All of the above

2. **You want to make a spray solution for use in sanitizing food-contact surfaces in the establishment. What should you do to ensure that you have made a proper sanitizing solution?**
   A. Compare the color of the solution to another solution of known strength.
   B. Try out the solution on a food-contact surface.
   C. Test the solution with a sanitizer test kit.
   D. Use very hot water when making the solution.

3. **What is the proper procedure for sanitizing a table that has been used to prepare food?**
   A. Spray it with a strong sanitizing solution, then wipe it dry.
   B. Wash it with a detergent, rinse it, then wipe it with a sanitizing solution.
   C. Wash it with a detergent, then wipe it dry.
   D. Wipe it with a dry cloth, then wipe it with a sanitizing solution.

4. **Which of the following items need to be both cleaned and sanitized?**
   A. Floors
   B. Walls
   C. Cutting boards
   D. Ceilings

Continued on next page...

Apply Your Knowledge

Multiple-Choice Study Questions *continued*

5. **If food-contact surfaces are in constant use, they must be cleaned and sanitized at**
   A. four-hour intervals.
   B. five-hour intervals.
   C. six-hour intervals.
   D. eight-hour intervals.

6. **Which one of the following steps is *incorrect* for cleaning and sanitizing a standing food mixer?**
   A. Clean the food and dirt from around the base of the mixer.
   B. Remove the detachable parts and wash them in the warewashing machine.
   C. Wash and rinse nondetachable food-contact surfaces. Wipe them with a sanitizing solution.
   D. Dry the detachable parts with a clean cloth and reassemble the machine.

7. **Your warewashing machine is not working and you must use your three-compartment sink to clean and sanitize tableware. What is the first thing you must do?**
   A. Fill the first sink with hot water and detergent.
   B. Clean and sanitize the sinks and drainboards.
   C. Prepare the sanitizing solution in the third sink.
   D. Gather clean towels for drying items.

8. **Which of the following is an improper method for storing clean and sanitized tableware and equipment?**
   A. Storing glasses and cups upside down
   B. Storing tableware six inches off the floor
   C. Storing flatware in containers with the handles down
   D. Storing utensils in a covered container until needed

Continued on next page...

Apply Your Knowledge

Multiple-Choice Study Questions *continued*

9. Managers should only purchase chemicals that
   A. are cost effective and do the job.
   B. will serve two purposes (clean inside and outside).
   C. are approved for use in a restaurant or foodservice establishment.
   D. are approved by OSHA.

10. When disposing of chemicals, managers should
    A. throw them away with other garbage.
    B. place them in a separate bag and then throw them away.
    C. pay someone to come and remove them from the establishment.
    D. follow the instructions on the label and any local regulations that may apply.

For answers, please turn to the Answer Key.

ADDITIONAL RESOURCES

Books and Periodicals

Cleaning house, safely. 2001. *Food Safety Illustrated.* 1 (3):14.

Hernandez, J. 1998. Keeping it clean. *Restaurant Hospitality.* 82 (9):118.

How to wash floors. 2003. *Food Safety Illustrated.* 3 (3):14.

Marriott, N. G. 1994. *Principles of food sanitation.* New York: Chapman & Hall.

National Restaurant Association. 2001. *Sanitation survival kit.* Washington, D.C.: National Restaurant Association.

Too steamy for print? Not when you're talking about dry vapor steam cleaners. 2003. *Food Safety Illustrated.* 3 (13):18.

Web Sites

Chlorine Chemistry Council®
www.c3.org
The Chlorine Chemistry Council, comprised of chlorine and chlorinated product manufacturers, is a business council of the American Chemistry Council. At this Web site, you will find the history of chlorine, its many uses at home and in the restaurant and in the foodservice industry, and chlorine in the news.

Colgate-Palmolive Company
www.colpalipd.com
A manufacturer of powerful, performance-tested cleaners and sanitizers, Colgate-Palmolive provides a Web site that contains useful information on its foodservice products and on food safety issues.

Ecolab, Inc.
www.ecolab.com
An excellent source of information on housekeeping and sanitation supplies for the restaurant and foodservice industry from the world's leading sanitation product supplier.

FDA Food Code
http://vm.cfsan.fda.gov/~dms/foodcode.html
As the basis for many local sanitation codes, as well as the basis for information in this textbook, the FDA Food Code, available at

this Web address, is a useful resource for information relating to food safety for the restaurant and foodservice industry.

KatchAll/San Jamar
www.sanjamar.com
KatchAll is an industry leader in innovative food safety products for the restaurant and foodservice industry. The Web site contains information on all KatchAll products that control time and temperature and help avoid cross-contamination.

MSDSonline
www.msdsonline.com
MSDSonline is a provider of electronic solutions for the outbound and inbound management of Material Safety Data Sheets (MSDS). From this site, your employees can access MSDS for all chemicals used in our operation.

National Restaurant Association
www.restaurant.org
The National Restaurant Association is the leading business association for the restaurant industry. Together with the National Restaurant Association Educational Foundation, the Association's mission is to represent, educate, and promote the rapidly growing restaurant and foodservice industry. This Web site should be your starting place for all issues and concerns related to your restaurant. This Web site has it all, from tips for running your establishment to vital data on your customers' spending habits.

Occupational Safety and Health Administration (OSHA)
www.osha.gov
OSHA's mission is to prevent work-related injuries, illnesses, and deaths. This Web site provides information related to workplace-safety conferences, technical links, and regulation and compliance documents.

Procter & Gamble
www.pg.com
Procter & Gamble produces cleaning and food and beverage products for the restaurant and foodservice industry. This Web site provides information on its products.

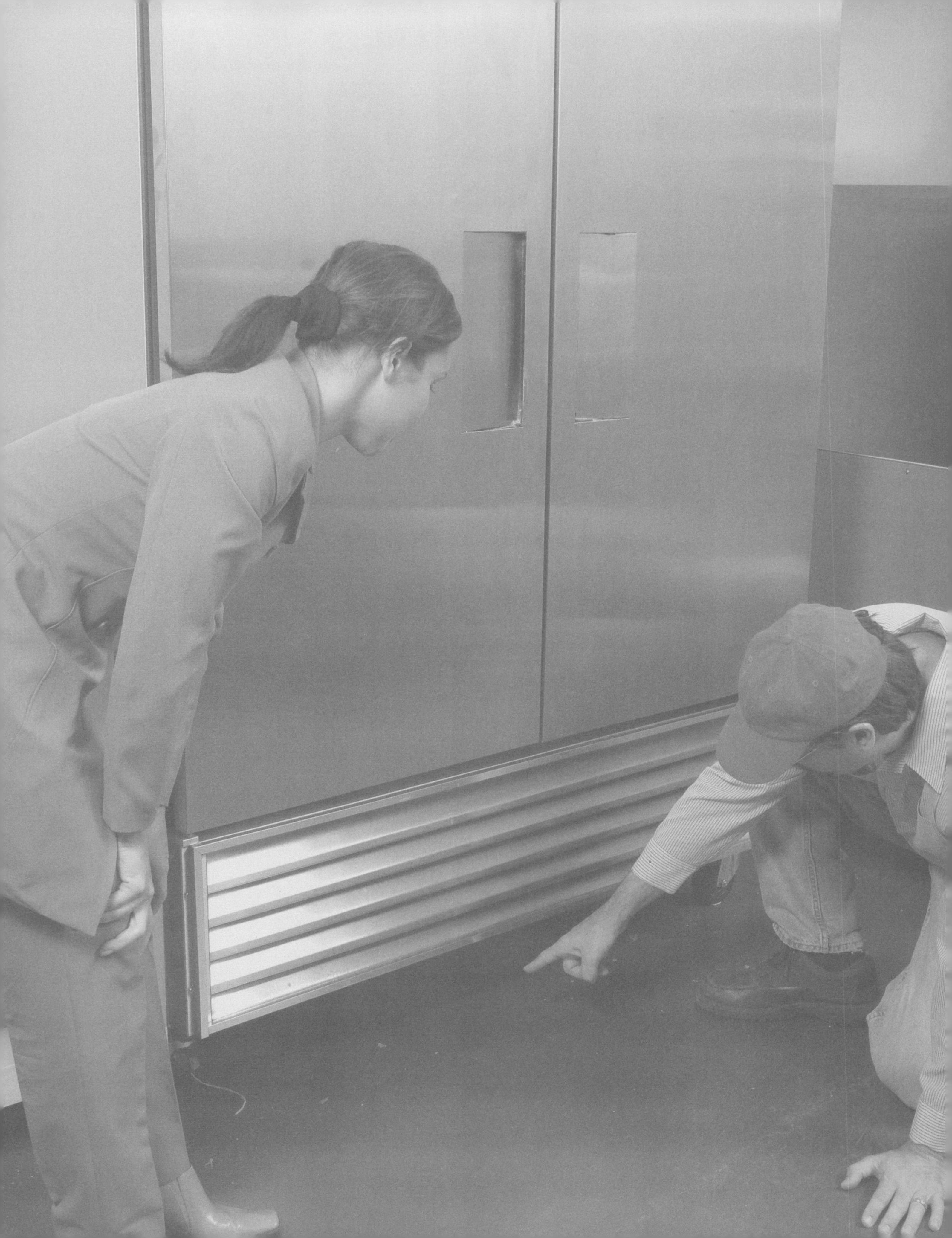

13

Integrated Pest Management

Inside this chapter:

- The Integrated Pest Management (IPM) Program
- Identifying Pests
- Working with a Pest Control Operator (PCO)
- Treatment
- Control Measures
- Using and Storing Pesticides

After completing this chapter, you should be able to:

- Identify requirements of an integrated pest management program.
- Differentiate between pest prevention and pest control.
- Identify ways to prevent pests from entering the facility.
- Identify the signs of pest infestation and/or activity.
- Identify requirements for applying pesticides.
- Identify proper storage requirements for pesticides and pest-application products.

Key Terms

Infestation
Integrated pest management (IFM)
Pest control operator (PCO)
Air curtains
Pesticide
Residual sprays
Contact sprays
Glue boards

Apply Your Knowledge

Check to see how much you know about the concepts in this chapter. Use the page references provided to explore the topic in each question.

Test Your Food Safety Knowledge

1. **True or False:** A strong oily odor may indicate the presence of roaches. *(See page 13-8.)*
2. **True or False:** The main purpose of an integrated pest management (IPM) program is to control pests once they have entered the establishment. *(See page 13-3.)*
3. **True or False:** Stationary equipment should not be covered before applying pesticides since it gives pests a place to hide. *(See page 13-18.)*
4. **True or False:** Glue traps are used to prevent roaches from entering the establishment. *(See page 13-8.)*
5. **True or False:** Pesticides can be stored in food-storage areas if they are labeled properly and closed tightly. *(See page 13-18.)*

For answers, please turn to the Answer Key.

INTRODUCTION

Pests such as insects and rodents can pose serious problems for restaurants and foodservice establishments. Not only are they unsightly to customers, they also damage food, supplies, and facilities. The greatest danger from pests comes from their ability to spread diseases, including foodborne illnesses.

THE INTEGRATED PEST MANAGEMENT (IPM) PROGRAM

Once pests have come into the facility in large numbers—an **infestation**—they can be very difficult to eliminate. Developing and implementing an **integrated pest management (IPM)** program is the key. An IPM program uses prevention measures to keep pests from entering the establishment and control measures to eliminate any pests that do infest it.

For your IPM program to be successful, it is best to work closely with a licensed **pest control operator (PCO).** These professionals use safe, up-to-date methods to prevent and control pests. Prevention is critical in pest control. If you wait until there is evidence of pests in your establishment, they may already be there in large numbers.

There are three basic rules of an IPM program:

1. Deny pests access to the establishment.
2. Deny pests food, water, and a hiding or nesting place.
3. Work with a licensed PCO to eliminate pests that do enter.

Deny Pests Access to the Establishment

Key Point

Check all deliveries before they enter the establishment, and refuse any shipment in which you find pests or signs of infestation.

Pests can enter an establishment in one of two ways. They either are brought inside with deliveries, or they enter through openings in the building itself. To prevent pests from entering your establishment in deliveries, start by using reputable suppliers. Check all deliveries before they enter your facility. Refuse any shipment in which you find pests or signs of infestation, such as egg cases and body parts (legs, wings, etc.).

Holes and cracks in the exterior of a building allow pests to get inside and hide, but there are many other openings pests can use to enter an establishment.

Doors, Windows, and Vents

- **Screen all windows and vents with at least sixteen mesh per square inch screening.** Anything larger might let in mosquitoes or flies. Check screens regularly, and clean and replace them as needed.
- **Install self-closing devices or door sweeps on all doors.** Repair gaps and cracks in door frames and thresholds. Use weather stripping on the bottoms of doors with no threshold.
- **Install air curtains (also called air doors or fly fans) above or alongside doors.** These devices blow a steady stream of air across the entryway, creating an air shield around open doors that insects avoid.
- **Keep drive-through windows closed when not in use.**
- **Keep all exterior openings closed tightly.**

Pipes

Mice, rats, and insects such as cockroaches use pipes as highways through a facility.

- **Use concrete to fill holes or sheet metal to cover openings around pipes.** (See *Exhibit 13a.*)
- **Install screens over ventilation pipes and ducts on the roof.**
- **Cover floor drains with hinged grates.** Rats are good swimmers and can enter buildings through drainpipes.

Floors and Walls

Rodents often burrow into buildings through decaying masonry or cracks in building foundations. They move through floors and walls the same way.

- **Seal all cracks in floors and walls.** Use a permanent sealant recommended by your PCO or local health department.
- **Properly seal spaces or cracks where stationary equipment is fitted to the floor.** Use an approved sealant or concrete, depending on the size of the spaces.

Exhibit 13a

Deny Entry to Pests

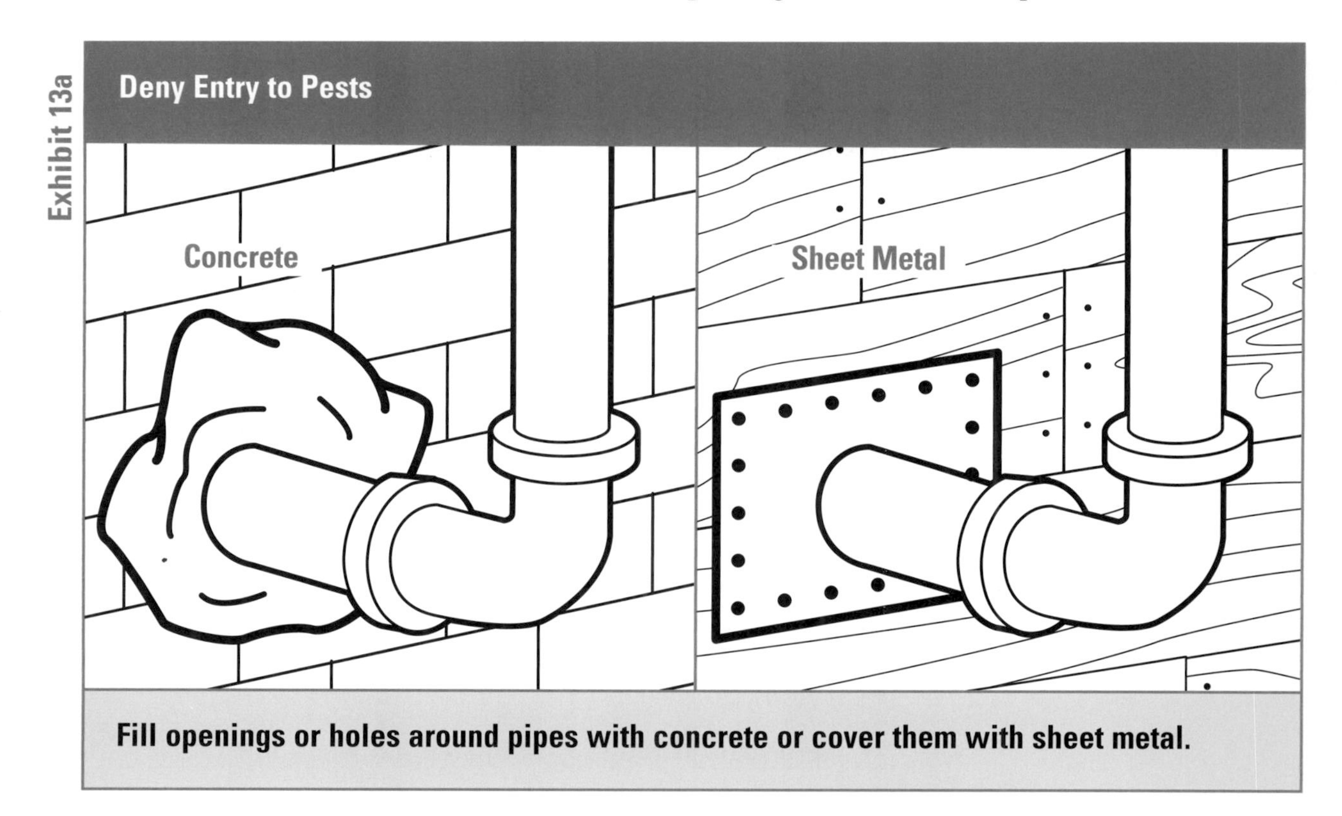

Fill openings or holes around pipes with concrete or cover them with sheet metal.

Deny Food and Shelter

Pests are usually attracted to damp, dark, dirty places. A clean and sanitary establishment offers them little in the way of food and shelter. The stray pest that might get in cannot thrive or multiply in a clean kitchen. Besides adhering to your master cleaning schedule, follow these additional guidelines:

Key Point

Garbage should be stored in clean, tightly covered containers that are in good condition. Dispose of garbage quickly so it will not attract pests.

- **Dispose of garbage quickly and correctly.** Garbage attracts pests and provides them a breeding ground. Keep garbage containers clean, in good condition, and tightly covered in all areas (indoors and outdoors). Clean up spills around garbage containers immediately. Wash and rinse containers regularly.
- **Store recyclables in clean, pest-proof containers, as far away from your building as local regulations allow.** Bottles, cans, paper, and packaging material provide shelter and food for pests.
- **Store all food and supplies properly and as quickly as possible.**
  - Keep all food and supplies away from walls and at least six inches off the floor.
  - When possible, keep humidity at fifty percent or lower. Low humidity helps prevent roach eggs from hatching. Use dehumidifiers as well as ventilation that forces air outside the building.
  - Refrigerate food such as powdered milk, cocoa, and nuts after opening. Most insects that might be attracted to this food become inactive at temperatures below 41°F (5°C).
  - Rotate products, so pests do not have time to settle into them and breed.

Key Point

Clean your establishment thoroughly. Careful cleaning eliminates a pest's food supply, destroys insect eggs, and reduces the number of places pests can take shelter.

- **Clean your establishment thoroughly.** Careful cleaning eliminates the food supply, destroys insect eggs, and reduces the number of places pests can safely take shelter.
  - Clean up food and beverage spills immediately, including crumbs and scraps.
  - Clean toilets and restrooms as often as necessary.

- Train employees to keep lockers and break areas clean. Food and dirty clothes should not be kept in or around lockers. Break rooms should be cleaned properly after use.
- Keep cleaning tools and supplies clean and dry. Store wet mops on hooks, rather than on the floor, since roaches frequently hide in them.
- Empty water from buckets to keep from attracting rodents.

Grounds and Outdoor Dining Areas

The popularity of outdoor dining presents different concerns. Birds, flies, bees, and wasps can be annoying and dangerous to the health of your customers. The key to protecting customers lies in minimizing these outdoor pests by denying them food and shelter. (See *Exhibit 13b.*)

- **Mow the grass, pull weeds, get rid of standing water, and pick up litter.**
- **Cover all outdoor garbage containers.**
- **Remove dirty dishes and uneaten food from tables, cleaning them as quickly as possible.**
- **Do not allow employees or customers to feed birds or wildlife on the grounds.**

- **Locate electronic insect eliminators ("zappers") away from food, customers, employees, serving areas, and combustible material.**
- **Call your PCO to remove hives and nests.**

IDENTIFYING PESTS

Pests may still get into your establishment even if you take careful preventive measures. Pests are good hitchhikers, hiding in delivery boxes and even coming in on employees' clothing or personal belongings. Learn how to spot signs of pests and identify what kind they are. If possible, record the time, date, and location of any pest sighting (or evidence of pests) and report this to your PCO. Early detection gives your PCO a chance to start treatment as soon as possible.

Cockroaches

Roaches often carry disease-causing microorganisms such as *Salmonella* spp., fungi, parasite eggs, and viruses. Research shows that many people are allergic to residue left by roaches on food and surfaces. Roaches reproduce quickly and can adapt to some pesticides, making it difficult to control them.

There are several different types of roaches. Most live and breed in dark, warm, moist, hard-to-clean places. You will typically find them in the following areas:

- Behind refrigerators, freezers, and stoves
- In sink and floor drains
- In spaces around hot-water pipes
- Inside equipment, often near motors and other electrical devices
- Under shelf liners and wallpaper
- Underneath rubber mats
- In delivery bags and boxes
- Behind unsealed coving (especially rubber-based)

Roaches generally feed in the dark. If you see a cockroach in daylight, you may have a major infestation. Only the weakest roaches come out in daylight.

If you suspect you have a roach problem, check for these signs:

- A strong oily odor
- Droppings (feces), which look like grains of black pepper
- Capsule-shaped egg cases that are brown, dark red, or black and may appear leathery, smooth, or shiny

You may have problems with more than one type of roach. Glue traps—containers with sticky glue on the bottom (see *Exhibit 13c*)—should be used to find out what type of roaches might be present. Work with your PCO to place the traps where roaches typically can be found. If possible, place them on the floor in the corner where two walls meet. Check the traps after twenty-four hours and show them to your PCO. The type and stage (nymph or adult) of roaches present will determine the type of treatment needed. Common types of roaches include American, Brown-banded, German, and Oriental. (See *Exhibit 13d.*)

Flies

The common housefly is also a great threat to human health. Because they feed on garbage and animal waste, flies transmit foodborne illnesses such as typhoid fever and dysentery. They also spread pathogens such as *Streptococcus* and *Staphylococcus*, which

Exhibit 13c

Roach Glue Trap

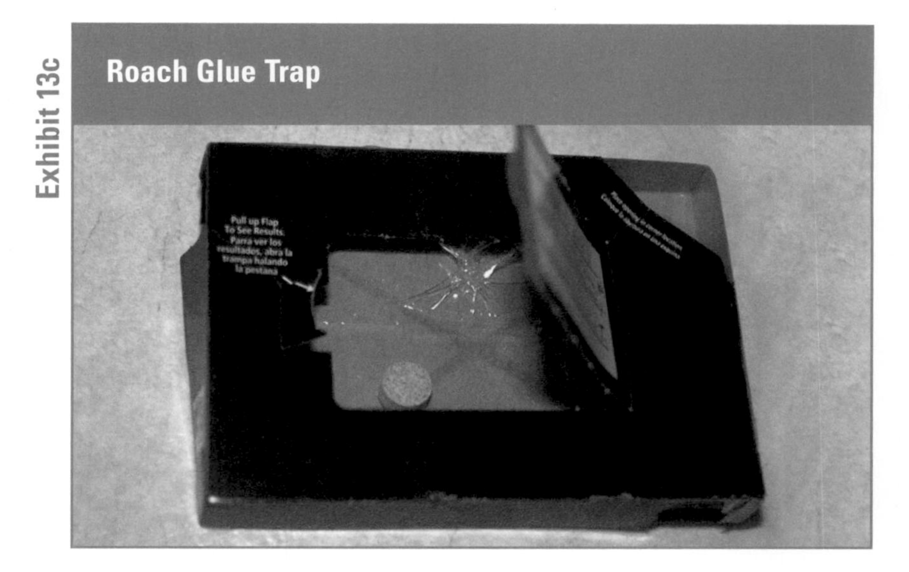

Glue traps can be placed in areas where roaches typically are found to monitor their presence.

Courtesy of the National Pest Management Association

Exhibit 13d

Common Cockroaches Found in Restaurants and Foodservice Establishments

American

German

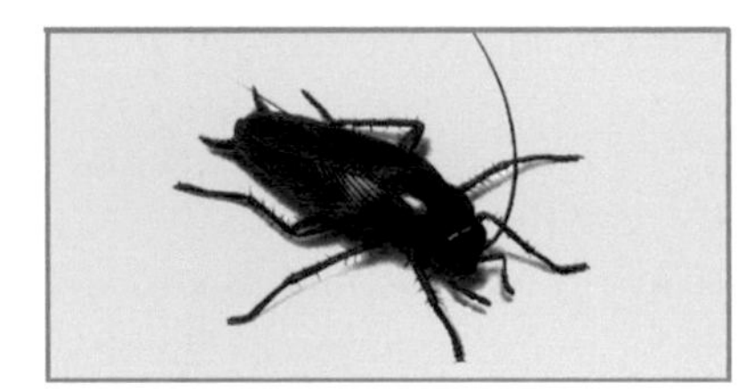

Brown-banded

Oriental

Courtesy of Orkin Commercial Services

can stick to their feet, hair, and mouths. Their feces and vomitus—swarming with bacteria—contaminate food. Flies have no teeth, so they only can eat liquids or dissolved foods. They inject a long spike and vomit into solid food, let it dissolve, and then eat it.

In general, houseflies have the following characteristics:

- They prefer calm air and the edges of objects, such as rims of garbage cans.
- To find food and lay their eggs, they are drawn to odors of decay, garbage, and animal waste.
- In warm weather, they reproduce rapidly (eggs can hatch in as few as thirty hours). For their eggs to hatch, they need warm, moist, decayed material located out of the sun. Eggs hatch into maggots, which can grow into adult flies in six days.

Other types of flies can be just as bothersome. Fruit flies are somewhat less harmful because they are primarily attracted to spoiled fruit, not animal waste. They can still transmit diseases, however. "Biting" flies, such as deer flies and horse flies, can be an additional nuisance to outdoor diners.

Other Insects

- **Beetles, weevils, and moths.** Flour-moth larvae and beetles are usually found in dry-storage areas. Look for insect bodies, wings, or webs, as well as clumped-together food and holes in food and packaging. To prevent these pests, cover food tightly, keep storage areas clean and sanitary, and practice FIFO.
- **Ants.** Ants often nest in walls and floors near stoves and hot-water pipes. They are drawn to grease and sweet food. Clean up all food scraps and spills to keep them out.
- **Termites and carpenter ants.** These insects can cause great structural damage, boring into wood and weakening walls, floors, and ceilings. They are rarely visible. Carpenter ants nest in wood, but forage for food elsewhere. Look for signs of sawdust that has fallen from the ceiling. Call your PCO at once if you see signs they are present.

- **Spiders.** Look for webs in corners and around fixtures. Spiders can be controlled by checking for and removing webs daily. The bites of some spiders—such as the black widow and brown recluse—are poisonous and can cause illness, but are rarely fatal. If you see either type, call your PCO.
- **Bees, wasps, and hornets.** Bees, wasps, and hornets can sting outdoor diners. Some people are allergic to their venom and can go into shock, or even die, from just one or two stings. In general, they are drawn to sweet food. Here are some suggestions for handling them:
  - Bees make hives in hollow trees, under eaves, or in other protected places. They usually will not bother people unless their hives are threatened. If you find a hive, call your PCO to remove it.
  - Wasps and hornets often build nests under eaves. Call your PCO to remove them.
  - Yellow jackets usually nest in the ground and are drawn to proteins in early summer and sweets in late summer. They aggressively defend their nests, so call a PCO to remove them.
- **Mosquitoes and gnats.** Mosquitoes carry diseases such as malaria, typhoid, yellow fever, and encephalitis. These members of the fly family feed on the blood of animals and humans. Mosquitoes are drawn to heat and are most active at dawn and dusk. Gnats are annoying to outdoor diners. To minimize their presence, keep all outdoor areas free of stagnant water and move lights away from the building.
- **Deer ticks.** Ticks can carry Lyme disease and cause other serious illnesses. Minimize risk by keeping the grass short around dining areas.

Rodents

Rodents are a serious health hazard. They eat and ruin food, damage property, and can spread disease. Most rodents have a simple digestive system. They urinate and defecate as they move around a facility. Their waste can fall into food and can contaminate surfaces.

Rats and mice are the most common types of rodent. (See *Exhibit 13e.*) They hide during the day and search for food at

Exhibit 13e

Common Types of Rodents

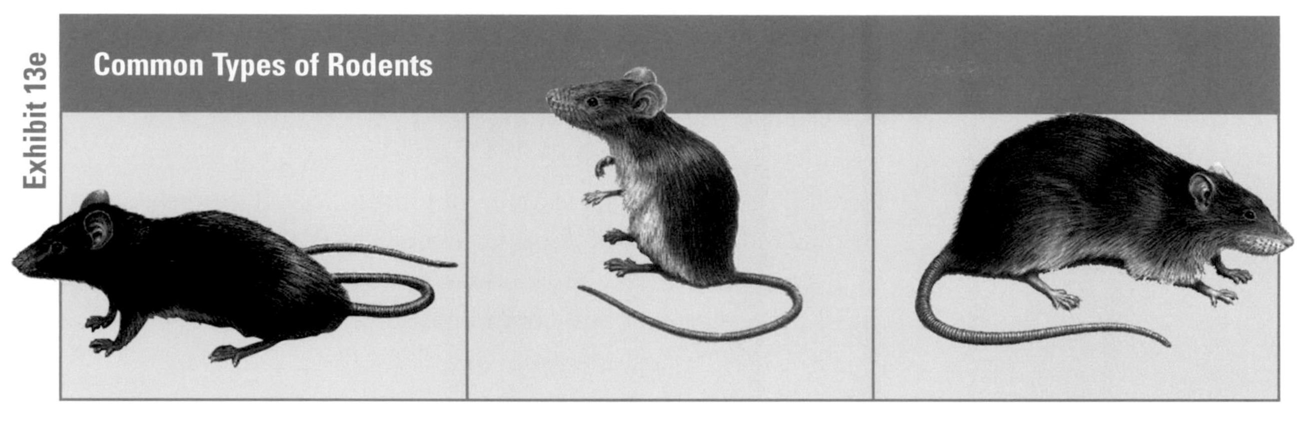

Roof Rat | Common House Mouse | Norway Rat

Courtesy of Orkin Commercial Services

night. Like other pests, they reproduce often. Typically, they do not travel far from their nests—rats travel only 100 to 150 feet, while mice travel only ten to thirty feet. Mice can squeeze through a hole the size of a dime to enter a facility, rats through quarter-sized holes. Rats can stretch to reach an item as high as eighteen inches, can jump three feet in the air, and can even climb straight up brick walls. They have very good senses of hearing, touch, and smell, and are smart enough to avoid poison bait and poorly laid traps. Effective control of rats and mice requires the knowledge and experience of professionals.

Exhibit 13f

Signs of Rodent Infestation

Rats and mice gnaw to reach food and to wear down their teeth.

Courtesy of the National Pest Management Association

A building can be infested with both rats and mice at the same time. Look for these signs:

- **Signs of gnawing.** Rats and mice gnaw to reach food and to wear down their teeth, which grow continuously. Rats' teeth are so strong they can gnaw through pipes, concrete, and wood. (See *Exhibit 13f.*)
- **Droppings.** Fresh droppings are shiny and black. Older droppings are gray.
- **Tracks.** Check dusty surfaces by shining a light across them at a low angle.
- **Nesting materials.** Mice use scraps of paper, cloth, hair, and other soft materials to build nests.
- **Holes.** Rats nest in burrows, usually in dirt, rock piles, or along foundations.

Birds

Bird droppings carry fungi and bacteria that can make people sick. They also may carry mites and microorganisms that can cause encephalitis and other diseases. Birds can be drawn to crumbs and food scraps in outdoor dining areas. Keep these areas clean, and remove food from tables quickly. Post signs asking customers not to feed birds. If birds become a problem, call a PCO specializing in bird-control measures.

Other Animals

Though less common, other animals such as bats, raccoons, and squirrels can infest your building too. Building damage, bites, and the possible spread of rabies are the biggest concerns. Prevent them from getting inside. If there is evidence any of these pests have entered your establishment, work with a PCO or your local animal control department to remove them.

- **Bats.** Bats will nest in high places that are warm, dark, and dry, often in eaves or attics. Bats are beneficial, since they eat insects, and in many places it is against the law to kill them. However, they will bite if cornered and can spread rabies. Established bat colonies can be hard to eliminate. At dusk, after bats have left to feed, block all crevices and entrances. Light the area for several days to keep them away until they find a new place to roost.
- **Raccoons.** Raccoons have adapted to humans and urban growth. They feed at night, but it is not unusual to see them during the day. They prefer wooded areas, but females will nest in dark places, such as attics and chimneys, to give birth. Raccoons can contract rabies, and they will bite if cornered.
- **Squirrels.** Like raccoons, squirrels will nest in attics if given the chance. Clear branches away from buildings and fill any holes or cracks around eaves and gutters.

WORKING WITH A PEST CONTROL OPERATOR (PCO)

Few pest problems are solved simply by spraying **pesticides**, chemical agents used to destroy pests. Although you can take most preventive measures yourself, most control measures should be carried out by professionals. Employ a licensed, certified PCO to handle pest

Key Point

Most pest control measures should be carried out by professional pest control operators (PCOs).

control. Working as a team, you and the PCO can prevent and/or eliminate pests and keep them from coming back. There are many advantages to working with a professional PCO, who will

- work with you to develop an integrated approach to pest management. This may include using a combination of chemical and nonchemical treatments to solve and prevent problems.
- stay up-to-date on new equipment and products.
- provide prompt service to address problems as they occur. Most contracts should provide regular visits plus immediate service when pests are spotted.
- take responsibility for keeping records on the use and safe handling of potentially hazardous chemicals, as well as all steps taken to prevent and control pests.

How to Choose a PCO

Key Point

When choosing a PCO, make sure he or she is licensed or certified, belongs to professional organizations, and is insured.

Hiring a PCO is like choosing any other supplier or service provider: you must do your homework. Use these guidelines to help make your decision.

- **Talk to other foodservice managers.** Find out what PCO they use and what their experiences have been. When you call a PCO, ask for references and check them thoroughly. Make sure the PCO has experience working with restaurants and other foodservice establishments.
- **Make sure the PCO is licensed or certified by your state, as required by federal law.** Certification means the PCO has passed a test on proper pesticide use and other control methods.
- **Ask the PCO if they belong to any professional organizations.** Membership in the National Pest Management Association (NPMA) or state or local groups usually means the PCO has up-to-date information on IPM.
- **Ask for proof of insurance.** Make sure the PCO has adequate coverage to protect you and your employees, customers, and facility.
- **Weigh all factors, not just price.** Make sure the PCO you choose has the expertise and resources you need to provide the service promised. A low bid may be costly in the long run.

Key Point

Always require a contract from your PCO. Service contracts should spell out the work to be performed and what is expected from both you and your PCO.

Service Contract

Always require a written service contract from your PCO. Service contracts outline the work to be performed and what is expected from both you and the PCO. Read your contract carefully and have your lawyer review it, if possible. A contract should include these items:

- A description of services to be provided, including an initial inspection, regular monitoring visits, follow-up visits, and emergency service
- A warranty for work to be done
- Legal liability of the PCO
- The period of service
- Your duties, including preventive measures and facility preparation before and after treatment
- Records to be kept by the PCO, such as:
    - Pests sighted and trapped, species, location, and actions taken
    - All chemicals used and Material Safety Data Sheets (MSDS) for each (Copies should be accessible to employees.)
    - Building and maintenance problems noted and fixed
    - Maps or photos of the facilities noting location of traps, bait, and problem spots
    - Schedule for checking and cleaning traps, replacing bait, and reapplying chemicals
    - Regular written summary reports from the PCO. Copies should be kept on file in the establishment for reference when planning improvements and assessing facility goals.

TREATMENT

Effective treatment starts with a thorough inspection of your facility and grounds. Give the PCO complete access to the building and cooperate fully during the inspection.

- Prepare employees to answer the PCO's questions.
- Provide building plans and equipment layouts.
- Point out possible trouble spots.

Key Point

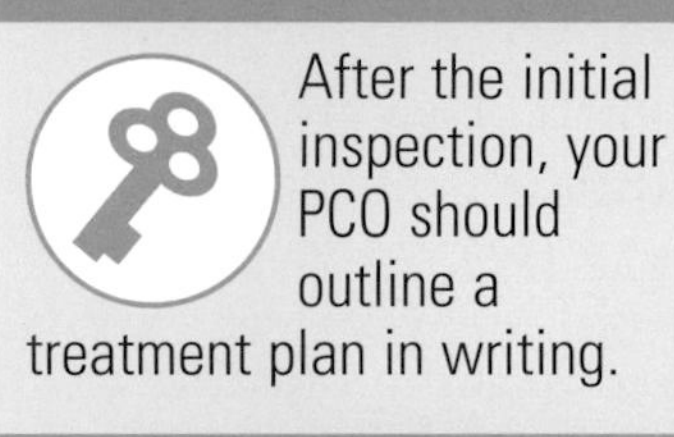

After the initial inspection, your PCO should outline a treatment plan in writing.

Courtesy of the National Pest Management Association

After the initial inspection, your PCO should outline—in writing—a treatment plan. In addition to price, the plan should contain the following information:

- **Exactly what treatment will be used for each area or problem and the potential risks involved**
- **Dates and times of each treatment.** The federal government requires a PCO to give you enough advance warning to prepare the facility properly. Employees must not be on-site during the treatment.
- **Steps you can take to control pests**
- **Building defects that may cause problems for prevention and control measures**
- **Timing of follow-up visits.** The PCO should review how well treatment is working and suggest alternate treatments if pests reappear.

CONTROL MEASURES

PCOs can use a variety of pest-control methods that are environmentally sound and safe for establishments. They are trained to know which techniques will work best to control different types of pests in your area. Since new technologies are being developed all the time, the more you know about each of these methods, the better you can evaluate how well your PCO is doing.

Controlling Insects

There are several methods your PCO can use to control insects, depending on the type of insect and degree of infestation.

- **Repellents.** Repellents are liquids, powders, or mists that keep insects away from an area, but do not kill them. Repellents are often used in hard-to-reach places, such as the spaces behind wallboards and plaster.
- **Sprays.** Chemical pesticide sprays are used to control insects. They include residual and contact sprays.
    - **Residual sprays** leave a film of insecticide that insects absorb as they crawl across it. Used in cracks and crevices

like those along baseboards, these sprays can be liquid or a dust, such as boric acid.

- **Contact sprays** kill insects on contact. They are usually used on groups of insects, such as clusters of roaches or a nest of ants.

- **Bait.** Chemical bait sometimes is used to control roaches or ants. The bait contains an attractant, and, when insects eat it, the chemical kills them. The advantage of using baits is that kitchen areas do not need to be prepped and people can remain on-site while they are set.
- **Traps.** There are several types of traps.
  - Light-only units simply entice insects to crawl inside where they find it hard to escape.
  - Electronic insect eliminators, or "zappers," use an electrically charged grid to kill insects attracted to the light.
  - Other units use both light and chemical attractants to lure insects onto a glue board.

Wasp and hornet traps are designed to hold nectar, which attracts these insects. Once inside, they cannot escape. Most fly traps use a light source—usually UV light—to attract insects to the trap. Tests have shown that different insects respond to different types or ranges of UV light. Make sure your PCO is using the most current technology.

The placement of traps is important. Never place them above or near food-preparation or storage areas or food-contact surfaces.

Controlling Rodents

Rats and mice tend to use the same routes through an establishment. Your PCO will choose the best method to eliminate these pests, which could include the following: (See *Exhibit 13g.*)

- **Traps.** Traps are a safe, effective way to kill rats and mice. If the infestation is large, however, traps will take time. Work with your PCO to set traps near or in rodent runways. Check traps often and remove dead rodents carefully. If a trapped rodent is still alive, have your PCO remove it.

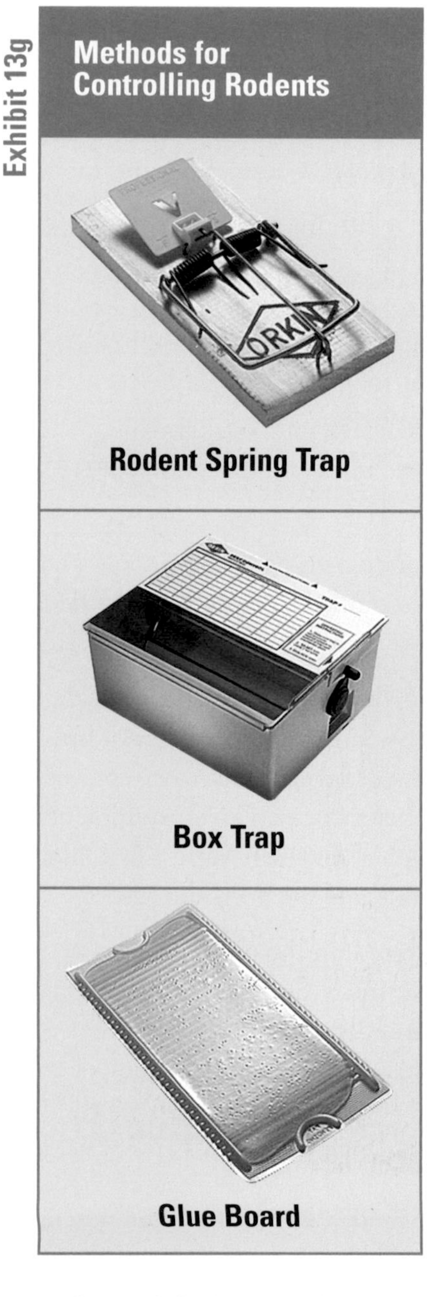

Several devices can be used to control rodents.

Courtesy of Orkin Commercial Services

- **Glue boards.** Glue boards kill mice. When these devices are placed in runways, the mice stick to the board and die in several hours—from exhaustion or lack of water or air. Check boards often and throw away any with trapped mice. Glue boards are not effective for controlling rats since these rodents are usually strong enough to escape.
- **Bait.** Chemical bait should only be used by a PCO in areas where it cannot contaminate food or food-contact surfaces. It is usually placed in special covered, locked containers near rodent runways and possible entry points. Your PCO may change the baits and their locations often until they work properly. Rats can easily detect chemical bait and often avoid it.

Controlling Birds

Birds can be a serious problem in outdoor dining areas. While there is no way to eliminate birds, your PCO can use several techniques to keep them from nesting and roosting on your building.

- **Repellents.** Chemical pastes are sometimes used on gutters and ledges to repel birds. Paste must be used carefully so it does not fall into food or onto tables.
- **Netting.** Fine-mesh wire netting is used to keep birds from roosting on statues and bas-relief carvings on buildings.
- **Wires.** A PCO might string wires across the roof to prevent birds from roosting. The wires can be electrified to deliver a mild shock.
- **Sound.** In some instances, birds can be frightened away by the sound of other birds in distress. Prerecorded tapes are played through loudspeakers near roosting sites, though birds might eventually ignore the sounds and return to roost.
- **Balloons.** Birds sometimes can be frightened away with helium-filled, mylar balloons. Ask your PCO about this technique.

USING AND STORING PESTICIDES

While it may seem more cost-effective to purchase and apply pesticides yourself, there are many reasons not to do so.

- Pesticides can be dangerous to employees, customers, and food.
- Applied improperly, they will be ineffective.
- Pests can develop resistance and an immunity to pesticides.
- Each region has its own pest-control problems, and some control measures are more effective than others.
- Pesticides are regulated by federal, state, and local laws, and some are not approved for use in restaurants and foodservice establishments.

Key Point

Pesticides should only be applied by a PCO. They are trained to determine the best pesticide for each pest, and how and where to apply it.

Courtesy of the National Pest Management Association

Rely on your PCO to decide if and when pesticides should be used in your establishment. They are trained to determine the best pesticide for each pest, and how and where to apply it.

To minimize the hazard to people, have your PCO use pesticides only when you are closed for business, and employees are not on-site. When pesticides will be applied, prepare the area to be sprayed by removing all food and food-contact surfaces. Cover equipment and food-contact surfaces that cannot be moved. Wash, rinse, and sanitize food-contact surfaces after the area has been sprayed.

Anytime pesticides are used or stored on the premises, you should have a corresponding MSDS, since they are hazardous materials. Your PCO should store and dispose of all pesticides used in your facility. If they are stored on the premises, follow these guidelines:

- **Keep pesticides in their original containers.**
- **Store pesticides in locked cabinets away from areas where food is stored or prepared.**
- **Store aerosol or pressurized spray cans in a cool place.** Exposure to temperatures higher than 120°F (49°C) could cause them to explode.
- **Check local regulations before disposing of pesticides.** Many are considered hazardous waste. Dispose of empty containers according to manufacturers' directions and local regulations.

SUMMARY

Pests can carry and spread a variety of diseases. Once they have infested a facility, it can be very difficult to eliminate them. Developing and implementing an integrated pest management (IPM) program is the key. An IPM program uses prevention measures to keep pests from entering the establishment and control measures to eliminate any pests that do get inside. To be successful, establishments must deny pests access to the facility and food, water, and shelter. Finally, you must work with a licensed pest control operator (PCO) to eliminate pests that do enter.

Pests can be brought inside the establishment with deliveries, or they can enter through openings in the building itself. To prevent them from getting inside, check deliveries before they enter your facility and refuse any shipment in which you find pests or signs of infestation. Screen all windows and vents, install self-closing doors and air curtains, and keep exterior openings closed when not in use. Fill or cover holes around pipes, and seal cracks in floors and walls.

Pests are usually attracted to damp, dark, dirty places. A clean and sanitary establishment offers them little food and shelter. Stick to your master cleaning schedule. Dispose of garbage quickly and keep containers clean and tightly covered in all areas. Store recyclables as far from your building as allowed. Keep food and supplies away from walls and at least six inches off the floor. Rotate products, so pests do not have time to settle into them and breed. Protect outdoor diners by denying pests food and shelter in these areas as well. Mow grass, remove standing water, and pick up litter. Cover outdoor garbage containers. Remove dirty dishes and uneaten food from tables, and clean them as quickly as possible. Do not allow employees or customers to feed birds or wildlife on the grounds.

Understanding pests is the key to controlling them. Roaches live and breed in dark, warm, and moist places. Check for a strong oily odor and droppings, which look like grains of black pepper, and egg cases. Use glue traps to find out what types of roaches are present. Rodents also are a serious health hazard. A building can be infested with both rats and mice at the same time. Look for droppings, signs of gnawing, tracks, nesting materials, and holes.

Although you can take most prevention measures yourself, most control measures should be carried out by professionals. Employ a licensed, certified PCO to handle pest control and require them to provide a written service contract outlining the work to be performed, as well as what is expected from both you and them. Working as a team, you can prevent and/or eliminate pests and keep them from coming back.

There are several methods your PCO can use to control insects, depending on the type of insect and degree of infestation. Control methods include repellents, sprays, chemical bait, and traps. Rodents can be controlled through the use of traps set near runways, glue boards (for mice), and chemical bait.

While it might seem cost-effective to apply pesticides yourself, there are many reasons for not doing so. Pesticides can be dangerous to your employees, customers, and food. Some pesticides are not approved for use in a restaurant or foodservice establishment, and pests can develop immunity to them. Rely on your PCO to decide if and when pesticides should be used in your establishment. They are trained to determine the best pesticide for each pest, and how and where to apply it.

Pesticides are hazardous materials. Any time they are used or stored on the premises, you should have corresponding Material Safety Data Sheets (MSDS). To minimize the hazard to people, have your PCO use pesticides only when you are closed for business and your employees are not on-site. Your PCO should store and dispose of all pesticides used in your facility. If they are stored on the premises, they should be kept in their original containers and stored in locked cabinets away from food-storage and food-preparation areas.

Apply Your Knowledge

1. What factors might Fred have overlooked in his recent remodeling that could have led to this infestation?
2. What should Fred do to eliminate the roaches?

For answers, please turn to the Answer Key.

A Case in Point

Now We're Cooking is a restaurant located in the middle of town on the first floor of a landmark, ninety-year-old building. The building is in a shopping area that includes several other restaurants. Fred, the manager, recently remodeled the restaurant. He chose materials that are easy to clean. All of the new foodservice equipment is designed for ease of cleaning as well. The cleaning procedures listed on Fred's master cleaning schedule are written out in detail and completed by the employees as scheduled, with frequent self-inspections. Spills are cleaned up immediately. FIFO is followed, and food is stored on metal racks, which keep it away from walls and six inches off the floor.

During a self-inspection two weeks after he finished remodeling, Fred found live cockroaches behind the sinks, in the vegetable-storage area, in the public restrooms, and near the garbage-storage area.

Apply Your Knowledge

Use these questions to review the concepts presented in this chapter.

Discussion Questions

1. What is the purpose of an integrated pest management program?
2. How can you prevent pests from entering the establishment?
3. How can you tell if your establishment has been infested with cockroaches or rodents?
4. What are the storage requirements for pesticides?
5. What precautions must be taken both before and after pesticides are applied in the establishment?

For answers, please turn to the Answer Key.

Apply Your Knowledge

Use these questions to test your knowledge of the concepts presented in this chapter.

Multiple-Choice Study Questions

1. **Which of the following is a sign you might have a problem with rodents?**
   A. You find capsule-shaped egg cases.
   B. You find scraps of paper and cloth gathered in the corner of a drawer.
   C. You see droppings that look like black grains of pepper.
   D. You smell a strong, oily odor.

2. **Cockroaches typically are found in places that are**
   A. cold, dry, and light.
   B. cold, moist, and dark.
   C. warm, moist, and dark.
   D. warm, dry, and light.

3. **Which of the following is a sign you might have a problem with cockroaches?**
   A. You find small holes burrowed through the storeroom wall.
   B. You find droppings that look like black grains of pepper underneath a refrigeration unit.
   C. You see small piles of sawdust that appear to have fallen from the ceiling.
   D. You find webs and wings in the dry-storage area.

4. **Why does careful cleaning help prevent insect infestations?**
   A. It eliminates their food supply.
   B. It destroys insect eggs.
   C. It reduces the number of places pests can take shelter.
   D. All of the above

5. **To prevent a pest infestation, you should store food at least _____ inch(es) off the floor.**
   A. one
   B. two
   C. four
   D. six

Continued on next page...

Apply Your Knowledge

Multiple-Choice Study Questions *continued*

6. All of the following are critical components of an integrated pest management program *except*
   A. denying pests access to the establishment.
   B. denying pests food, water, and a hiding or nesting place.
   C. working with a licensed PCO to eliminate pests that do enter.
   D. notifying the EPA that pesticides are being used in the establishment.

7. Which of the following is *not* used to control roaches?
   A. Repellent
   B. Contact spray
   C. Glue traps
   D. Chemical bait

8. When pesticides are applied in the establishment, you must do all of the following *except*
   A. leave stationary equipment uncovered.
   B. remove all movable, food-contact surfaces.
   C. wash, rinse, and sanitize food-contact surfaces that have been sprayed.
   D. make a corresponding MSDS available to employees for the pesticide used.

9. Anytime pesticides are stored on the premises,
   A. they must be transferred from the original container to smaller containers.
   B. they must be stored away from food-preparation areas.
   C. they must be kept in dry-storage areas.
   D. they must be registered with the local regulatory agency.

For answers, please turn to the Answer Key.

ADDITIONAL RESOURCES

Books and Periodicals

Pest Control. 2002. *Food Safety Illustrated.* 2 (3):8–10.

Pest management in restaurants: An integrated approach. 1997. Washington, D.C.: National Restaurant Association, Technical Services, Public Health and Safety Department.

Web Sites

Ecolab, Inc.
www.ecolab.com
An excellent source of information on housekeeping and sanitation supplies for the restaurant and foodservice industry from the world's leading sanitation product supplier.

FDA Food Code
http://vm.cfsan.fda.gov/~dms/foodcode.html
As the basis for many local sanitation codes, as well as the basis for information in this textbook, the FDA Food Code, available at this Web address, is a useful resource for information relating to food safety for the restaurant and foodservice industry.

National Integrated Pest Management Network (NIPMN)
www.ipmcenters.org
The National IPM Network is an evolving and cooperating group of universities, government agencies, and other organizations coming together to provide up-to-date, accurate information for pest management. This Web site has a wealth of information on pest control, arranged by commodity, pest, region, and tactics.

National Pest Management Association
www.pestworld.org
The mission of the National Pest Management Association is to communicate their role as protectors of food, health, property, and the environment, and to affect the success of its members through education and advocacy. This Web site provides information on how to become a member, key conferences, sales leads, and a pest control operator (PCO) locator.

National Restaurant Association
www.restaurant.org
The National Restaurant Association is the leading business association for the restaurant industry. Together with the National Restaurant Association Educational Foundation, the Association's mission is to represent, educate, and promote the rapidly growing restaurant and foodservice industry. This Web site should be your starting place for all issues and concerns related to your restaurant. This Web site has it all, from tips for running your establishment to vital data on your customers' spending habits.

Occupational Safety and Health Administration (OSHA)
www.osha.gov
OSHA's mission is to prevent work-related injuries, illnesses, and deaths. This Web site provides information related to workplace-safety conferences, technical links, and regulation and compliance documents.

Orkin Commercial
www.orkin.com/commercial
From the leader in pest control, Orkin's Web site contains useful information on how to prevent and control pests in your foodservice establishment. It also contains a bug guide providing all the information you will want to know about ants, rodents, flies, and birds.

Apply Your Knowledge

Notes

Food Establishment Inspection Report
As Governed by State Code Section XXX.XXX
Do Good County
12344 Any Street, Our Town, State, 11111
FOODBORNE RISK FACTORS AND PUBLIC HEALTH INTERVENTIONS
Compliance Status
Demonstration of Knowledge
Employee Health
Good Hygienic Practices
Approved Source
Protection from Contamination
Potentially Hazardous Food Time/Temperature
Consumer Advisory
Highly Susceptible Populations
Chemical
Conformance with Approved Procedures
GOOD RETAIL PRACTICES (X = not in compliance)
Safe Food and Water
Food Temperature Control
Food Identification
Food Protection from Contamination
Proper Use of Utensils
Utensils, Equipment and Vending
Physical Facilities
Person in Charge (Signature)
Inspector (Signature)
Follow-up: YES NO

Unit 4

Sanitation Management

Food Safety Regulation and Standards

Inside this chapter:

- Objectives of a Foodservice Inspection Program
- Government Regulatory System for Food
- The Food Code
- Foodservice Inspection Process
- Federal Regulatory Agencies
- Voluntary Controls Within the Industry

After completing this chapter, you should be able to:

- Identify the principles and procedures needed to comply with food safety regulations.
- Identify state and local regulatory agencies and regulations that require food safety compliance.
- Prepare for a regulatory inspection.
- Identify the proper procedures for guiding a health inspector through the establishment.

Key Terms

U.S. Department of Agriculture (USDA)
Food and Drug Administration (FDA)
Regulations
Health inspector
Centers for Disease Control and Prevention (CDC)
Environmental Protection Agency (EPA)
National Marine Fisheries Service (NMFS)

Apply Your Knowledge

Check to see how much you know about the concepts in this chapter. Use the page references provided to explore the topic in each question.

Test Your Food Safety Knowledge

1. **True or False:** The FDA writes the food regulations that must be followed by each establishment. *(See page 14-4.)*
2. **True or False:** Health inspectors are generally employees of the Centers for Disease Control and Prevention (CDC). *(See page 14-4.)*
3. **True or False:** You should ask to accompany the health inspector during the inspection of your establishment. *(See page 14-7.)*
4. **True or False:** Critical violations noted during a health inspection should be corrected within one week of the inspection. *(See page 14-8.)*
5. **True or False:** A HACCP-based inspection focuses on the flow of food in an establishment as opposed to the sanitary appearance of the facility. *(See page 14-12.)*

For answers, please turn to the Answer Key.

INTRODUCTION

There are several reasons why it is important to have a foodservice inspection program. Most important is that failure to ensure food safety can jeopardize the health of your customers and could cost you your business. All establishments—including quick-service and fine-dining restaurants, delis, hospitals, nursing homes, and schools—must follow standard food safety practices critical to the safety and quality of the food served. An inspection system lets the establishment know how well it is following these practices.

OBJECTIVES OF A FOODSERVICE INSPECTION PROGRAM

All establishments serving the public must provide safe food and are subject to inspection. It does not matter whether there is a charge for the food or whether the food is consumed on or off premises.

The purpose of an inspection program is to

- evaluate whether the establishment is meeting minimum sanitation and food safety standards.
- protect the public's health by requiring establishments to provide food that is safe, uncontaminated, and presented properly.
- convey new food safety information to the establishment.
- provide an establishment with a written report, noting deficiencies, so it can be brought into compliance with safe food practices.

Key Point

Recommendations for restaurant and foodservice regulations are written at the federal level, regulations are made at the state level, and enforcement is carried out at the local level.

Federal Level
Regulations recommended

▼

State Level
Regulations written

▼

Local Level
Regulations enforced

GOVERNMENT REGULATORY SYSTEM FOR FOOD

Every country has a history of government involvement in the development of health laws. Today, government control of food in the U.S. is exercised at three levels: federal, state, and local.

At the federal level, the **U.S. Department of Agriculture (USDA)**, and the **Food and Drug Administration (FDA)** are directly involved in the inspection process.

The USDA is responsible for inspection and quality grading of meat, meat products, poultry, dairy products, eggs and egg products, and fruit and vegetables shipped across state boundaries. The USDA provides these services through the Food Safety and Inspection Service (FSIS) agency.

The FDA is the agency that writes the Food Code. *(See page 14-4.)* In addition, it inspects foodservice operations that cross state borders (interstate establishments such as those on planes and trains, as well as food manufacturers and processors) because they overlap the jurisdictions of two or more states. The FDA shares responsibility with the USDA for inspecting food-processing plants to ensure standards of purity, wholesomeness, and compliance with labeling requirements.

In the U.S., most food **regulations** affecting restaurant and foodservice operations are written at the state level (except regulations for interstate or international establishments, which are determined at the federal level). Each state decides whether to adopt the Food Code or some modified form of it.

State regulations may be enforced by local (city or county) or state health departments. In a large city, the city health department will probably be responsible for enforcing health codes. In smaller cities or in rural areas, a county or state health department may be responsible for enforcement. In any case, the manager must be familiar with the local agencies and their enforcement system. City, county, or state **health inspectors** (also called sanitarians, health officials, or environmental health specialists) conduct restaurant and foodservice inspections in most states. They generally are trained in food safety, sanitation, and public health principles and methods.

THE FOOD CODE

Key Point

The FDA writes the Food Code, which provides recommendations for restaurant and foodservice regulations.

The Food Code is written by the FDA based on input from the Conference for Food Protection (CFP). The Food Code lists the government's recommendations for foodservice regulations. Currently, these recommendations are updated every two years to reflect developments in the restaurant and foodservice industry and the field of food safety.

The Food Code is intended to assist state health departments in developing regulations for a foodservice inspection program. It is not an actual law. Although the FDA recommends adoption by the states, it cannot require it. Rather, the Food Code represents the FDA's best advice for a uniform system of regulation to ensure food safety. Some states use the Food Code as a basis for developing their own codes, instead of adopting it in its entirety. Food and sanitation codes are written very broadly and generally cover the following areas:

- **Foodhandling and preparation:** sources, receiving, storage, display, service, transportation
- **Personnel:** health, personal cleanliness, clothing, practices

- **Equipment and utensils:** materials, design, installation, storage
- **Cleaning and sanitizing:** facility, equipment
- **Utilities and services:** water, sewage, plumbing, restrooms, waste disposal, integrated pest management (IPM)
- **Construction and maintenance:** floors, walls, ceilings, lighting, ventilation, dressing rooms, locker areas, storage areas
- **Foodservice units:** mobile, temporary
- **Compliance procedures:** restaurant and foodservice inspections, enforcement actions

Key Point

Managers must contact local health departments to find out which regulations apply to their operations.

Adoption and interpretation of food safety standards may vary widely from one state to another or, in some cases, from one locality to another. Therefore, restaurant and foodservice managers must consult with local health departments to find out which specific regulations apply to their operations.

Currently, when a state adopts, develops, or amends its food code, it is required to provide time for public input and comment. This is the time when any interested person or association can comment on how the proposed changes will impact their establishment(s). If you are interested in participating in this process, make sure you keep in contact with your health department or State Restaurant Association.

The lack of uniformity in local food codes can frustrate efforts by the restaurant industry to establish uniform food safety standards. For example, some jurisdictions require refrigerated food temperatures to be 45°F (7°C) or lower, while others require 41°F (5°C) or lower. States may also differ in the recommended inspection frequency. Some states require inspections for restaurant and foodservice establishments at least every six months, while others schedule inspections more or less frequently. Some states may even differ as to which establishments should be inspected; for example, some states do not inspect convenience food stores. Some states might have separate regulations or agencies for different types of foodservice operations (such as vending machines or delis in grocery stores).

Although all inspectors focus on food safety practices, the areas emphasized during the inspection can vary among jurisdictions or individual inspectors. For example, some inspectors are more concerned with refrigeration temperatures while others may focus on the physical appearance of the facility. One inspector may examine the flow of food while another may focus primarily on the personal hygiene of employees. It is the responsibility of the manager to keep food safe and wholesome throughout the establishment at all times, regardless of the inspector or the inspection process.

FOODSERVICE INSPECTION PROCESS

Key Point

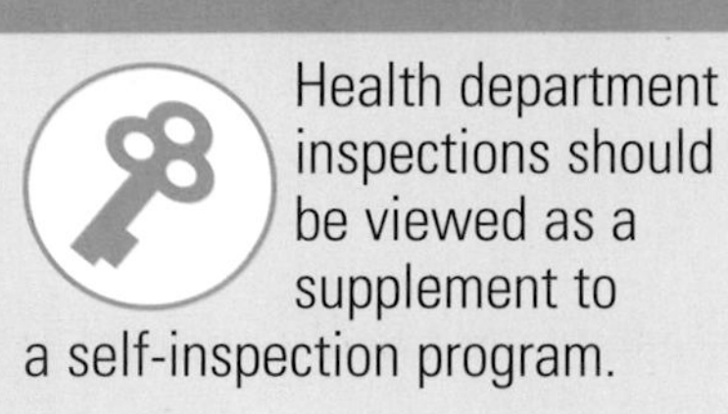

Health department inspections should be viewed as a supplement to a self-inspection program.

Well-managed establishments will perform continuous self-inspections to keep food safe, in addition to the regular inspections performed by the health department. Establishments with high standards for sanitation and food safety consider health department inspections only a supplement to their self-inspection programs.

There are many benefits that come from establishing a good self-inspection program. A good program can result in safer food, improved food quality, a clean environment for employees and customers, and higher inspection scores. Strive to exceed the expectations of both your customers and the health department. The higher you set your standards, the more likely you are to do well on your health department inspection. In addition, your customers will notice your commitment to providing them a safe, sanitary dining experience.

During health department inspections, the local health code serves as the inspector's guide. You should keep a current copy of your local or state sanitation regulations and be familiar with them. Regularly compare the code to procedures at your establishment, but remember that code requirements are only minimum standards to keep food safe.

Keep in mind that changes continue to occur in the inspection process. Some areas use traditional, scored inspection systems with a number or letter grade, while others use HACCP-based inspections, or a combination of the two.

Key Point

The frequency of regulatory inspections will vary based on several factors, including complexity of the establishment's menu, the size of the operation, and its clientele.

Traditional Inspection System

Some health departments are required to conduct inspections at least every six months. However, the frequency will vary depending on the area, type of establishment, or food served. Many health departments use a risk-based approach to inspection frequency. Determining factors can include:

- **Size and complexity of the operation.** Larger operations offering a considerable number of potentially hazardous food items might be inspected more frequently.
- **Inspection history of the establishment.** Establishments with a history of low sanitation scores or consecutive violations might be inspected more frequently.
- **Clientele's susceptibility to foodborne illness.** Nursing homes, schools, day-care centers, and hospitals might receive more frequent inspections.
- **Workload of the local health department and the number of inspectors available.**

In most cases, an inspector will arrive without prior warning and ask for the manager or the person in charge of the operation. In your absence, make sure your employees know who is in charge. Some companies have specific policies on how to handle an inspection. Make sure you are aware of those policies. Establishments should not refuse entry. In some jurisdictions, inspectors could have the authority to gain access or to revoke the establishment's permit for refusal to allow an inspection.

Key Point

Accompany the sanitarian during the inspection so you can answer any questions.

It is best to accompany the sanitarian during the inspection. This allows you to answer any questions and, possibly, correct deficiencies immediately. Accompanying the inspector will also give you the opportunity to learn from the inspector's comments and suggestions, and to gain sanitation advice.

After the inspection, the inspector will discuss the results and the score (if a score is given) and arrange for any follow-up, if necessary. Managers will be asked to sign the inspection report to acknowledge they have received it. Follow your company's policy regarding this issue. A copy of the report is then given to the manager or person in charge at the time of the inspection. All deficiencies noted on the report should be acted upon. Critical deficiencies should be corrected immediately or within forty-eight hours of the inspection. All other deficiencies should be corrected as soon as possible. Copies of all reports should be kept on file in the establishment and referred to when planning improvements and assessing facility goals. Copies of reports are kept on file at the health department. These are considered public documents and may be made available to the public upon request.

The following steps will help managers and operators get the most out of sanitation inspections:

1. **Ask for identification.** Many inspectors will volunteer their credentials. Do not let anyone enter the back of the facility without proper identification. Ask the purpose of the visit; make sure you know whether it is a routine inspection, the result of a customer complaint, or for some other purpose.

2. **Cooperate.** Most inspectors have learned to expect some defensiveness and resentment from restaurant and foodservice operators. This can be interpreted as having something to hide. Answer all of the inspector's questions to the best of your ability. Instruct employees to do the same. Explain to the inspector that you wish to accompany him or her during the inspection. This will encourage open communication and a good working relationship. If a deficiency can be corrected quickly, do so, or tell the inspector when it can be corrected.

3. **Take notes.** As you accompany the inspector, make a note of any problem pointed out. Make it clear you are willing to correct problems. Taking your own notes will help you remember exactly what was said. If you believe the inspector is incorrect about something, note what was commented upon. Then ask the inspector's supervisor for a second opinion.

4. **Keep the relationship professional.** Do not offer food or drink before, during, or after an inspection. This could be viewed as bribery.

5. **Be prepared to provide records requested by the inspector**. You might ask the inspector why these records are needed. If a request appears inappropriate, you can check with the inspector's supervisor or with your lawyer about limits on confidential information. Records you provide to the inspector will become part of the public record. Inspectors might ask for

records of purchases to verify that food is received from an approved source, records of integrated pest management (IPM) treatments, and a list of all chemicals used in the facility. HACCP records could be requested in some cases. In fact, HACCP records very well could be an important part of an inspection because they document the establishment's efforts to ensure food safety.

6. **Discuss violations and time frames for correction with the inspector.** The inspection report should be studied closely. Deficiencies and comments should be discussed in detail with the inspector. If any deficiencies were corrected on the spot, make sure they are noted. In order to make complete and permanent corrections, you will need to know the exact nature of the violation, how it impacts food safety, how to correct it, and whether or not the inspector will follow up. The inspector sees many operations and may offer expert advice on how to correct deficiencies.

7. **Follow up.** Take the inspection report with you through your facility and correct the problems. Determine why each problem occurred by evaluating sanitation procedures, the master cleaning schedule, and employee foodhandling practices. Establish new procedures or revise existing ones to correct the problem permanently, if necessary. (See *Exhibit 14a.*)

Exhibit 14a

Inspection Follow-Up

Determine why each violation occurred by evaluating sanitation procedures, the master cleaning schedule, and employee foodhandling practices. Establish new procedures or revise existing ones.

Many health departments use a traditional inspection system using a demerit scoring scale. (See *Exhibit 14b.*) Usually the highest possible score is one hundred points. For every violation, between one and five points are subtracted from one hundred to get the final score. Noncritical violations worth one or two points must be corrected by the time of the next routine inspection. Critical violations worth four to five points must be corrected within a time frame specified by the inspector. Other health departments use a letter to score the establishment. Whatever scoring system is used, make sure you understand it, as well as your options for improving an unsatisfactory score.

Establishments are generally given a short amount of time (forty-eight hours or less) to correct critical violations and improve their overall score. If a low score is received upon reinspection, the establishment may be fined or even closed.

Key Point

Establishments can be closed when the health department feels they pose an immediate and substantial health hazard to the public.

In some states, if the inspector determines a facility poses an immediate and substantial health hazard to the public, he or she may ask for a voluntary closure, or issue an immediate suspension of the permit to operate. A suspension requires the approval of the local health offices and means that operations at the establishment must cease immediately. Examples of hazards calling for closure include:

- Significant lack of refrigeration
- Backup of sewage into the establishment or its water supply
- Emergency, such as a building fire or flood
- Serious infestation of insects or rodents
- Long interruption of electrical or water service

The suspension order may be posted at a public entrance to the establishment; however, this is not required if the establishment closes voluntarily. Regulations vary from one jurisdiction to another, but to reinstate a permit to operate, the establishment must eliminate the hazards causing the suspension, then pass a reinspection.

The health department can request a hearing if it feels that an establishment must correct a hazard, but it does not wish to suspend the establishment's permit. The establishment can also

Traditional Inspection Form

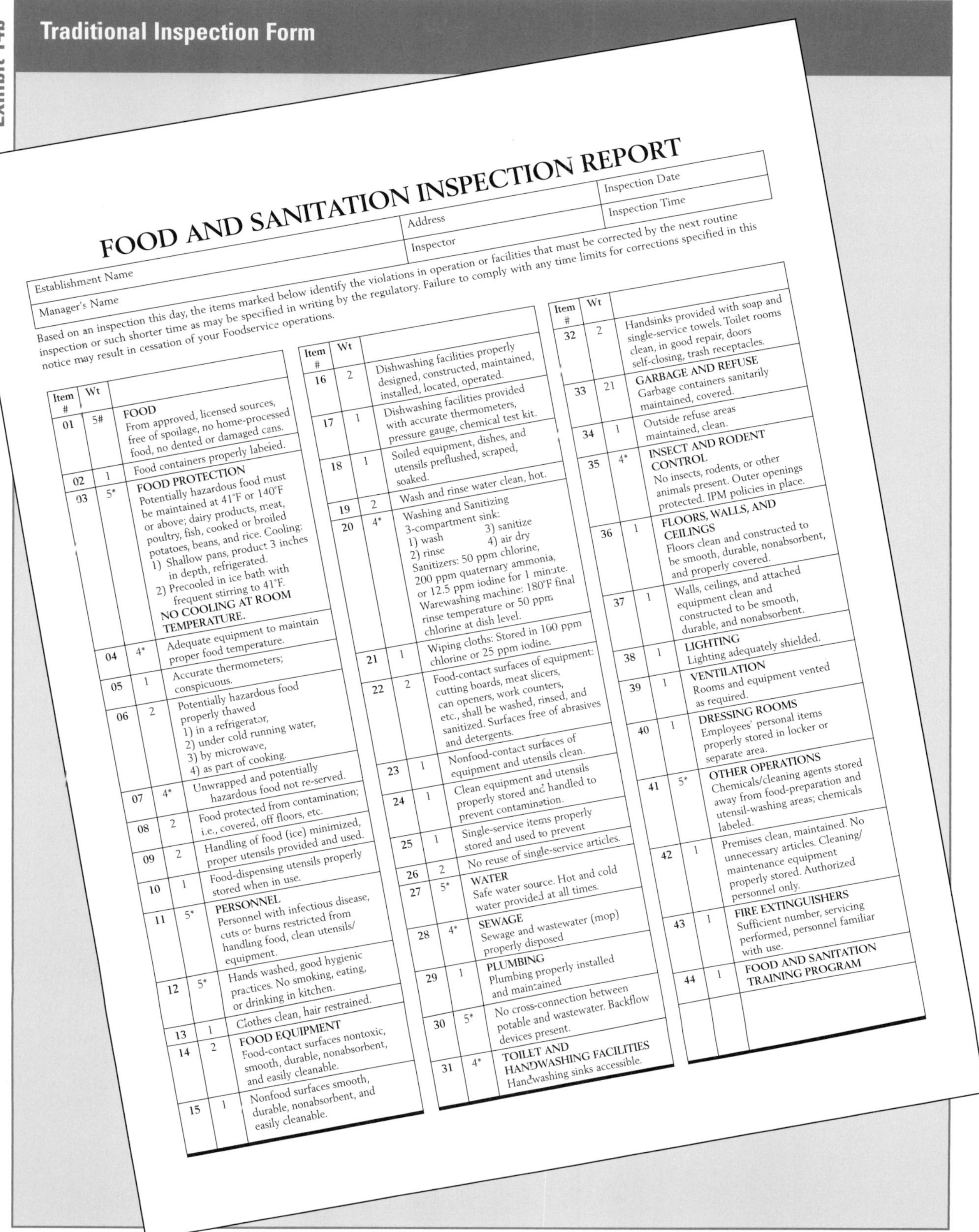

FOOD AND SANITATION INSPECTION REPORT

Establishment Name	Address	Inspection Date
Manager's Name	Inspector	Inspection Time

Based on an inspection this day, the items marked below identify the violations in operation or facilities that must be corrected by the next routine inspection or such shorter time as may be specified in writing by the regulatory. Failure to comply with any time limits for corrections specified in this notice may result in cessation of your Foodservice operations.

Item #	Wt	
01	5#	**FOOD** From approved, licensed sources, free of spoilage, no home-processed food, no dented or damaged cans.
02	1	Food containers properly labeled.
03	5*	**FOOD PROTECTION** Potentially hazardous food must be maintained at 41°F or 140°F or above; dairy products, meat, poultry, fish, cooked or broiled potatoes, beans, and rice. Cooling: potatoes, beans, and rice. Cooling: 1) Shallow pans, product 3 inches in depth, refrigerated. 2) Precooled in ice bath with frequent stirring to 41°F. NO COOLING AT ROOM TEMPERATURE.
04	4*	Adequate equipment to maintain proper food temperature.
05	1	Accurate thermometers; conspicuous.
06	2	Potentially hazardous food properly thawed 1) in a refrigerator, 2) under cold running water, 3) by microwave, 4) as part of cooking.
07	4*	Unwrapped and potentially hazardous food not re-served.
08	2	Food protected from contamination; i.e., covered, off floors, etc.
09	2	Handling of food (ice) minimized, proper utensils provided and used.
10	1	Food-dispensing utensils properly stored when in use.
11	5*	**PERSONNEL** Personnel with infectious disease, cuts or burns restricted from handling food, clean utensils/ equipment.
12	5*	Hands washed, good hygienic practices. No smoking, eating, or drinking in kitchen.
13	1	Clothes clean, hair restrained.
14	2	**FOOD EQUIPMENT** Food-contact surfaces nontoxic, smooth, durable, nonabsorbent, and easily cleanable.
15	1	Nonfood surfaces smooth, durable, nonabsorbent, and easily cleanable.
16	2	Dishwashing facilities properly designed, constructed, maintained, installed, located, operated.
17	1	Dishwashing facilities provided with accurate thermometers, pressure gauge, chemical test kit.
18	1	Soiled equipment, dishes, and utensils preflushed, scraped, soaked.
19	2	Wash and rinse water clean, hot.
20	4*	Washing and Sanitizing 3-compartment sink: 1) wash 2) rinse 3) sanitize 4) air dry Sanitizers: 50 ppm chlorine, 200 ppm quaternary ammonia, or 12.5 ppm iodine for 1 minute. Warewashing machine: 180°F final rinse temperature or 50 ppm chlorine at dish level.
21	1	Wiping cloths: Stored in 100 ppm chlorine or 25 ppm iodine.
22	2	Food-contact surfaces of equipment: cutting boards, meat slicers, can openers, work counters, etc., shall be washed, rinsed, and sanitized. Surfaces free of abrasives and detergents.
23	1	Nonfood-contact surfaces of equipment and utensils clean.
24	1	Clean equipment and utensils properly stored and handled to prevent contamination.
25	1	Single-service items properly stored and used to prevent
26	2	No reuse of single-service articles.
27	5*	**WATER** Safe water source. Hot and cold water provided at all times.
28	4*	**SEWAGE** Sewage and wastewater (mop) properly disposed
29	1	**PLUMBING** Plumbing properly installed and maintained
30	5*	No cross-connection between potable and wastewater. Backflow devices present.
31	4*	**TOILET AND HANDWASHING FACILITIES** Handwashing sinks accessible.
32	2	Handsinks provided with soap and single-service towels. Toilet rooms clean, in good repair, doors self-closing, trash receptacles.
33	21	**GARBAGE AND REFUSE** Garbage containers sanitarily maintained, covered.
34	1	Outside refuse areas maintained, clean.
35	4*	**INSECT AND RODENT CONTROL** No insects, rodents, or other animals present. Outer openings protected. IPM policies in place.
36	1	**FLOORS, WALLS, AND CEILINGS** Floors clean and constructed to be smooth, durable, nonabsorbent, and properly covered.
37	1	Walls, ceilings, and attached equipment clean and constructed to be smooth, durable, and nonabsorbent.
38	1	**LIGHTING** Lighting adequately shielded.
39	1	**VENTILATION** Rooms and equipment vented as required.
40	1	**DRESSING ROOMS** Employees' personal items properly stored in locker or separate area.
41	5*	**OTHER OPERATIONS** Chemicals/cleaning agents stored away from food-preparation and utensil-washing areas; chemicals labeled.
42	1	Premises clean, maintained. No unnecessary articles. Cleaning/ maintenance equipment properly stored. Authorized personnel only.
43	1	**FIRE EXTINGUISHERS** Sufficient number, servicing performed, personnel familiar with use.
44	1	**FOOD AND SANITATION TRAINING PROGRAM**

request a hearing if it believes an action by the health department is unjustified. There is often a time limit to request such a hearing (usually within five to ten days after an inspection). Check local regulations to determine these limits.

One limitation of the traditional inspection system centers on the fact that multiple violations of the same type do not cause any more demerits than a single violation. For example, leaving one potentially hazardous food at room temperature incurs the same number of demerits as leaving all food at room temperature. Clearly, leaving all food at room temperature is a greater health concern.

As a result, a new inspection system has been implemented in some areas. In this type of inspection system, a narrative format is used to report violations rather than a demerit system. When inspectors see problems with a facility, they write up the problems—as well as any recommended corrections—in paragraph form. Reference numbers to the health code and repeated violations are noted, with multiple violations of the same type receiving greater emphasis. In the old scoring system, violations were considered critical or noncritical. The new system allows inspectors to use their professional judgment regarding some of the violations. It also provides establishments with more feedback. See *Exhibit 14c* for an example of the new inspection report form recommended by the FDA.

HACCP-Based Inspections

Some health departments use HACCP-based inspections, focusing on the flow of food rather than on the sanitary appearance of the facility. Inspectors can observe the way an establishment receives, stores, prepares, cooks, holds, cools, reheats, and serves food, and assess whether the critical control points identified are actually in control. Since this type of inspection could be viewed as complex and time-consuming, it might only be performed under special circumstances. A health department might perform a HACCP-based inspection to do the following:

- Trace the source of contamination following a report of foodborne illness.
- Evaluate an establishment with significant hazards.
- Assist an establishment in developing or implementing a HACCP system.

Exhibit 14c

New Inspection Form Recommended by the FDA

DEPARTMENT OF HEALTH AND HUMAN SERVICES
PUBLIC HEALTH SERVICE
FOOD AND DRUG ADMINISTRATION

FDA

FOOD ESTABLISHMENT INSPECTION REPORT

Violations cited in this report shall be corrected within the time frames specified below, but within a period not to exceed 10 calendar days for critical items (§ 8–405.11) or 90 days for noncritical items (§ 8–406.11)

VIOLATIONS: CRITICAL ______ NONCRITICAL ______

		PERMIT NUMBER:	DATE:
ESTABLISHMENT:			
	CITY:	STATE:	ZIP:
ADDRESS:		TELEPHONE:	
PERSON IN CHARGE / TITLE:			
INSPECTOR / TITLE:			TIME:
INSPECTION TYPE: ROUTINE	FOLLOW-UP	COMPLAINT	OTHER:

Critical (X)	Repeat (X)	Code Reference	Violation Description / Remarks / Corrections

Food Establishment Inspection Report Page ___ of ___

Exhibit 14d

HACCP-Based Inspection Form for Charting the Flow of Food

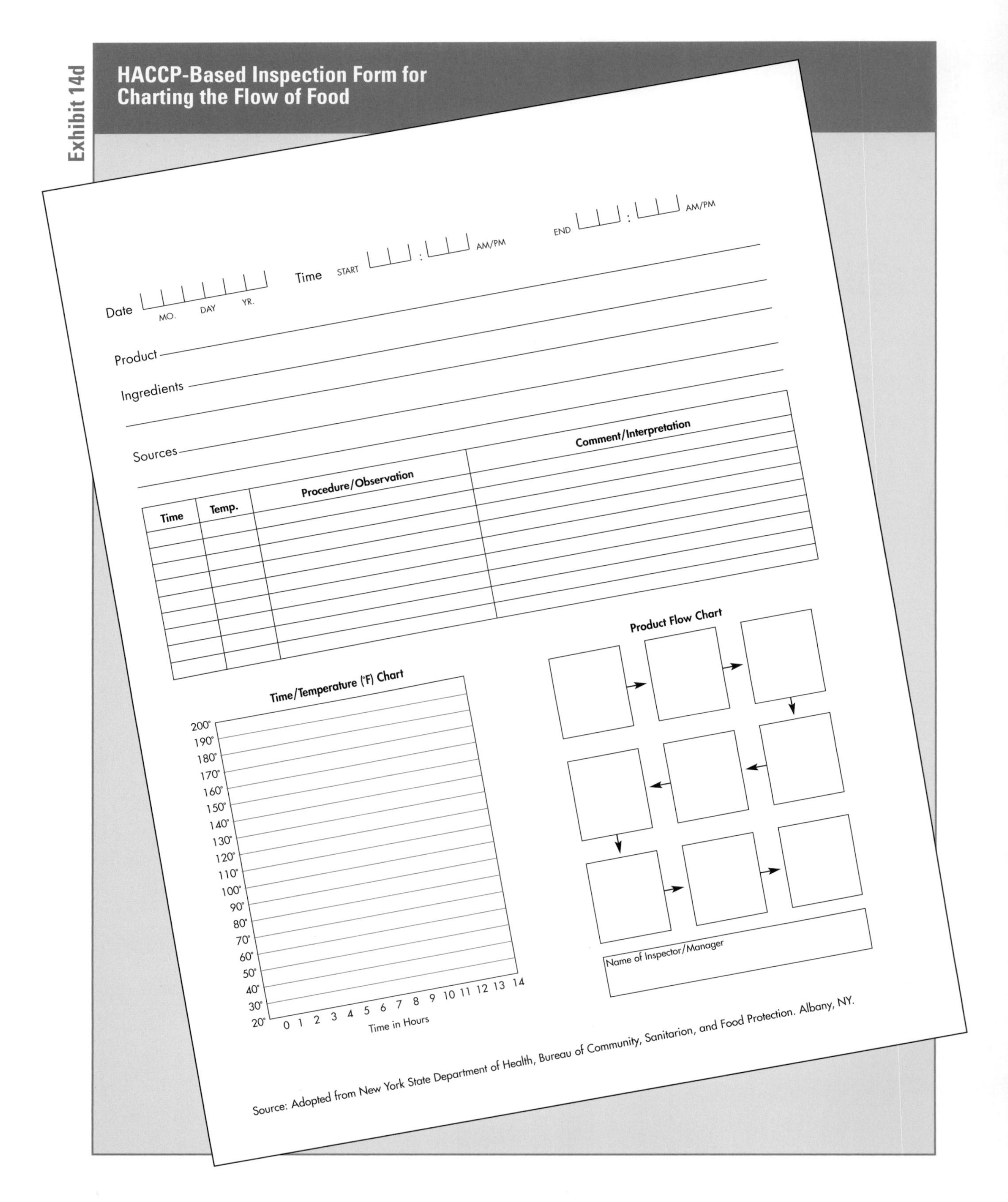
Date MO. DAY YR.

Time START : AM/PM END : AM/PM

Product

Ingredients

Sources

Time	Temp.	Procedure/Observation	Comment/Interpretation

Source: Adopted from New York State Department of Health, Bureau of Community, Sanitarion, and Food Protection. Albany, NY.

Other health agencies are using the HACCP-based approach to decide how often different establishments should be inspected, or to determine how the inspections should be conducted. With this approach, scoring is done differently. Instead of using a point system with demerits, HACCP-based inspections determine critical violations. Many of these critical violations are similar to the four- and five-point items identified in the traditional inspection system. *Exhibit 14d* shows one type of form used during a HACCP-based inspection.

FEDERAL REGULATORY AGENCIES

Several federal government agencies work to protect the sanitary quality of food in an establishment. Earlier in this chapter, the roles of the FDA, and USDA, were discussed. A few other agencies concerned with food safety should also be mentioned.

The **Centers for Disease Control and Prevention (CDC),** located in Atlanta, Georgia, are agencies of the U.S. Department of Health and Human Services. The Centers provide the following services:

- Investigate outbreaks of foodborne illness
- Study the causes and control of disease
- Publish statistical data and case studies in the *Morbidity and Mortality Weekly Report (MMWR)*
- Provide educational services in the field of sanitation
- Conduct the Vessel Sanitation Program—an inspection program for cruise ships

The **Environmental Protection Agency (EPA)** sets air- and water-quality standards and regulates the use of pesticides (including sanitizers) and the handling of wastes.

The **National Marine Fisheries Service (NMFS)** of the U.S. Department of Commerce implements a voluntary inspection program that includes product standards and sanitary requirements for fish-processing operations.

VOLUNTARY CONTROLS WITHIN THE INDUSTRY

Few industries have devoted as much effort to regulating themselves as the restaurant and foodservice and food-processing industries. Scientific and trade associations, manufacturing firms, and foodservice corporations have vigorously pursued programs to raise the standards of the industry through research, education, and cooperation with government. Although participation in these programs is voluntary, these organizations have actively promoted professional standards, recommended legislative policy, sponsored uniform enforcement procedures, and provided educational opportunities. The overall results in food safety have included:

- Increased understanding of foodborne illness and its prevention
- Improvements in the sanitary design of equipment and facilities
- Industry-wide efforts to maintain the sanitary quality of food during processing, shipment, storage, and service
- Efforts to make foodservice laws more science-based, practical, and uniform

While many organizations have contributed to these endeavors, those having the most relevance for the foodservice operator and manager include:

Key Point

The National Restaurant Association works closely with government agencies to develop and implement proper guidelines and regulations regarding the food supply.

- **The National Restaurant Association,** founded in 1919, is the leading business association for the restaurant and foodservice industry. The Association's mission is to represent, educate, and promote the rapidly growing industry. The Association is a strong voice advocating proper food safety practices and science-based regulations. The Association's Health and Safety Regulatory Affairs Department provides information to the industry on topics such as sanitation, pest control, nutrition, ergonomics, and the Americans with Disabilities Act.
  - In addition, it works closely with government agencies to develop and implement proper guidelines and regulations regarding the food supply. Through its excellent Research Department, the Association generates both operator and consumer research on a variety of issues and trends, including the annual *Restaurant Industry Operations Report* and the *Restaurant Industry Forecast.* The Association's

successful lobbying arm represents the interests of the industry by being in constant contact with legislators.

- Each May, the National Restaurant Association holds the Restaurant, Hotel-Motel Show, one of the largest trade shows in the U.S. More than one hundred thousand people come to Chicago, Illinois, to see everything from tablecloths and plates to industrial warewashers and food safety systems.
- For more information about the National Restaurant Association, its services, and member benefits, call 800.424.5156 or 202.331.5900, or visit their Web site at www.restaurant.org.

Key Point

The National Restaurant Association Educational Foundation is dedicated to developing, promoting, and providing educational and training solutions for the restaurant and foodservice industry.

- **The National Restaurant Association Educational Foundation (NRAEF)** is a not-for-profit organization dedicated to fulfilling the educational mission of the National Restaurant Association. Focusing on three key strategies of risk management, recruitment, and retention, the NRAEF is the restaurant and foodservice industry's premier provider of educational resources, materials, and programs. Sales from all NRAEF products and services benefit the industry by directly supporting the NRAEF's educational initiatives.
  - As the educational resource for the industry, the NRAEF administers the largest scholarship program of its kind for the restaurant and hospitality sector. Whether students are beginning their restaurant and hospitality education or continuing a successful career, the NRAEF offers scholarships to help them. Scholarships are available for senior high school students, undergraduate and graduate students, industry professionals, and educators/administrators.
  - For more information about the NRAEF and its products and services, call 800.765.2122, ext. 701 (outside the Chicago area), or 312.715.1010, ext. 701 (within the Chicago area). For information or applications for the Scholarships Program, call 312.715.1010, ext. 733. You can also visit www.nraef.org.

- **The National Environmental Health Association (NEHA)** is an organization of environmental specialists and professionals, including those responsible for food-inspection services and environmental health programs. NEHA's Registered Sanitarian (RS) program has established national standards for education, experience, and testing for health inspectors. For more information, visit their Web site at www.neha.org.
- **The International Association for Food Protection (IAFP)** was a pioneer in the highly successful U.S. milk-sanitation program. It publishes information on food safety and on generally accepted standard procedures for investigating a foodborne illness. For more information, visit their Web site at www.foodprotection.org.
- **The Institute of Food Technologists (IFT)** is a multidisciplinary scientific society with expertise in technology, research, education, manufacturing, and the safety of food in the foodservice industries. IFT sponsors food and restaurant industry research in a variety of areas, such as food quality, safety, and process controls. For more information, visit their Web site at www.ift.org.
- **The National Society of Professional Sanitarians (NSPS)** and the American Academy of Sanitarians (AAS) are composed of food industry professionals and sanitarians concerned with environmental health protection policies at the national, state, and local levels.
- **The Association of Food and Drug Officials (AFDO)** develops and publishes food sanitation codes and encourages food protection through the adoption of uniform legislation and enforcement procedures. For more information, visit their Web site at www.afdo.org.
- **The Council of Hotel and Restaurant Trainers (CHART)** is an association of restaurant and foodservice trainers and human resource professionals. CHART was formed in 1971 to provide members with a forum in which to grow professionally and increase their effectiveness as trainers. For more information, visit their Web site at www.chart.org.

- **The National Pest Management Association (NPMA)** consists of licensed and certified Pest Control Operators (PCOs) throughout the U.S. NPMA provides guidelines and training materials for IPM treatment programs, hazard communications, and other topics relating to safety and sanitation. For more information, visit their Web site at www.pestworld.org.
- **NSF International** evaluates and lists foodservice equipment meeting its standards. It also issues the NSF International mark. NSF International develops and publishes widely accepted standards for equipment design, construction, and installation, updated every five years. A listing of equipment meeting these standards is updated every six months. NSF's approach demonstrates the kind of progress the foodservice industry can achieve through voluntary programs. For more information, visit their Web site at www.nsf.org.
- **Underwriters Laboratories, Inc. (UL)** performs a service similar to NSF International, listing equipment that meets NSF International standards. UL also lists products complying with their own published environmental and public health standards. In addition, UL lists electrical equipment that passes its own safety requirements. For more information, visit their Web site at www.ul.org.

SUMMARY

Public and private organizations and agencies can offer valuable assistance to restaurant and foodservice managers in meeting their commitment to food safety. It is up to the manager to use available help and to maintain a sanitary establishment.

Federal governmental agencies create standards affecting the establishment directly or indirectly. The Food Code, developed by the FDA, serves as a guideline for many state and local regulations. In addition, the FDA regulates the purity and safety of food in interstate commerce.

While state and local health codes vary throughout the nation, virtually all contain provisions governing food safety, personal hygiene, sanitary facilities, equipment and utensils, safe operating practices, training, and enforcement procedures.

The inspector—a representative from the state or local health department—is a professional in sanitation and public health.

To keep food safe in your establishment, implement a food safety management system and follow a self-inspection program. The effectiveness of this system will be reflected in inspection reports. HACCP-based inspections—a departure from the traditional emphasis on sanitary facilities—are being adopted by some regulatory agencies. During this type of inspection, inspectors observe the way an establishment receives, stores, prepares, cooks, holds, cools, reheats, and serves food, and the critical control points identified.

Professional and trade organizations in foodservice and public health recommend guidelines for restaurant and foodservice sanitation, investigate sanitation problems, and promote best practices. These associations also develop educational programs and conduct research into the causes and prevention of foodborne illness.

Apply Your Knowledge

1. Did Jerry handle the inspection correctly?
2. What does he need to do following this inspection?

For answers, please turn to the Answer Key.

A Case in Point

Carolyn, the inspector from the city's public health department, was standing at the door of Jerry's Diner. Jerry greeted her, and they both walked to the kitchen. Carolyn pulled out a HACCP worksheet. While Jerry explained that the cook was preparing stir-fried chicken and vegetables, Carolyn noted the ingredients and their sources. Then she observed the preparation procedures and noted times and temperatures, which she plotted on a time-temperature graph. She also filled in a product flowchart and indicated the critical control points for the stir-fried chicken. Jerry told her what corrective actions his employees were trained to carry out if critical control points were not met.

Next, Carolyn checked the concentration of the sanitizing solution in the three-compartment sink the dish washer used to clean, rinse, and sanitize equipment. She also checked the handwashing station.

Jerry and Carolyn went to Jerry's office, where they discussed the report. Jerry compared his flowchart for the stir-fried chicken and vegetables with the one Carolyn had done, determining where changes could be made to improve monitoring.

Apply Your Knowledge

Use these questions to review the concepts presented in this chapter.

Discussion Questions

1. What is the role of the FDA regarding food protection?
2. What are the roles of federal, state, and local agencies regarding the regulation of food safety in establishments?
3. What should a manager do during and after an inspection?
4. What are some factors that determine the frequency of a health inspection in an establishment?
5. What are some of the significant industry organizations that help managers deal with employee training and sanitation regulations and standards?

For answers, please turn to the Answer Key.

Apply Your Knowledge

Use these questions to test your knowledge of the concepts presented in this chapter.

Multiple-Choice Study Questions

1. **An establishment can be closed for all of the following reasons *except***
   - A. significant lack of refrigeration.
   - B. backup of sewage.
   - C. serious infestation of insects or rodents.
   - D. minor violations not corrected within twenty-four hours.

2. **Which of the following is a goal of the food safety inspection program?**
   - A. To evaluate whether an establishment is meeting minimum food safety and sanitation standards
   - B. To protect the public's health
   - C. To convey new food safety information to establishments
   - D. All of the above

3. **Which operation would most likely be subject to a food safety inspection by a federal agency?**
   - A. Hospital
   - B. Cruise ship crossing international waters
   - C. Local ice cream store with a history of safety violations
   - D. Food kitchen run by church volunteers

4. **A person shows up at a restaurant claiming to be a health inspector. What should the manager do?**
   - A. Ask to see identification.
   - B. Ask to see an inspection warrant.
   - C. Ask for a hearing to determine if the inspection is necessary.
   - D. Ask for a one-day postponement to prepare for the inspection.

Continued on next page...

Apply Your Knowledge

Multiple-Choice Study Questions *continued*

5. **How does a HACCP-based food safety inspection differ from an ordinary inspection?**
   A. It focuses more on the sanitary appearance of the facility.
   B. It needs to be done more frequently.
   C. It focuses more on the flow of food through the establishment.
   D. It uses a point system with demerits.

6. **Which of the following agencies enforce food safety in a restaurant?**
   A. FDA
   B. CDC
   C. State or local health departments
   D. USDA

7. **Violations noted on the health inspection report should be**
   A. discussed in detail with the inspector.
   B. corrected within forty-eight hours or less if they are critical.
   C. explored to determine why they occurred.
   D. All of the above

8. **The responsibility for keeping food safe in an establishment rests with the**
   A. state health department.
   B. manager/operator.
   C. health inspector.
   D. FDA.

9. **Food codes developed by state agencies are**
   A. minimum standards necessary to ensure food safety.
   B. maximum standards necessary to ensure food safety.
   C. voluntary guidelines for establishments to follow.
   D. inspection practices for grading meats and meat products.

For answers, please turn to the Answer Key.

ADDITIONAL RESOURCES

Books and Periodicals

Frable, F. 1997. Industry needs central resource for code, regulation information. *Nation's Restaurant News.* 31 (20):44.

Hertneky, P. B. 1996. You and your health inspector. *Restaurant Hospitality.* 80 (6):57.

Marriott, N. G. 1994. *Principles of food sanitation.* New York: Chapman & Hall.

Murray, J. 1994. The model food code. *FoodService Director.* 7 (4):82.

Penner, K. 1992. *Exploring public policy options: Food safety in food service.* Manhattan, KS: Kansas State University Cooperative Extension Service.

Solis, O. C. 1997. Partner or adversary? *Food & Service.* 58 (3):16.

Web Sites

Centers for Disease Control and Prevention (CDC)

www.cdc.gov

The mission of the CDC is to promote health and quality of life by preventing and controlling disease, injury, and disability. To prevent and control foodborne illness, the CDC collect data on outbreaks. This Web site provides general information on foodborne illnesses and their prevention.

Code of Federal Regulations (CFR)

http://access.gpo.gov/nara/cfr

The CFR provides electronic access to the general and permanent rules published in the Federal Register by the executive departments and agencies of the federal government. The codes of interest to the restaurant and foodservice industry include Chapters 7 and 21.

Conference for Food Protection (CFP)

www.foodprotect.org

The CFP is a nonprofit organization providing a representative and equitable partnership among regulators, industry, academia, professional organizations, and consumers to identify problems, formulate recommendations, and develop and implement practices ensuring food safety. This Web site houses previous Conference issues and proceedings, its constitution and by-laws, and current Conference activities.

Department of Health and Human Services (HHS)
www.hhs.gov
HHS is the federal government's principal agency for protecting the health of all Americans and providing essential human services. The department includes more than three hundred programs, covering a wide spectrum of activities, including the FDA and the CDC. The various food safety programs the Department oversees can be accessed from this site.

Environmental Protection Agency (EPA)
www.epa.gov
The EPA endeavors to abate and control pollution systematically, by proper integration of a variety of research, monitoring, standard setting, and enforcement activities. This Web site provides information on all areas of pollution prevention and control.

Food and Drug Administration (FDA)
www.fda.gov
FDA's mission is to promote and protect the public's health by helping safe and effective products reach the market in a timely manner, as well as to monitor products for continued safety after they are in use. This Web site provides a wealth of information on all areas they regulate, including food and product recalls, enforcement activities, all field programs and their inspectional references, and projects to keep industry and consumers aware of health-related topics.

FDA Enforcement Report Index
www.fda.gov/opacom/enforce.html
The FDA Enforcement Report, accessible through this Web site, is published weekly by the FDA and contains information on actions taken in connection with agency regulatory activities.

FDA Food Code
http://vm.cfsan.fda.gov/~dms/foodcode.html
As the basis for many local sanitation codes, as well as the basis for information in this textbook, the FDA Food Code, available at this Web address, is a useful resource for information relating to food safety for the restaurant and foodservice industry.

FDA Center for Food Safety and Applied Nutrition (CFSAN)

www.cfsan.fda.gov

As the center within the FDA responsible for food safety and nutrition, CFSAN promotes and protects public health by researching and implementing guidelines, policies, and standards to ensure that food is safe, nutritious, wholesome, and properly labeled. This Web site provides a wealth of information on food safety and sanitation, including corresponding guidelines, policies, and standards.

Food and Drug Law Institute

www.fdli.org

The Food and Drug Law Institute is a nonprofit institute dedicated to advancing the public health by providing a neutral forum for critical examination of the laws, regulations, and policies related to drugs, medical devices, other healthcare technologies, and foods. This Web site contains information on pertinent publications, conferences, and sources for experts in this field.

Food Marketing Institute (FMI)

www.fmi.org

The FMI is a nonprofit association conducting programs in research, education, industry relations, and public affairs on behalf of its members, which include large multistore chains, small regional firms, and independent supermarkets.

FoodSafety.gov

www.foodsafety.gov

FoodSafety.gov is a gateway Web site providing links to selected information related to government food safety. This Web site is a great starting point for accessing all government-related food safety information.

Grocery Manufacturers of America (GMA)

www.gmabrands.com

GMA is the world's largest association of food, beverage, and consumer-product companies. With U.S. sales of more than $460 billion, GMA members employ more than 2.5 million workers in all fifty states. The organization applies legal, scientific, and political expertise from its member companies to vital food, nutrition, and public-policy issues affecting the industry. This Web site contains information on public policy and industry affairs, products and services, and facts and figures concerning the grocery industry.

International Dairy-Deli-Bakery Association (IDDBA)
www.iddba.org
IDDBA is a resource for information and services across all food channels for the dairy, deli, and bakery industries. This Web site provides information relevant to all people who work in the dairy, deli, and bakery industries.

National Association of Convenience Stores (NACS)
www.nacsonline.com
NACS is a proactive organization representing the convenience store segment. This informative Web site provides up-to-date industry happenings and contains reports outlining vital industry statistics.

National Automatic Merchandising Association (NAMA)
www.vending.org
NAMA is the national trade association of the merchandising vending and contract-foodservices management business. This Web site provides up-to-date information on all issues concerning this segment of the foodservice industry, as well as provides a resource for conferences and education for those working in this segment.

National Food Processors Association (NFPA)
www.nfpa-food.org
The NFPA is the voice of the food processing industry on scientific and public policy issues involving food safety, nutrition, technical and regulatory matters, and consumer affairs. Their Web site offers many resources on food safety and food processing for the entire food industry.

National Frozen Food Association (NFFA)
www.nffa.org
NFFA's mission is to promote the sales and consumption of frozen food through education, training, research, sales planning, and menu development, and to provide a forum for industry dialogue. Publications and information on frozen foods and how to market them to consumers are available at this Web site.

National Marine Fisheries Service

www.nmfs.noaa.gov

This organization provides a voluntary inspection service to the industry, assuring compliance with all applicable food regulations. Product-quality evaluation, grading, and certification services on a product-lot basis are also provided. Benefits include the ability to apply official marks—such as the U.S. Grade A, Processed Under Federal Inspection (PUFI), and lot-inspection marks. This Web site provides information on National Oceanic and Atmospheric Administration (NOAA) fisheries and their initiatives, as well as information on the voluntary inspection service.

U.S. Department of Agriculture (USDA)

www.usda.gov

The USDA touches many topic areas involved with land and its use in the U.S. These areas include ensuring a safe, affordable, nutritious, and accessible food supply; caring for agricultural, forest, and range lands; providing economic opportunities for farm and rural residents; and working to reduce hunger in the U.S. and throughout the world. This Web site provides information on all of these topics and particularly the safety of food from animal and plant farms.

USDA–FSIS

www.fsis.usda.gov

This branch of the USDA focuses on the safety of food produced on farms through the regulation and inspection of meat, poultry, and egg products. This Web site provides a wealth of information on food safety practices and procedures at all levels for meat, poultry and eggs.

Employee Food Safety Training

Inside this chapter:

- Purpose of Food Safety Training
- Key Elements of Effective Training
- Developing the Training Program
- Conducting the Training Session
- Food Safety Certification

After completing this chapter, you should be able to:

- Assess the training needs of employees.
- Design, implement, and reinforce a training program.
- Evaluate the success of a training program.
- Identify when and where food safety training should take place.
- Recognize the importance of food safety certification.

Key Terms

Presentation
Feedback
Application
Training program
Training need
Objective
Demonstration
Lecture
Role-plays
Job aids
Technology-based training
Evaluation

Apply Your Knowledge

Check to see how much you know about the concepts in this chapter. Use the page references provided to explore the topic in each question.

Test Your Food Safety Knowledge

1. **True or False:** Practicing a task without feedback will not be effective. *(See page 15-4.)*
2. **True or False:** A training need is a gap between what employees know and what they need to know to do their job. *(See page 15-5.)*
3. **True or False:** The ideal length for a training session is forty to sixty minutes. *(See page 15-14.)*
4. **True or False:** An objective states what the learner will be able to do after training. *(See page 15-6.)*
5. **True or False:** A written test is the only way to determine if training objectives have been met. *(See page 15-18.)*

For answers, please turn to the Answer Key.

INTRODUCTION

Training means teaching employees how to do a job properly. This sounds simple, but training is actually more involved. In your establishment, it is up to you to train all employees in a wide range of tasks and to include food safety training as a key element in your training program.

Training programs must be established for both new and current employees. For new employees, food safety training must be mandatory. It is dangerous to assume they know the safe and sanitary procedures used in your establishment. While employees already on staff might know the correct procedures, they might not always follow them since they are often in a hurry, forget, or lack motivation. For these employees, it might be important to schedule short retraining sessions, update meetings on new procedures, or motivational sessions that reinforce methods and practices.

In this chapter, the following aspects of food safety training for employees will be explored:

- Assessing and analyzing food safety training needs
- Planning, executing, and reinforcing the training program
- Evaluating the success of the training effort

PURPOSE OF FOOD SAFETY TRAINING

Food safety training provides employees with the knowledge and skills needed to handle food safely in your establishment. The final responsibility for food safety rests with the manager.

Employee training is an important factor in every operation's financial statement. At first glance, training appears costly. The manager, of course, must evaluate every facet of the business for cost and might wonder how training will affect the bottom line. It is true that food safety training might require time away from regular tasks for both employees and managers. It might also require the use of Web-based programs, the services of professional trainers, and the selection and use of training materials, such as videos, slides, books, and CD-ROMs, to reinforce safe practices. However, staff training will have a positive return on investment in the long run. Benefits from food safety training include:

- **Avoiding the costs associated with a foodborne-illness outbreak.** These costs may include legal fees and medical bills.
- **Preventing the loss of revenue and reputation incurred when an establishment is forced to close due to a foodborne-illness outbreak.**
- **Improving employee morale and reducing turnover.** Most employees want to do their jobs right and expect to receive training, which helps instill employee confidence.
- **Increasing customer satisfaction.** When customers see an establishment is committed to serving safe food, satisfaction will be higher.

A successful training program has these essential elements:

- Clearly defined and measurable objectives
- Training that supports the objectives
- Evaluation to ensure the objectives have been achieved
- A work climate that reinforces training
- Management support

In order for the training program to be effective, employees must see that the commitment to food safety comes from the top down. Management should lead by example. If managers show a commitment to food safety by behavior and attitude, employees are likely to follow.

KEY ELEMENTS OF EFFECTIVE TRAINING

Key Point

As a general rule, one-third of the time spent training should be devoted to the presentation of content, while the remaining two-thirds should be spent allowing trainees to apply the information learned, with feedback.

Successful training involves three key elements: presentation, feedback, and application. **Presentation** is the delivery of content to the learner, which can be accomplished through a variety of methods. Once the content is presented, the learner must have the opportunity to practice, apply, or respond to the content in order to retain it. While learners are practicing or applying the content, they must receive **feedback** (positive or negative reinforcement) on their performance. The feedback received must be specific and immediate. **Application** or practice without feedback will be ineffective.

As a general rule, one-third of the time spent training should be devoted to the presentation of content, while the remaining two-thirds should be devoted to activities that allow trainees to apply what they have learned, with feedback.

An effective trainer will have good presentation skills and be knowledgeable in the science of food safety. In addition, a good trainer recognizes that each trainee learns at a different rate and through different media.

DEVELOPING THE TRAINING PROGRAM

A **training program** is a structured sequence of events that leads to learning. To be effective, the training program must be well organized. Some large establishments set up a training department and have a training professional on staff to develop and deliver their programs. Other establishments rely on the manager to develop and deliver training. A foodservice manager can develop an effective training program through careful planning and the use of good support tools.

There are several steps in the development and delivery of an effective training program. These include:

1. Assessing training needs
2. Establishing learning objectives
3. Choosing training delivery methods
4. Selecting an instructor (if applicable)
5. Selecting training materials
6. Scheduling a training session
7. Selecting a training area
8. Preparing the trainer

Assessing Training Needs

Key Point

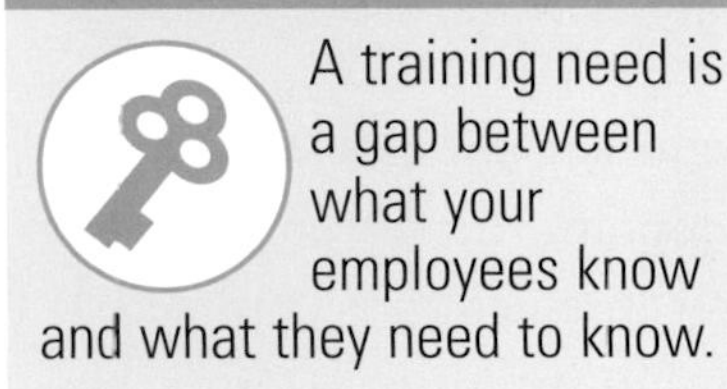

A training need is a gap between what your employees know and what they need to know.

The first task in developing a training program is to assess the training needs in your establishment. A **training need** is a gap between what your employees are required to know to perform their job and what they actually know. For new hires, the need might be apparent. For employees already on staff, the need is not always as obvious. Determining the gap in performance may require work. To identify food safety training gaps, a manager or trainer can do the following:

- Test employees' food safety knowledge.
- Observe employee job performance.
- Question or survey employees to identify areas of weakness.

Your staff needs the correct food safety knowledge and skills. From their first day on the job, new foodhandlers will require training in the following areas:

- **Overview:** The importance of food safety
- **Personal hygiene:** Health, personal cleanliness, proper work attire, and hygienic practices, including handwashing
- **Food preparation:** Time-temperature control, the prevention of cross-contamination, and safe practices for preparing, cooking, holding, serving, cooling, and reheating food
- **Cleaning and sanitizing:** Procedures for cleaning and sanitizing food-contact surfaces
- **Chemicals:** Procedures for safely handling the chemicals used in the establishment
- **Pests:** Pest identification and prevention measures

Training needs will differ for each job within the establishment. All employees will need general information. Other information will be job specific. For example, all employees need to know the proper way to wash their hands, but only cooks need to know the required minimum internal cooking temperature for chicken.

Regardless of the position, employees need to be retrained periodically on the food safety practices previously listed. It is the manager's responsibility to keep employees informed of the changes in the science of food safety and best practices in the industry.

Establishing Learning Objectives

Key Point

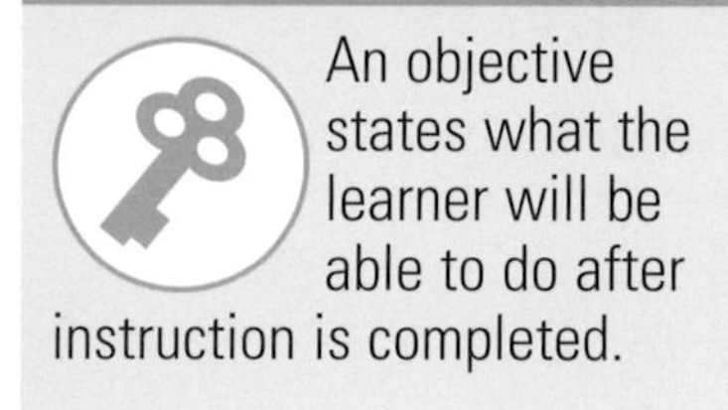

An objective states what the learner will be able to do after instruction is completed.

Once training needs have been identified, the next step is to define the objectives of the training program. An **objective** states what the learner will be able to do after instruction is completed. Objectives need to be stated clearly and in measurable terms. Use action verbs such as *operate, demonstrate, or practice,* rather than *be aware, observe, understand, or notice.* Examples of clearly stated objectives include:

- Demonstrate the proper procedure for calibrating a bimetallic stemmed thermometer using the ice-point method.

- List the internal cooking temperature for meat, seafood, and poultry.

Choosing Training Delivery Methods

There are many methods for delivering content to the learner. Some delivery methods are better suited to one-on-one learning, and others to group learning. No single delivery method is best for training at all levels. Using several delivery methods will result in more effective learning. Some methods for delivering training include:

- Demonstration
- Lecture
- Role-play
- Job aids
- One-on-one training (on-the-job training)
- Technology-based training
- Group training

Key Point

The delivery method chosen should allow trainees to apply what they have learned, with feedback on their performance.

Remember that the delivery method chosen should allow trainees to apply what they have learned while they receive feedback on their performance. However, the choice will also depend upon the number of people needing training, the cost, and—most importantly—how the trainees will learn best.

Every participant learns differently. Ask yourself if the delivery methods you have chosen allow learners to do one or more of the following:

- Talk to each other
- Reflect on the content and determine how it applies to their job
- See the task being performed
- Be physically active and perform the task
- Hear the instructions spoken
- Read the materials and take notes
- Reason through real-life situations

Key Point

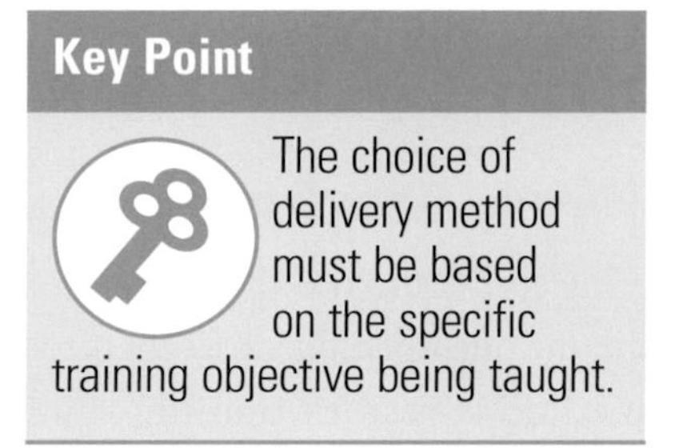

The choice of delivery method must be based on the specific training objective being taught.

The delivery method chosen must also be based on the specific training objective being taught. For example, if a manager wants to train an employee to wash, rinse, and sanitize the salad-preparation counter properly, the best method could be to demonstrate the process, then have the employee perform the task. Demonstration allows immediate feedback. The manager can explain key points of a task while performing it, then observe progress as the employee practices the same task.

Demonstration

Demonstration is the process of illustrating a skill or task before another person or group. When conducting a demonstration, remember the following steps:

1. Preface the demonstration with an explanation of what trainees should look for when the skill or task is demonstrated.
2. Emphasize key points as you demonstrate the task or skill.
3. Explain how each step fits into the task sequence. Trainees need to know where they are, step by step, in the process of performing the task.
4. Demonstrate the skill or task slowly, so trainees can see what is happening. Then repeat the skill or task at normal speed.
5. Before giving trainees an opportunity to perform the same task, ask them to explain each of the steps in sequence.
6. Ask trainees to demonstrate the skill or task. Provide appropriate feedback, correcting any errors as they occur.

Lecture

Key Point

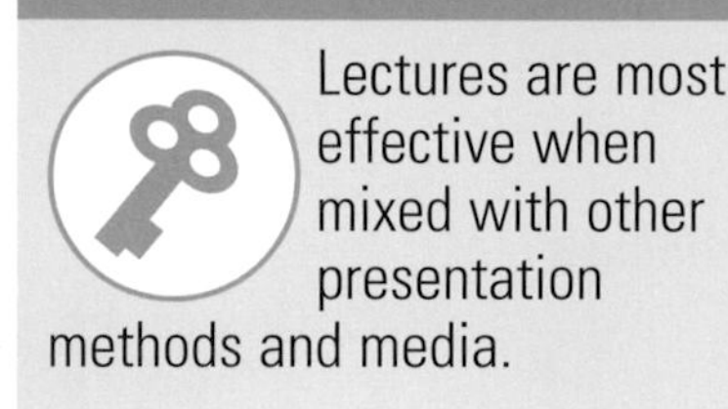

Lectures are most effective when mixed with other presentation methods and media.

A **lecture** is a prepared oral presentation used to deliver content to a group of participants. Lectures are most effective when mixed with other presentation methods and media, and are better received and accepted when the following techniques are used:

- Start with an interesting statement, observation, quotation, or question.
- Use relevant humor where appropriate.
- Use interesting and relevant examples, anecdotes, analogies, and statistics.

- Ask frequent questions to solicit audience participation.
- Use frequent small-group discussions and activities.
- Build in a review.

Only one-third of the training time should be spent lecturing, while two-thirds of the time should be spent allowing participants to apply the information learned, with feedback.

Role-Play

In **role-plays,** trainees enact a situation in order to try out new skills or apply what has been learned. Different types of role-plays are suitable for different types of learning situations. They usually are set up so a trainee is confronted by another trainee and must answer questions, handle problems, provide satisfaction, solve a complaint, etc. For example, a role-play can be used to teach a manager how to deal with a health inspector during an inspection.

When using a role-play, you should

- make certain new content is understood before beginning.
- provide trainees with detailed instructions.
- explain and model the situation before trainees begin.
- keep the role-play simple.

Job Aids

Key Point

Job aids should be used when tasks are performed infrequently, are complex, when the consequences of making a mistake are severe, or when safety is a concern.

Job aids can also be used to deliver content to employees. They may include worksheets, checklists, samples, flowcharts, procedural guides, glossaries, diagrams, decision tables, etc. Other job aids might include visual reminders, such as posters illustrating the proper steps in handwashing, manual warewashing, or other tasks. Employees can then use these job aids once they return to their jobs. Job aids are particularly useful when

- tasks are performed infrequently.
- tasks are complex.
- sequence of performance is critical.
- consequences of making a mistake are severe.
- safety is a concern.

One-on-One Training

One-on-one training has the following advantages:

- Takes into consideration the special needs of individual employees
- Can take place on the job, eliminating the need for a separate training location
- Enables the manager to monitor employee progress
- Allows for immediate feedback
- Offers the opportunity to apply information that has been learned

One-on-one training does have some disadvantages, however. Most importantly, its effectiveness depends upon the ability of the person delivering the training. Therefore, the trainer must be selected very carefully. Many establishments certify trainers or validate that they have the appropriate skills before allowing them to train.

Technology-Based Training

Web-based training, interactive CD-ROMs, and other **technology-based training** programs offer yet another way to deliver training. The advantages of technology-based delivery methods include:

- **Standardized delivery.** The training is delivered the same way every time.
- **Standardized feedback.** Each time a trainee responds to a situation, the program can provide standardized feedback.
- **Customizable instruction.** Each trainee can choose their own learning path.
- **Increased performance practice.** Trainees are allowed to practice a skill until they are proficient.

Exhibit 15a

Group Training

Group training can be cost effective and might offer more uniform content delivery.

Group Training

If several employees require food safety training, group sessions might be more practical than one-on-one training. (See *Exhibit 15a*.) Group training can be delivered through a number of methods, including those discussed earlier in this chapter. It has several advantages:

- It is more cost effective.
- Training is more uniform.
- You know precisely what employees have been taught.

One disadvantage of group training is that it is designed to meet the overall needs of a group of employees rather than the needs of each individual. Slower learners, less-skilled employees, or those with limited proficiency in English may not understand all the material presented. A trainer must remember that it is not enough simply to cover the material. Trainees must be involved in the training process too. If they are involved, they will retain the information.

Key Point

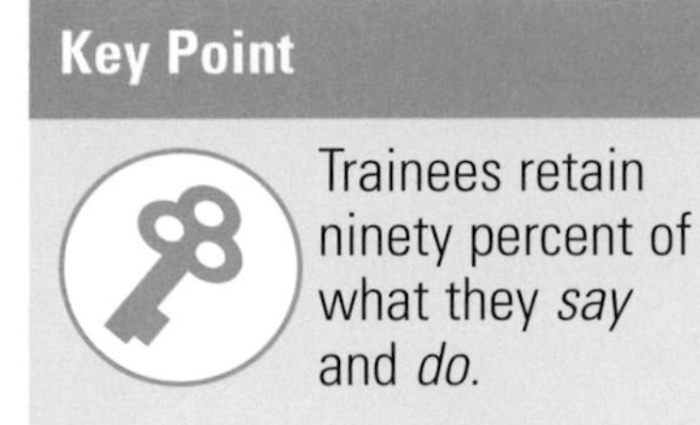

Trainees retain ninety percent of what they *say* and *do*.

Studies on training effectiveness show that trainees retain

- ten percent of what they read.
- twenty percent of what they hear.
- thirty percent of what they see.
- fifty percent of what they hear and see.
- seventy percent of what they say.
- ninety percent of what they say and do.

Selecting an Instructor

The manager often is the most likely person to teach food safety to their staff. He or she knows the operation and is responsible for food safety practices within the establishment. If the manager chooses not to conduct the training personally, an instructor should be selected based on the following criteria:

- Knowledge of food safety practices
- Understanding of the operation's food safety challenges
- Demonstrated skill teaching others
- Good communication skills

Individuals who may be able to conduct food safety training include:

- **Immediate supervisors.** Often these people are a logical choice because of their working relationship with the employee.
- **Staff trainer.** In large establishments, a training professional is often on staff to deliver training and support the training needs of the operation.
- **Representative of the health department.** The local sanitarian may agree to teach a food safety training session. This fosters a cooperative working relationship with his or her agency.

Key Point

Contact your local health department for help in teaching food safety in your establishment.

- **Representative from a professional or educational organization.** These organizations often provide trainer-preparation courses for operators and also have staff instructors who conduct courses on-site or at other locations. Check with suppliers and your State Restaurant Association.

Very often, you will experience greater results when you use a combination of these resources.

Selecting Training Materials

Key Point

To be useful, training materials must be accurate, appropriate, and attractive.

Training materials save time, add interest, help participants learn and retain information, and make the trainer's job easier.

The manager or trainer should be guided by the "Three A's" when selecting materials. To be useful, training materials must be *accurate, appropriate,* and *attractive.*

Accurate. Materials must be factual, up-to-date, and complete. To ensure that the materials are accurate, they should meet the learning objectives of the training program. This may be accomplished by developing them internally or purchasing them from professional or educational organizations, industry suppliers, or from an authority in foodservice training.

Appropriate. Materials must be suitable for the purpose they are to serve. Written materials must be matched to the reading-comprehension levels of trainees. For employees with limited English proficiency, training materials in other languages are available. In addition, materials should suit the abilities of the trainer. Limitations imposed by the training location will influence your choice of training materials. Reliable equipment must be available in order to use videotapes, slides, CD-ROMs, Web-based programs, overhead transparencies, and other audiovisual or technological tools.

Attractive. In order to teach people, you must first get their attention. When teaching subjects that generally do not have a wide appeal, make information exciting and memorable with eye-catching materials, whenever possible.

Materials Used in a Typical Lesson Plan

The manager or trainer should start each training session by clearly stating the objectives of the session. It is a good idea to provide handouts to students with these objectives clearly stated, or to write them on a flipchart, easel, or chalkboard.

Sometimes a pretest or quiz is given to assess the trainees' prior knowledge. The new information is then presented, using audiovisual materials to supplement the presentation. Afterwards, a brief written, oral, or practical exam is given to test the trainees' knowledge.

The trainer can use written examinations prepared by an educational or training organization, or they can write examinations themselves. The advantage of using prepared exams is that they are usually validated for content and evaluated for reading-level comprehension. Any mistakes made by the trainee should be discussed and corrected immediately after the exam is scored.

To reinforce the training, use signs, posters, bulletin boards, reminders in pay envelopes, and other materials.

Scheduling Training Sessions

Key Point

Training content should ideally be broken up into segments twenty to thirty minutes long.

Developing and implementing a master training schedule at your establishment can be a useful method to determine training priorities and show your commitment to training. Both orientation and ongoing training should be included in this schedule. For legal reasons, it is important to document the training that took place, even if the training session was brief.

Training sessions should not be too long. The ideal length is probably about twenty to thirty minutes per segment. Successful training sessions, however, may be as long as an hour and a half, if well planned. The ideal length depends on the type of learning activity used. Training sessions might need to be given more than once during the day in order to cover all employee shifts.

Training sessions should be scheduled during slow times for the establishment: before opening, after closing, or on a day when the establishment is normally closed. If there is a separate area in the facility for training purposes, scheduling can be more flexible.

Selecting a Training Area

Conduct training in a comfortable location. An on-site location allows for demonstrations and provides the opportunity to relate instruction to your establishment. Employee workstations can be used as training areas, particularly for one-on-one sessions. In most establishments, an employee break room, an executive office, or a section of the facility not in use can also be used.

Consider the following when choosing an area for group training:

- The area should be an appropriate size; no one should feel crowded.
- Tables or desks should be available so trainees can take notes.
- Seating should be adequate, comfortable, and arranged to encourage open discussion.
- A blackboard or flip chart should be provided.
- The room should have enough electrical outlets for projectors, sound equipment, videotape players, computers, and other electronics.
- If audiovisual materials are to be used, it must be possible to darken the room so trainees can see the visuals adequately.
- The area should be free of distractions.

Preparing the Trainer

Training employees requires excellent communication and organizational skills. The following suggestions will help you conduct a successful training session.

- **Make sure you are knowledgeable in all areas of food safety.** Feeling comfortable with the subject is important, in case trainees ask questions. If you are uncertain of your subject, trainees will know.
- **Prepare for the presentation.** Make an outline of the presentation. Practice it until it feels natural. Familiarize yourself with the training room and check all audiovisual equipment to make sure it is in working order.

- **Maintain eye contact with trainees during your presentation.**
- **Keep your delivery conversational and informal.** This makes the atmosphere more comfortable. Vary the tone of your voice for emphasis and do not speak too quickly. A moderate rate will allow trainees to take notes and ask questions.
- **Use simple language.** Using unnecessary technical terms will cause trainees to lose interest or become lost. Pause often and ask questions to make sure your audience understands the material.
- **Treat all questions and comments made by the trainees seriously.** Answer them in a straightforward way. You can give trainees positive feedback by saying, "That's a good question."
- **Look for cues indicating employees do not understand the information or are bored and losing interest.** These cues could include fidgeting, looking at watches, doodling, and not asking or answering questions. Vary the way in which you present the information to hold the trainees' attention.
- **Keep the sessions short.**
- **Keep the training as practical as possible and relate information to the employee's job.** If the subject is too general, employees will tend to lose interest. Give specific examples of how the training content relates to the job they do.

CONDUCTING THE TRAINING SESSION

Key Point

Participants want to know how the training will enable them to do their jobs faster, easier, or better.

For training to be effective, make sure trainees know how the material will benefit them. Answer the question, "What's In It For Me (WIIFM)?" Trainees want to know what they will gain from participating in the training session. They need to know how the training will enable them to do their jobs faster, easier, or better.

It is the trainer's job to capture their attention and to help them remember the material. Here are some guidelines to help you achieve this.

Key Point

Present no more than five to nine major points in a short training session.

Keep It Short and Simple

It is important to take a simple, straightforward approach. Because time is often limited, focus on five to nine major points in the training session.

Individualize Training

Make trainees feel you are talking to them individually and that you have a real interest in them. Encourage employee participation during sessions and show you are interested in what they have to say.

Recognize Employee Achievement in the Training Program

Wall charts tracking individual progress through the training course, as well as certificates of completion, can serve as incentives.

Be Creative

Put questions and answers in a game-show format. Create teams and have trainees compete to give correct answers for points or small prizes. Ask one trainee to draw a sanitation procedure on a chalkboard or flip chart while the rest of the group guesses what they are drawing. If the group is not interested in drawing, have them act out the words.

A variety of unit-level training tools and games are available through your State Restaurant Association or by calling the NRAEF at 800.765.2122, ext. 701, or by visiting www.nraef.org.

Provide Feedback

Praise and positive feedback are part of all good learning experiences, and they should be part of the training process in your establishment. Praise indicates to employees that they are performing well and have been successful in the training program. Generous praise reinforces employees' positive attitudes about their jobs. When they see that management values their best effort, they will continue their good practices.

Evaluate

The training program is not complete until it has been evaluated. The **evaluation** process is important because it tells you if training has provided the participants with the knowledge and skills needed to do their job. To evaluate the training program, the manager must carefully judge the trainee's performance against the learning objectives.

When evaluating effectiveness, ask the following questions: Did the training produce results on the job? If the intended results were not produced, why not?

If training was ineffective, several factors might be involved. For example, the trainee could have acquired the knowledge, but is simply not applying it. Or perhaps equipment used during the training session is different from the equipment used on the job. Perhaps there are negative consequences for doing the job the way the employee was trained to do it.

In some cases, an employee has learned the material, but is simply applying it incorrectly. There may be several reasons for this. Perhaps the employee has been improperly trained and his or her performance is consistent with the bad practices taught. Perhaps the employee has learned the skill, but it is not being reinforced on the job. Management intervention is required to ensure that the proper skills are being taught, practiced, and reinforced.

Key Point

Evaluation should always be based on the training objectives.

Evaluation should always be based on the training objectives. There are several ways in which objectives can be measured. Most commonly, objectives are measured through written or oral tests. Test results can help a manager determine if a trainee needs to review the content. Objectives can also be measured by evaluating an employee while he or she performs a task or skill required by the objective. Evaluation works best when a combination of written and performance-based tests are used.

FOOD SAFETY CERTIFICATION

The National Restaurant Association and federal and state regulatory officials recommend food safety training and certification, particularly for managers and supervisors. Certification demonstrates that a person comprehends basic food safety principles and recommended food safety practices that prevent foodborne illness.

Manager certification is already a requirement in some states. In other states, some cities and counties require certification although the state as a whole does not require it. The National Restaurant Association Educational Foundation Web site provides a jurisdictional summary of training and certification requirements for the entire country. *(See page 15-28.)*

Regardless of state regulations, many proactive establishments and managers have made a strong commitment to food safety by making food safety training and education an integral part of their establishments. You, too, should make a commitment to training, as well as certification, when possible, even if your state or city does not require it. As a conscientious foodservice manager, you should train your employees, monitor their practices, and make food safety a part of everyone's job description. In this way, you will be able to ensure that the food you serve is safe.

SUMMARY

Food safety training provides employees with the knowledge and skills needed to handle food safely in your establishment. Some of the benefits include preventing a foodborne-illness outbreak, avoiding the loss of revenue and reputation following an outbreak, improving employee morale, and increasing customer satisfaction.

Successful training involves presentation, application, and feedback. Once the content is presented, the learner must have the opportunity to practice and apply it, with feedback, in order to retain it.

There are several steps in the development and delivery of an effective training program.

- First, you must assess the training needs in your establishment. This is the gap between what your employees are required to know and what they actually know. To identify food safety-training needs, you can observe employee performance, review past health inspection reports, and test your employees' knowledge.

- Once training needs have been identified, clear and measurable objectives need to be defined stating specifically what the learner will be able to do after instruction is complete. The trainer then must determine how content will be delivered to the trainees.
- Content can be delivered through demonstration, lecture, role-play, job aids, and technology-based programs. Using several delivery methods will result in more effective learning. However, whatever method is chosen, it should allow trainees to apply what they have learned with feedback on their performance.
- Instructors and training materials need to be selected. Instructors should be knowledgeable, must have the ability to teach others, and must have good communication skills. Training materials must be accurate, appropriate, and attractive.
- A master training schedule should be developed to schedule around training priorities and to show your commitment to training. Training sessions should be short—from twenty to thirty minutes—and should be scheduled when the operation is slow.
- Comfortable training areas, free of distractions, should be identified within the establishment.
- Finally, the trainer should be prepared before the training session begins. This includes making sure you are knowledgeable in the areas you are teaching and practicing your presentation.
- For training to be effective, make sure you let trainees know what they will gain from participating in the training session. They should be able to see how the training will help them do their job faster, easier, or better. Keep the training short and simple—focusing on five to nine major points in the session.
- Encourage employee participation during sessions and recognize employee achievement during the program. Praise and positive feedback are part of all good learning experiences and should be a part of the training process in your

establishment. Finally, evaluate the training program to determine if employees can do what you trained them to do.

The National Restaurant Association and federal and state regulatory officials recommend food safety training and certification, particularly for managers. Certification shows that a person comprehends basic food safety principles and recommended food safety practices that prevent foodborne illnesses.

Apply Your Knowledge

1. Do you think Paul's training program for the cook and server will be effective?
2. What else should Paul be doing?

For answers, please turn to the Answer Key.

A Case in Point

Paul has two employees who need food safety training. One is an assistant cook, Albert, who used to be a buser and needs further training. The other, Maria, is an inexperienced server. Albert received general training in the fundamentals of food safety about three months ago, but Maria has had no training. Both Albert and Maria require specialized training. Paul maintains an ongoing food safety training program, so he has already trained the rest of his employees and evaluated the effectiveness of his original program.

Paul decides to have the kitchen supervisor train Albert using the one-on-one method, so the supervisor can demonstrate procedures and give immediate feedback to Albert as he practices them. By relying on the supervisor to train the cook, Paul can have both the cook and server trained at the same time.

Paul gives Maria a copy of the job description and learning objectives and sets up a food safety videotape in his office for her to watch. While the videotape is playing, Paul goes to the back of the facility to talk with a clerk about an expected shipment. After half an hour, Paul returns to his office to give Maria a written test. While Maria is completing the test, Paul goes to the kitchen to observe the supervisor training Albert, the cook. Noting that everything looks good in the kitchen, Paul goes to the dining room.

Apply Your Knowledge

Use these questions to review the concepts presented in this chapter.

Discussion Questions

1. What are the benefits of teaching food safety in an establishment?
2. What are the three key elements of successful training? How do they fit together?
3. How can an establishment determine its food safety training needs?
4. What is an objective? Give an example of a good objective.
5. What are some methods that can be used to deliver training?

For answers, please turn to the Answer Key.

Apply Your Knowledge

Multiple-Choice Study Questions

1. **Mary has hired college students to run her lakeside hot-dog stand for the summer. She would like to develop a brief training program to prepare them for the job. What should Mary do first?**
   A. Write training objectives.
   B. Choose training delivery methods.
   C. Assess her training needs.
   D. Evaluate her training program.

2. **The purpose of implementing a training program for food safety is to**
   A. avoid the costs associated with a foodborne-illness outbreak.
   B. prevent the loss of reputation associated with a foodborne-illness outbreak.
   C. improve employee morale.
   D. All of the above

3. **The ideal length for a training session is**
   A. five to ten minutes.
   B. twenty to thirty minutes.
   C. forty to sixty minutes.
   D. one to two hours.

4. **Which of the following is not a good way for a restaurant or foodservice manager to identify food safety-training needs in their establishment?**
   A. Reading the comments on the health inspection report
   B. Testing employees to find out what they know
   C. Observing employees while they perform their jobs
   D. Reading a textbook on training restaurant employees

5. **During the training session, a trainer should focus on five to ____ major points.**
   A. nine
   B. ten
   C. twelve
   D. fifteen

Continued on next page...

Multiple-Choice Study Questions *continued*

6. As a general rule, one-third of the time in a training session should be spent presenting content while the remaining two-thirds of the time should be spent
   A. measuring knowledge with a test.
   B. allowing employees to apply what was learned, with feedback.
   C. determining the training need.
   D. determining a delivery method.

7. A WIIFM shows trainees all of the following *except*
   A. what's in it for them.
   B. how the training will help them do their job faster, easier, or better.
   C. how the training will benefit them.
   D. how the training will benefit the establishment.

8. An instructor should use a delivery method that
   A. supports the training objectives.
   B. allows trainees to apply what they have learned.
   C. complements how trainees learn.
   D. All of the above

9. Which of the following objectives is stated clearly and in measurable terms? You will...
   A. be able to demonstrate the proper procedure for calibrating a thermometer.
   B. understand how to calibrate a thermometer.
   C. be aware of how to calibrate a thermometer.
   D. know how to calibrate a thermometer.

10. Which of the following statements about training objectives is *not* true?
    A. They should describe the end result that the training is intended to produce.
    B. They should be stated in measurable terms.
    C. They should state what the learner will be able to do prior to training.
    D. They should be phrased using action verbs.

For answers, please turn to the Answer Key.

ADDITIONAL RESOURCES

Books and Periodicals

Bax, B. 1997. *Handbook for safe food service management.* Upper Saddle River, NJ: Prentice-Hall.

Craig, R. L., ed. 1987. *Training and development handbook: A guide to human resource development,* 3rd ed. New York: McGraw-Hill. (Sponsored by the American Society for Training and Development.)

Culture Shock: The language of food safety must speak to many cultures in today's foodservice kitchens. 2003. *Food Safety Illustrated.* 3 (3):8–12.

National Restaurant Association Educational Foundation. 2001. *The manager's guide to employee-level training.* Chicago, IL: National Restaurant Association Educational Foundation.

National Restaurant Association Educational Foundation. 2001. *ServSafe® train-the-trainer program.* Chicago, IL: National Restaurant Association Educational Foundation.

Pike, R. W. 1994. *Creative training techniques handbook,* 2nd ed. Minneapolis MN: Lakewood Books.

Rothwell, W. J., and H. C. Kazanas 1998. *Mastering the instructional design process: A systematic approach,* 2nd ed. San Francisco: Jossey-Bass.

Training in the bilingual kitchen. 2001. *Food Safety Illustrated.* 1 (3):8–10.

Web Sites

Council of Hotel and Restaurant Trainers (CHART)

www.chart.org

CHART is one of the oldest and largest organizations dedicated to training in the hospitality industry, with a mission to help hospitality-training professionals improve operational performance by developing people. This Web site provides information on how to become a member of CHART, information on conferences, and a sample of their newsletter.

FDA Food Code
http://vm.cfsan.fda.gov/~dms/foodcode.html
As the basis for many local sanitation codes, as well as the basis for information in this textbook, the FDA Food Code, available at this Web address, is a useful resource for information relating to food safety for the restaurant and foodservice industry.

Foodservice Consultants Society International (FCSI)
www.fcsi.org
The FCSI links many professionals working in the foodservice industry, helping consultants advance their careers, providing all-important professional recognition, and supplying numerous networking opportunities with colleagues. This Web site allows you to become a member, take continuing education classes, and even find a job.

International Council on Hotel, Restaurant, and Institutional Education (CHRIE)
www.chrie.org
CHRIE is the global advocate of hospitality and tourism education for schools, colleges, and universities, offering programs in hotel and restaurant management, foodservice management, and culinary arts. This Web site provides a place for exchanging information, ideas, research, and products and services related to education, training, and resource development for the hospitality and tourism industry.

International Food Safety Council
www.nraef.org/ifsc
In 1993, the National Restaurant Association Educational Foundation recognized the need for food safety awareness and created the International Food Safety Council as its food safety awareness initiative. The Council's mission is to heighten the awareness of the importance of food safety education throughout the restaurant and foodservice industry. This Web site is an excellent resource center on the industry's food safety issues.

National Agricultural Library (NAL)
www.nalusda.gov
NAL has a mission to increase the availability and utilization of agricultural information for researchers, educators, policymakers, consumers of agricultural products, and the public. Use this Web site to locate valuable food safety and HACCP information.

National Association of College and University Food Services (NACUFS)

www.nacufs.org

NACUFS is the trade association for foodservice professionals at over 650 institutions of higher education in the U.S., Canada, and abroad. This Web site provides a full range of educational programs, industry-specific publications, management services, and networking opportunities.

National Restaurant Association

www.restaurant.org

The National Restaurant Association is the leading business association for the restaurant industry. Together with the National Restaurant Association Educational Foundation, the Association's mission is to represent, educate, and promote the rapidly growing restaurant and foodservice industry. This Web site should be your starting place for all issues and concerns related to your restaurant. This Web site has it all, from tips for running your establishment to vital data on your customers' spending habits.

National Restaurant Association Educational Foundation (NRAEF)

www.nraef.org

The NRAEF's Web site is a comprehensive overview of the initiatives, products, and services offered by the organization, and an effective tool to help you become engaged in NRAEF initiatives, as well as its education and training opportunities.

Resource Center for Workforce Solutions™

www.nraef.org/solutions

The National Restaurant Association Educational Foundation and Coca-Cola North America Foodservice & Hospitality Division (CCNA) have formed a dynamic alliance to build a comprehensive resource of actionable workforce solutions for the restaurant and foodservice industry. This growing portfolio of tools from a variety of industry-leading sources is designed to help operators attract, engage, and retain their employees.

Apply Your Knowledge

Notes

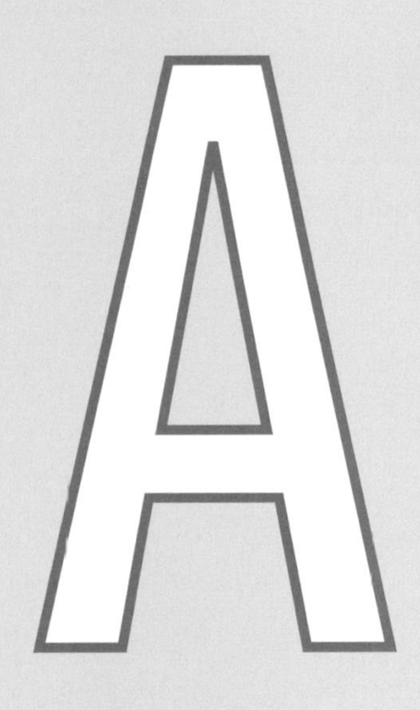

Answer Key

Chapter 1

Page Activity

1-2 Test Your Food Safety Knowledge

1. False 2. True 3. True 4. False 5. False

1-17 A Case in Point

1. André tried to ensure that all of his employees went through initial food safety training.
2. He failed to periodically retrain his employees on food safety practices. Without the refresher training, his employees were likely to forget proper food safety practices and procedures, which led to the lower inspection scores and a potential foodborne illness.
3. In the future, André should ensure that employees receive initial food safety training and refresher training on an ongoing basis. He should also create or utilize a tool that assesses his employees' knowledge and compares it to what they should know. Any gap that is identified should be filled using follow-up training.

1-18 Discussion Questions

1. A *foodborne illness* is a disease carried or transmitted to people by food. A *foodborne-illness outbreak*, according to the CDC, is an incident in which two or more people experience the same illness after eating the same food.
2. The potential costs of a foodborne-illness outbreak include the following:
   - Loss of customers and sales
   - Loss of prestige and reputation
   - Embarrassment
   - Lawsuits resulting in legal fees
   - Increased insurance premiums
   - Lowered employee morale
   - Employee absenteeism
   - Need for retraining employees
   - Food costs that result from discarding food supplies that may or may not be contaminated
   - Fees for testing food supplies and employees
3. The elderly are at higher risk for contracting a foodborne illness because their immune systems and resistance to illness may have weakened with age. Also, as people age, their senses of smell and taste are diminished, so they may be less likely to detect "off" odors or tastes, which indicate that food may be spoiled.
4. The three major types of hazards to food safety are biological hazards, chemical hazards, and physical hazards.
5. This is not an acceptable practice because the chef failed to clean and sanitize the knife and cutting board properly after cutting up the salmon. Microorganisms, which may have been present on the salmon, could be transferred to the knife and cutting board and then to the parsley. This is a classic example of cross-contamination.

1-19 Multiple-Choice Study Questions

1. B 2. B 3. B 4. B 5. D 6. A

Chapter 2

Page	Activity

2-2 Test Your Food Safety Knowledge

1. True 2. False 3. False 4. True 5. True

2-28 A Case in Point

1. The illness was caused by bacteria.
2. *Bacillus cereus* was the microorganism responsible for the outbreak.
3. Given the rapid onset and the symptoms, the illness was most likely an intoxication.

2-29 Discussion Questions

1. If potentially hazardous food is left in the temperature danger zone for more than four hours, bacteria can grow to levels high enough to make someone ill.
2. *A foodborne infection* results when a person eats food containing pathogens, which then grow in the intestine and cause illness. Typically, symptoms of a foodborne infection do not appear immediately. A *foodborne intoxication* results when a person eats food containing toxins that cause illness. The toxin may have been produced by pathogens found on the food or may be the result of a chemical contamination. The toxin might also be a natural part of a plant or animal consumed. Typically, symptoms of foodborne intoxication appear quickly, within a few hours. A *foodborne toxin-mediated infection* results when a person eats food containing pathogens, which then produce illness-causing toxins in the intestine.
3. An outbreak of salmonellosis can be prevented by the following:
   - ▷ Cooking food to its required minimum internal temperature
   - ▷ Avoiding cross-contamination
   - ▷ Properly refrigerating food
   - ▷ Properly cooling cooked meat and meat products
   - ▷ Properly handling and cooking eggs
   - ▷ Ensuring that employees practice good personal hygiene
4. The following are basic characteristics of viruses:
   - ▷ Unlike bacteria, they rely on a living cell to reproduce.
   - ▷ They are not complete cells.
   - ▷ Unlike bacteria, they do not reproduce in food.
   - ▷ Some may survive freezing and cooking.
   - ▷ They can be transmitted from person to person, from people to food, and from people to food-contact surfaces.
   - ▷ They usually contaminate food through a foodhandler's improper personal hygiene.
   - ▷ They can contaminate both food and water supplies.
5. The two FAT TOM requirements for growth that are easiest for an establishment to control are time and temperature.

   To control time:
   - ▷ Limit the amount of time food spends in the temperature danger zone.
   - ▷ Prepare food in small batches, as close to service as possible.

   To control temperature:
   - ▷ Cook food to the proper temperature.
   - ▷ Freeze food properly.
   - ▷ Refrigerate food to 41°F (5°C) or lower.

2-30 Multiple-Choice Study Questions

1. A	3. D	5. D	7. B	9. A
2. C	4. B	6. D	8. D	10. A

Chapter 3

Page Activity

3-2 Test Your Food Safety Knowledge

1. False 2. False 3. True 4. False 5. True

3-16 A Case in Point 1

1. Two factors contributed to the outbreak.
   A. When the mahi-mahi (a scombroid species of fish) was time-temperature abused, the bacteria associated with the fish produced the toxin histamine.
   B. Since cooking does not destroy this toxin, the consumption of the fish resulted in a scombroid poisoning (intoxication).

3-16 A Case in Point 2

1. Mark did what he was asked to do.
2. There are several problems.
   A. Only a licensed pest control operator (PCO) should apply pesticides in an establishment.
   B. Pesticides and other chemicals should be stored away from food, utensils, and equipment used for food.

3-17 Discussion Questions

1. The scombroid toxin is produced when the scombroid species of fish is time-temperature abused, while the toxins associated with other types of seafood are either a natural part of the animal or occur as a result of the animal's diet.
2. To guard against a seafood-specific foodborne illness, establishments should do the following:
   - Purchase seafood from reputable suppliers who maintain strict time-temperature controls and harvest from safe waters.
   - Refuse fish that show evidence of thawing and refreezing.
   - Refuse fresh fish that have not been received at 41°F (5°C) or lower.
   - Thaw frozen fish at refrigerator temperatures of 41°F (5°C) or lower.
3. The ciguatera toxin can occur in certain predatory tropical reef fish such as red snapper. For this reason, it should only be purchased from approved sources. Toxins are a natural part of some varieties of wild mushrooms. However, foodborne-illness outbreaks associated with mushrooms are almost always caused by the consumption of wild mushrooms collected by amateur mushroom hunters. For this reason, all mushrooms should be purchased from approved sources.
4. The enamel on the pitcher may be chipped or cracked, exposing toxic metals. When the acidic orange juice comes in contact with these metals, it may leach them from the pitcher into the juice, possibly causing illness.
5. A glass should never be used to scoop ice because it may chip or break into the ice, creating a physical hazard.

3-18 Multiple-Choice Study Questions

1. C	3. D	5. C	7. B	9. C
2. C	4. B	6. A	8. B	10. B

Chapter 4

Page Activity

4-2 Test Your Food Safety Knowledge

1. False 2. True 3. True 4. True 5. True

4-17 A Case in Point 1

1. Yes, this situation represents a threat to food safety.
2. Chris did several things correctly.
   A. She properly washed her hands and put on a new pair of single-use gloves at the beginning of her shift.
   B. She stepped away from the food-preparation area to sneeze.
3. Chris did several things wrong.
   A. After using the tissue, Chris should have taken off her gloves, washed her hands, and put on a new pair of gloves, since she may carry pathogens such as *Staphylococcus aureus* in her nose and mouth.
   B. She should have informed her manager of her condition.
   C. While some jurisdictions allow foodhandlers to drink from a covered container with a straw, Chris was drinking from an open container in a food-preparation area.
   D. Chris should have placed her medication inside a covered, leak-proof container that was clearly labeled before placing it inside the walk-in refrigerator.

4-17 A Case in Point 2

1. Marty had shigellosis. Because he was in a hurry, he failed to wash and dry his hands properly. Even though he used tongs to handle the food, Marty must have made direct contact with the food or food-contact surfaces, which resulted in a foodborne-illness outbreak.
2. Marty should have informed his manager that he was ill. However, his frequent and hurried trips to the restroom should have indicated this to the manager. If, after exploring the problem, the manager found that Marty was suffering from diarrhea, he should have restricted Marty from working with or around food.

4-18 Discussion Questions

1. Employees must meet the following work attire requirements:
   - Wear a clean hat or other hair restraint.
   - Wear clean clothing daily.
   - Remove aprons when leaving food-preparation areas.
   - Wear clean, closed-toe shoes with a sensible, nonslip sole.
   - Remove jewelry prior to preparing or serving food or while around food-preparation areas.
2. The following personal behaviors can contaminate food:
   - Nose picking
   - Rubbing an ear
   - Scratching the scalp
   - Touching a pimple or an open sore
   - Running fingers through the hair
3. All cuts and infected wounds should be covered with a clean, dry bandage. Disposable gloves or finger cots should be worn over bandaged cuts on hands.

4. Foodhandlers must follow these procedures when using gloves:
   - Gloves must never be used in place of handwashing.
   - Hands must be washed before putting on gloves and when changing to a fresh pair.
   - Gloves used to handle food are for single use only and should never be washed and re-used.
   - Gloves should be removed by grasping them at the cuff and peeling them off inside-out over the fingers while avoiding contact with the palm and fingers.
   - Gloves should be changed
     - as soon as they become soiled or torn.
     - before beginning a different task.
     - at least every four hours during continual use, and more often when necessary.
     - after handling raw meat and before handling cooked or ready-to-eat food.
5. According to the FDA Food Code, managers must *exclude* from the establishment foodhandlers who have been diagnosed with a foodborne illness, and they must report employee illnesses to the local regulatory agency resulting from the following pathogens:
   - *Salmonella* typhi
   - *Shigella* spp.
   - Shiga toxin-producing *E. coli*
   - Hepatitis A virus

   Managers must *restrict* foodhandlers from working with or around food if they have the following symptoms:
   - Fever
   - Diarrhea
   - Vomiting
   - Sore throat with fever
   - Jaundice

   If a foodhandler has any one of the above symptoms and the establishment primarily serves a high-risk population, the foodhandler must be excluded from the establishment.

4-19 Multiple-Choice Study Questions

1. D	5. B	9. A	13. D
2. D	6. C	10. A	14. C
3. C	7. B	11. D	15. D
4. D	8. A	12. B	16. C

Chapter 5

Page Activity

5-2 Test Your Food Safety Knowledge

1. True 2. False 3. False 4. False 5. False

5-14 Discussion Questions

1. Food can be time-temperature abused when it is not
   - ▷ cooked to its required minimum internal temperature.
   - ▷ cooled properly.
   - ▷ reheated properly.
   - ▷ held at the proper temperature.
2. Cross-contamination can be prevented in an establishment by:
   - ▷ assigning specific equipment to each type of food product prepared.
   - ▷ cleaning and sanitizing all work surfaces, equipment, and utensils after each task.
   - ▷ preparing raw and ready-to-eat food at different times.
   - ▷ purchasing ingredients that require minimal preparation.
3. The steps for calibrating a thermometer using the ice-point method are:
   - ❶ Fill a large container with crushed ice. Add clean tap water until the container is full. Stir the mixture.
   - ❷ Put the thermometer stem or probe into the ice water so the sensing area is completely submerged. Wait thirty seconds, or until the indicator stops moving. Do not let the stem or probe touch the sides or bottom of the container. Keep the stem or probe in the ice water.
   - ❸ Hold the calibration nut securely with a wrench or other tool and rotate the head of the thermometer until it reads 32°F (0°C). On some thermocouples or thermistors, it may be possible to press a reset button adjust the readout.

5-15 Multiple-Choice Study Questions

1. A
2. C
3. C
4. B
5. D
6. C
7. D

Chapter 6

Page Activity

6-2 Test Your Food Safety Knowledge

1. True
2. False
3. True
4. True
5. True

6-24 A Case in Point 1

1. Ed should have asked the delivery driver to come back later. Products must be delivered when employees have adequate time to inspect them. Deliveries must be inspected immediately. By putting the deliveries away without inspecting them, Sunnydale missed an important opportunity to identify food that
   - A. was not delivered at the proper temperature.
   - B. was damaged or mishandled.
   - C. had been thawed and refrozen.
   - D. had expired code dates.
   - E. showed signs of an insect infestation.

6-25 A Case in Point 2

1. John had good intentions and did most things correctly. However, he did make mistakes that could result in a foodborne illness. John should have done the following:
   - A. Notified the kitchen manager that the shipment had arrived.
   - B. Made sure that the bimetallic stemmed thermometer he took from the kitchen had been properly calibrated, as well as cleaned and sanitized before using it.
   - C. Cleaned and sanitized the thermometer after checking the temperature of each product. He should not have wiped the thermometer on his apron.
   - D. Inserted the thermometer stem into the middle of the bucket of live oysters between the shellfish for an ambient reading instead of trying to judge how cold they were with his hand. The temperature of the oysters should have been 45°F (7°C) or lower.
   - E. Marked the delivery date on the shellstock identification tags attached to the shipment of oysters and mussels. He should not have dumped the remaining shellfish from the previous shipment into the new containers, mixing one shipment with another.

6-26 Discussion Questions

1. General guidelines for receiving food safely include the following:
   - ▷ Training employees to inspect deliveries properly
   - ▷ Planning ahead for shipments
   - ▷ Scheduling deliveries for off-peak hours
   - ▷ Planning a back-up menu in case you have to return some food items
   - ▷ Receiving only one delivery at a time
   - ▷ Having the right information available
   - ▷ Inspecting deliveries immediately
   - ▷ Correcting mistakes immediately
   - ▷ Putting products away as quickly as possible
   - ▷ Keeping the receiving area clean and well lit to discourage pests
2. When checking the temperature of:
   - ▷ **Poultry:** Insert the thermometer stem or probe into the thickest part of the product. The temperature should be 41°F (5°C) or lower.
   - ▷ **Bulk milk:** Fold the bag or pouch around the thermometer stem or probe, being careful not to puncture the bag. The temperature should be 41°F (5°C) or lower.
3. Fresh poultry should be rejected for all of the following conditions:
   - ▷ Purple or green discoloration around the neck
   - ▷ Dark wing tips (red wing tips are acceptable)
   - ▷ Stickiness under wings or around joints
   - ▷ Abnormal unpleasant odor
4. Four types of external damage to cans that are cause for rejection are:
   - ▷ Swollen ends
   - ▷ Leaks and flawed seals
   - ▷ Rust
   - ▷ Dents

6-27 Multiple-Choice Study Questions

1. D	4. C	7. A	10. C
2. C	5. B	8. D	11. B
3. D	6. B	9. B	12. B

Chapter 7

Page Activity

7-2 Test Your Food Safety Knowledge

1. False 2. False 3. True 4. True 5. False

7-15 A Case in Point 1

1. Pete and Angie made several errors.
   A. Pete should not have placed the hot stockpot of soup into the refrigerator to cool. Not only is this an unsafe way to cool hot, potentially hazardous food, but the hot pot could warm up the interior of the refrigerator enough to put other stored food in the temperature danger zone. Pete should have cooled the soup properly *(see Chapter 8)* before storing it in the refrigerator.
   B. Angie should not have placed the uncovered pan of raw chicken on the top shelf in the refrigerator. Nor should she have stored the carrot cake below the raw chicken breasts, since juices could have dripped onto the cake, contaminating it. Cooked or ready-to-eat food must be stored above raw meat, poultry, and fish if these items are stored in the same unit. Ideally, raw product like fresh meat, fish, and poultry should be stored in a separate unit from cooked and ready-to-eat food. All stored food should also be wrapped properly to prevent cross-contamination.
2. Unfortunately, their mistakes could affect all of the food stored in the refrigerator.

7-16 A Case in Point 2

1. Here is what Ed did wrong:
   A. He failed to check use-by or expiration dates prior to storing food in the storeroom. He should not have placed the new case of tomato sauce in front of the existing case unless the cans in the new case had earlier use-by or expiration dates. He made the same mistake when storing the soup and pasta.
   B. He should not have stored the case of crackers on the floor. All items must be stored at least six inches off the floor.
   C. He should have cleaned up the flour that he spilled immediately to keep from attracting rodents.
   D. He should not have stored the cleaning supplies in the storeroom with food. Cleaning tools and chemicals should be kept in their own storage area away from food and food-contact surfaces.
   E. He should not have stored the produce underneath the dining room stairs, since this can lead to contamination. Food must only be stored in designated areas.

7-17 Discussion Questions

1. The recommended top-to-bottom order for storing the items in the same refrigerator is:
   - Raw trout
   - Uncooked beef roast
   - Raw ground beef
   - Raw chicken

2. If food is removed from its original packaging, it should be placed in a clean and sanitized container and covered. The new container must be labeled with the name of the food being stored and its original use-by or expiration date.
3. Frozen dairy products such as ice cream and frozen yogurt can be stored at 6°F to 10°F (–14°C to –12°C).
4. To hold food at a specific internal temperature, refrigerator air temperature usually must be at least 2°F (1°C) lower than the desired temperature. For example, to hold food at an internal temperature of 41°F (5°C), the air temperature in the refrigerator should be at least 39°F (4°C).
5. The first-in, first out (FIFO) method is commonly used to ensure that refrigerated, frozen, and dry products are properly rotated during storage. By this method, a product's use-by, expiration, or preparation date is first identified. The products are then stored to ensure that the oldest are used first. One way to do this is to train employees to store products with the earliest use-by or expiration dates in front of products with later dates. Once shelved, those stored in front are used first.

7-18 Multiple-Choice Study Questions

1. D
2. C
3. A
4. B
5. D
6. A
7. C
8. D
9. C
10. B

Chapter 8

Page Activity

8-2 Test Your Food Safety Knowledge

1. False
2. False
3. False
4. False
5. False

8-23 A Case in Point 1

1. Here is what John did wrong:
   - A. He failed to wash his hands before starting work.
   - B. He failed to thaw the shrimp properly. The shrimp should have been thawed in the refrigerator at 41°F (5°C) or lower. This, however, would have required advance planning. John also could have thawed the shrimp by submerging it under running, potable water at a temperature of 70°F (21°C) or lower.
   - C. He took out more whole fish from the walk-in refrigerator than he could prepare in a short period of time, unnecessarily subjecting the fish to time-temperature abuse.
   - D. He failed to clean and sanitize the boning knife, cutting board, and worktable properly after cleaning and filleting the fish. Microorganisms that may have been present on the fish could have been transferred to the shrimp that John prepared with the contaminated knife and cutting board.

8-24 A Case in Point 2

1. Here is what Angie did wrong:
   - A. She cooled the leftover chicken breasts improperly. Food should never be left out to cool at room temperature. Angie could have divided the chicken into smaller portions and refrigerated them, or used a blast chiller.
   - B. She unnecessarily subjected the chicken salad ingredients to time-temperature abuse by leaving them out on the counter while she performed other duties. Angie should have left the ingredients in the refrigerator until she was ready to prepare the salad.
   - C. She used raw shell eggs in a nursing home where she serves a population at high risk for foodborne illness. She should not have done this, especially since she was cooking the eggs to hold them for later service. Angie should be using pasteurized shell eggs or egg products.

D. She failed to handle the large number of pooled eggs properly. She left a bowl of eggs near a warm stove in the temperature danger zone. She should have pooled a smaller number of eggs and kept them in an ice bath away from the stove. Also, she failed to check the temperature of the eggs prior to placing them on the steam table. Eggs that will be cooked and held for later service must be cooked to at least 155°F (68°C) for fifteen seconds.

E. She failed to wash her hands before starting to prepare food and did not wash her hands between foodhandling tasks. If there were any microorganisms on her hands, she would have transferred them to the food and food-contact surfaces she handled. Also, Angie wiped her hands on her apron. Wiping hands is not an adequate substitute for proper handwashing.

8-25 Discussion Questions

1. The required minimum internal cooking temperatures are:

   Poultry: 165°F (74°C) for fifteen seconds

   Fish: 145°F (63°C) for fifteen seconds

   Pork: 145°F (63°C) for fifteen seconds (roasts for four minutes)

   Ground beef: 155°F (68°C) for fifteen seconds

2. The four proper methods for thawing food are:
   - Thaw it in the refrigerator at 41°F (5°C) or lower.
   - Submerge it under running, potable water at a temperature of 70°F (21°C) or lower.
   - Thaw it in a microwave oven if it will be cooked immediately afterward.
   - Thaw it as part of the cooking process as long as the product reaches the required minimum internal cooking temperature.

3. There are a number of methods that can be used to cool food, including:
   - Reducing the quantity or size of the food being cooled
   - Using ice-water baths
   - Using blast chillers
   - Stirring food as it cools
   - Adding ice or cool water as an ingredient
   - Using properly equipped, steam-jacketed kettles by running cold water through the jacket

4. Meat, poultry, and fish cooked in a microwave must be heated to 165°F (74°C) or higher. The steps for properly cooking food in a microwave include:

   1. Cover food to prevent the surface from drying out.
   2. Rotate or stir food halfway through the cooking process to distribute heat more evenly.
   3. Let food stand for at least two minutes after cooking to let product temperature equalize.

8-26 Multiple-Choice Study Questions

1. B	3. D	5. D	7. C	9. B
2. C	4. D	6. A	8. D	10. C

Chapter 9

Page Activity

9-2 Test Your Food Safety Knowledge

1. True 2. False 3. False 4. True 5. True

9-16 A Case in Point 1

1. Here is what Jill did wrong:
   A. She packed the deliveries in cardboard boxes instead of rigid, insulated carriers.
   B. She used the wrong utensil to fill the soup baine.
   C. She failed to make sure that the internal temperature of the food on the steam table was checked at least every four hours. This would have alerted her to the fact that the steam table was not maintaining the proper temperature and that the casserole was in the temperature danger zone.
2. The following is what Jill should have done:
   A. She should have kept the delivery meals in a hot-holding cabinet or left the food in a steam table until suitable containers were found or the driver arrived.
   B. She should have used a long-handled ladle, which would have kept her hands away from the soup, preventing possible contamination, as she ladled it out.
   C. She should have discarded the casserole and any other food that was not at the right temperature, since she did not know how long the food was in the zone.
   D. She should have made sure that an employee was assigned to monitor the food bar to ensure that customers, such as the children, followed proper etiquette.

9-17 A Case in Point 2

1. Megan made the following errors:
   A. She tasted the food on the customer's plate.
   B. She failed to wash her hands after clearing the dirty dishes from the table.
   C. She failed to clean the table properly after busing it. Meagan should not have wiped the table with the cloth she kept in her apron.
   D. She improperly scooped ice into glassware. Megan should not have used the glass itself to retrieve ice from the bin. Using a glass this way could cause it to chip or break in the ice.
   E. She re-served bread and butter that had been previously served to a customer. Uneaten bread or rolls should never be re-served to other customers.
   F. She failed to wash her hands after scratching a sore. By scratching it and not washing her hands afterwards, she could easily have contaminated everything else she touched.
2. Here is what Megan should have done before beginning her shift:
   A. She should have washed her hands after clearing the table and before she touched the water glasses.
   B. When cleaning tables between guest seatings, Megan should wipe up spills with a disposable, dry cloth. The table should then be cleaned with a clean cloth stored in a sanitizer solution *(see Chapter 11)*.
   C. She should have used tongs or an ice scoop to get ice.
   D. She should have served a fresh basket of bread.
   E. She should have washed her hands immediately after scratching the sore.

9-18 Discussion Questions

1. The following practices can minimize contamination in self-service areas:
   - ▷ Protect food on display with sneeze guards or food shields.
   - ▷ Assign an employee to replenish food-bar items and to hand out fresh plates and silverware for return visits.
   - ▷ Identify all food items on display.
   - ▷ Keep raw meat, fish, and poultry separate from cooked and ready-to-eat food.
2. When transporting food, protect it from contamination and time-temperature abuse during transport by doing the following:
   - ▷ Use rigid, insulated storage containers capable of maintaining food temperatures of 135°F (57°C) or higher or 41°F (5°C) or lower.
   - ▷ Clean and sanitize the inside of delivery vehicles.
   - ▷ Check internal food temperatures regularly.
   - ▷ Make sure employees practice good personal hygiene.
   - ▷ Keep raw and ready-to-eat products separate during delivery and storage.
3. Ready-to-eat, potentially hazardous food can be displayed or held for consumption without temperature control for up to four hours under the following conditions:
   - ▷ Prior to removing the food from temperature control, it has been held at 41°F (5°C) or lower, or 135°F (57°C) or higher.
   - ▷ The food contains a label that specifies when the item must be discarded.
   - ▷ The food is sold, served, or discarded within four hours.
4. When serving food off-site, it must be protected from contamination and time-temperature abuse, and facilities and equipment used to prepare food must be clean and sanitary. In addition:
   - ▷ Make sure employees practice good personal hygiene.
   - ▷ Ensure there is safe drinking water for cooking, warewashing, and handwashing.
   - ▷ Check internal food temperatures regularly.
   - ▷ Label food with storage, shelf-life, and reheating instructions for employees at off-site locations.
   - ▷ Provide food safety guidelines for consumers.
   - ▷ Ensure there is adequate power for holding and cooking equipment.
   - ▷ Provide adequate garbage storage and disposal.

9-19 Multiple-Choice Study Questions

1. A
2. B
3. C
4. B
5. D
6. A
7. B

Chapter 10

Page Activity

10-2 Test Your Food Safety Knowledge

1. False
2. True
3. True
4. True
5. True

10-17 Discussion Questions

1. Active managerial control is a proactive approach for addressing the five most common risk factors responsible for foodborne illness as identified by the CDC. By continuously monitoring and verifying procedures responsible for preventing these risks, establishments can ensure they are being controlled.

2. Active managerial control focuses on controlling the five most common risk factors responsible for foodborne illness as identified by the CDC. HACCP focuses on identifying significant biological, chemical, or physical hazards at specific points within a product's flow through the operation, so they can be prevented, eliminated, or reduced to safe levels.
3. The five foodborne illness risk factors identified by the CDC are:
   - ▷ Purchasing food from unsafe sources
   - ▷ Failing to cook food adequately
   - ▷ Holding food at improper temperatures
   - ▷ Using contaminated equipment
   - ▷ Poor personal hygiene
4. A critical limit is a minimum or maximum limit that a critical control point must meet to prevent, eliminate, or reduce an identified hazard to a safe level. Cooking ground beef to 155°F (68°C) is an example of a critical limit.
5. An establishment is required to have a HACCP plan in place if they perform the following activities:
   - ▷ Smoke or cure food as a method of food preservation
   - ▷ Use food additives as a method of food preservation
   - ▷ Package food using a reduced-oxygen packaging method
   - ▷ Offer live, molluscan shellfish from a display tank
   - ▷ Custom-process animals for personal use
   - ▷ Package unpasteurized juice for sale to the consumer without a warning label

10-18 Multiple-Choice Study Questions

1. B 3. C 5. C 7. B 9. D
2. B 4. B 6. A 8. C

Chapter 11

Page Activity

11-2 Test Your Food Safety Knowledge

1. True 2. False 3. True 4. False 5. True

11-32 A Case in Point

1. The people who had iced drinks at the bar became ill from the chemical drain cleaner used to clean the glasswasher drain. It is likely that the grease trap on the kitchen sink was blocked with grease. This caused waste water (and drain cleaner) to back up into the glasswasher drain and the drain for the icemaker since this equipment shared the same piping system. Carlos had failed to install an air gap between the icemaker and the floor drain, which resulted in waste water backing up into the icemaker, contaminating the ice. Carlos should not have tried to install and maintain the plumbing in the establishment himself. Establishments must rely on professionals to do this.
2. Aside from handling a potential lawsuit, Carlos will require the services of a professional to fix the problem and will have to clean and sanitize the icemaker storage bin thoroughly before it can be used again.

11-33 Discussion Questions

1. One of the most important factors to consider when selecting flooring for food-preparation areas is the material's porosity, or the extent to which it can become saturated with liquids. The FDA Food Code recommends the use of nonporous (nonabsorbent) flooring in food-preparation areas.

2. A backup of raw sewage in an establishment is cause for immediate closure, correction of the problem, and thorough cleaning.
3. To prevent backflow in an establishment:
   - ▷ Install vacuum breakers or other approved backflow-prevention devices on threaded faucets and connections between two piping systems.
   - ▷ Install air gaps wherever practical and possible. (This is the only completely reliable method.) An air gap is a space used to separate a water supply outlet from any potentially contaminated source.
4. Sources of potable water include:
   - ▷ Approved public water mains
   - ▷ Private water sources regularly maintained and tested
   - ▷ Bottled drinking water
   - ▷ Closed portable water containers filled with potable water
   - ▷ On-site water storage tanks
   - ▷ Properly maintained water transport vehicles

   If an establishment uses a private water supply, such as a well, rather than an approved public source, they should check with their local regulatory agency for information on inspections, testing, and other requirements. Generally, nonpublic water systems should be tested at least annually and the report kept on file in the establishment.
5. The requirements of a handwashing station include:
   - ▷ **Hot and cold running water.** Supplied through a mixing valve or combination faucet at a temperature of at least 100°F (38°C)
   - ▷ **Soap.** Liquid, bar, or powder
   - ▷ **Means to dry hands.** Most local codes require establishments to supply disposable paper towels in handwashing stations.
   - ▷ **Waste container.** Required if disposable paper towels are provided
   - ▷ **Signage indicating employees are required to wash their hands before returning to work.**

   Handwashing stations are required in food-preparation areas, service areas, warewashing areas, and restrooms.
6. When installing stationary equipment:
   - ▷ It must be mounted on legs at least six inches off the floor or it must be sealed to a masonry base.
   - ▷ Stationary tabletop equipment should be mounted on legs, providing a minimum clearance of four inches between the base of the equipment and the tabletop. Alternatively, the equipment should be tiltable, or it should be sealed to the countertop with a nontoxic, food-grade sealant.

11-34 Multiple-Choice Study Questions

1. A
2. D
3. D
4. B
5. C
6. B
7. C
8. D
9. A
10. D
11. C
12. A
13. D
14. B

Chapter 12

Page Activity

12-2 Test Your Food Safety Knowledge

1. False
2. False
3. False
4. False
5. False

12-29 A Case in Point 1

1. Schedules do not clean dining rooms, people do. Tim had made his schedule too rigid and had failed to monitor it. He should have made the necessary adjustments to take late-night banquets into account.
2. Tim should enlist the cooperation of Norman, the night shift manager, to make sure the cleaning program is followed. Norman should bring problems to Tim's attention. Shift supervisors, employees, and managers must communicate effectively with one another.
3. Shifts should be scheduled to include all cleaning duties. If the banquet room closes at 1:00 A.M., the shift should extend beyond the closing time to take cleaning into account. Tim should also encourage his employees to follow a clean-as-you-go approach. In this way, the soiled tableware no longer being used by the banquet attendees could have been brought to the dishwasher before midnight.

12-29 A Case in Point 2

1. The tableware can be washed, rinsed, and sanitized in a three-compartment sink until the machine has been repaired. Before cleaning and sanitizing the items, clean and sanitize each compartment and all work surfaces. Then follow these steps:

   ❶ Rinse, scrape, or soak the items.

   ❷ Wash the items in the first sink in a detergent solution at least 110°F (43°C).

   ❸ Immerse or spray-rinse the items in the second sink, using water at least 110°F (43°C).

   ❹ Immerse the items in the third sink in hot water or a chemical sanitizing solution. If hot-water immersion is used, the water must be at least 171°F (77°C). (Some jurisdictions require 180°F [82°C].) If chemical sanitizing is used, the sanitizer must be mixed at the proper concentration and tested with a sanitizer test kit.

   ❺ Air-dry the items.

12-30 Discussion Questions

1. Food-contact surfaces must be cleaned and sanitized
   - after each use.
   - anytime you begin working with another type of food.
   - anytime you are interrupted during a task and the tools or items you have been working with may have been contaminated.
   - at four-hour intervals, if the items are in constant use.
2. *Cleaning* is the process of removing food and other types of soil from a surface, while *sanitizing* is the process of reducing the number of microorganisms on that surface to safe levels.
3. When cleaning and sanitizing items in a three-compartment sink, follow these steps:

   ❶ Rinse, scrape, or soak the items.

   ❷ Wash the items in the first sink in a detergent solution at least 110°F (43°C).

   ❸ Immerse or spray-rinse the items in the second sink, using water at least 110°F (43°C).

   ❹ Immerse the items in the third sink in hot-water or a chemical sanitizing solution. If hot water immersion is used, the water must be at least 171°F (77°C). (Some jurisdictions require 180°F [82°C].) If chemical sanitizing is used, the sanitizer must be mixed at the proper concentration and tested with a sanitizer test kit.

   ❺ Air-dry the items.

4. To store clean and sanitized tableware, utensils, and equipment:
   - Store tableware and utensils at least six inches off the floor. Keep them covered or otherwise protected from dirt and condensation.
   - Clean and sanitize drawers and shelves before clean items are stored.
   - Clean and sanitize trays and carts used to carry clean tableware and utensils. Do this daily or as often as necessary.
   - Store glasses and cups upside down. Store flatware and utensils with handles up so employees can pick them up without touching food-contact surfaces.
   - Keep the food-contact surfaces of clean-in-place equipment covered until ready for use.
5. Factors that affect the efficiency of sanitizers include the following:
   - **Contact time.** For a sanitizer to kill microorganisms, it must make contact with the object for a specific amount of time.
   - **Temperature.** Generally, sanitizers work best at temperatures between 55°F (13°C) and 120°F (49°C).
   - **Concentration.** Concentrations below those recommended could fail to sanitize objects, while concentrations higher than recommended can be unsafe, and might corrode metals.
6. Cleaning tools and supplies should be cleaned and sanitized before being stored in a locked area away from food and food-preparation areas. When storing cleaning materials:
   - Air-dry wiping cloths overnight.
   - Hang mops, brooms, and brushes on hooks to air-dry.
   - Clean, rinse, and sanitize buckets.

12-31 Multiple-Choice Study Questions

1. D
2. C
3. B
4. C
5. A
6. D
7. B
8. C
9. C
10. D

Chapter 13

Page Activity

13-2 Test Your Food Safety Knowledge

1. True
2. False
3. False
4. False
5. False

13-21 A Case in Point

1. Fred should have been working with a licensed PCO and had an IPM program in place prior to his discovery of the roach infestation. While it sounds as if Fred and his staff are doing a good job keeping the establishment clean, some other measures he could have taken to prevent the infestation include:
   A. Screening windows and vents
   B. Installing self-closing doors and door sweeps
   C. Keeping exterior openings closed tightly
   D. Filling holes around pipes
   E. Sealing cracks in floors and walls
   F. Disposing of garbage quickly and correctly
   G. Making sure that shipments are inspected for signs of pest infestation
2. Fred needs the help of a licensed PCO to help eliminate the roach infestation. The PCO may use repellents, sprays, bait, and/or traps to eliminate the roaches.

13-22 Discussion Questions

1. The purpose of an integrated pest management program is to do the following:
   - ▷ Prevent pests from entering the establishment.
   - ▷ Deny pests food, water, and a hiding or nesting place.
   - ▷ Work with a licensed PCO to eliminate any pests that do infest it.
2. To prevent pests from entering an establishment:
   - ▷ Screen all windows and vents with at least sixteen mesh per square inch screening.
   - ▷ Install self-closing devices or door sweeps on all doors.
   - ▷ Install air curtains above or alongside doors.
   - ▷ Keep drive-through windows closed when not in use.
   - ▷ Keep all exterior openings closed tightly.
   - ▷ Use concrete to fill holes or sheet metal to cover openings around pipes.
   - ▷ Install screens over ventilation pipes and ducts on the roof.
   - ▷ Cover floor drains with hinged grates.
   - ▷ Seal all cracks in floors and walls.
   - ▷ Properly seal spaces or cracks where stationary equipment is fitted to the floor.
3. Signs of a cockroach infestation include:
   - ▷ A strong oily odor
   - ▷ Droppings (feces), which look like grains of black pepper
   - ▷ Capsule-shaped egg cases that are brown, dark red, or black and may appear leathery, smooth, or shiny

   Signs of a rodent infestation include:
   - ▷ Signs of gnawing
   - ▷ Droppings that are shiny and black (fresh) or gray (older)
   - ▷ Tracks across dusty surfaces
   - ▷ Nesting materials, such as scraps of paper, cloth, hair, and other soft materials
   - ▷ Holes in dirt, rock piles, or along foundations
4. Your PCO should store and dispose of all pesticides used in the facility. If they are stored on the premises, follow these guidelines:
   - ▷ Keep them in their original container.
   - ▷ Store them in locked cabinets away from areas where food is stored and prepared.
   - ▷ Check local regulations before disposing of pesticides.
5. To minimize the hazard to people, have your PCO use pesticides only when you are closed for business, and employees are not on-site. When pesticides will be applied, prepare the area to be sprayed by removing all food and food-contact surfaces. Cover equipment and food-contact surfaces that cannot be moved. Wash, rinse, and sanitize food-contact surfaces after the area has been sprayed. Anytime pesticides are used or stored on the premises, you should have a corresponding MSDS since they are hazardous materials.

13-23 Multiple-Choice Study Questions

1. B
2. C
3. B
4. D
5. D
6. D
7. C
8. A
9. B

Chapter 14

Page Activity

14-2 Test Your Food Safety Knowledge

1. False 2. False 3. True 4. False 5. True

14-21 A Case in Point

1. Jerry did a pretty good job handling the inspection. He was cooperative and professional, answering the inspector's questions to the best of his ability. However, he should have asked the inspector for identification and inquired about the purpose of the visit. He also should have taken notes during the inspection. This would have helped him remember later exactly what was said.
2. Jerry did discuss violations with the inspector and now needs to follow up on the findings. He should walk through his facility and determine why each problem occurred. Then he should establish new procedures or revise existing ones to permanently correct problems.

14-22 Discussion Questions

1. The FDA writes the Food Code. In addition, it inspects foodservice operations that cross state borders (interstate establishments such as those on planes and trains, as well as food manufacturers and processors) because they overlap the jurisdictions of two or more states. The FDA shares responsibility with the U.S. Department of Agriculture (USDA) for inspecting food-processing plants to ensure standards of purity, wholesomeness, and compliance with labeling requirements.
2. Recommendations for restaurant and foodservice regulations are written at the federal level, regulations are made at the state level, and enforcement is carried out at the local level.
3. During an inspection, a manager should do the following:
   - Ask for identification.
   - Cooperate.
   - Take notes.
   - Keep the relationship professional.
   - Be prepared to provide records requested by the inspector.

   After the inspection, the manager should do the following:
   - Discuss violations and time frames for correction with the inspector.
   - Follow up by determining why each problem occurred, and then establish new procedures or revise existing ones.
4. Factors that determine the frequency of a health inspection include:
   - **Size and complexity of the operation.** Larger operations offering a large number of potentially hazardous food items might be inspected more frequently.
   - **Inspection history of the establishment.** Establishments with a history of low sanitation scores or consecutive violations might be inspected more frequently.
   - **Clientele's susceptibility to foodborne illness.** Nursing homes, schools, day-care centers, and hospitals might receive more frequent inspections.
   - **Workload of the local health department and the number of inspectors available.**

5. Industry organizations that help managers deal with employee training and sanitation regulations and standards include the following:
   - National Restaurant Association (NRA)
   - National Restaurant Association Educational Foundation (NRAEF)
   - National Environmental Health Association (NEHA)
   - International Association for Food Protection (IAFP)
   - Institute of Food Technologists (IFT)
   - National Society of Professional Sanitarians (NSPS)
   - Association of Food and Drug Officials (AFDO)
   - Council of Hotel and Restaurant Trainers (CHART)
   - National Pest Management Association (NPMA)
   - NSF International
   - Underwriters Laboratories (UL)

14-23 Multiple-Choice Study Questions

1. D
2. D
3. B
4. A
5. C
6. C
7. D
8. B
9. A

Chapter 15

Page	Activity

15-2 Test Your Food Safety Knowledge

1. True
2. True
3. False
4. True
5. False

15-22 A Case in Point

1. Ongoing supervision and feedback throughout the training process corrects performance problems and assures employees that they are doing a good job. While Paul gets off to a good start, he does not follow up on his efforts. The videotape on sanitation that Maria watched may be packed with information, but if Paul does not stop the tape to explain concepts or terms or to allow Maria to ask questions, she may not understand or retain the information presented. Paul should know that learning is better accomplished through interaction. The test he gives Maria can be a good tool to measure the level of learning, but it will be more effective if he corrects it as soon as she is done, and then discusses each question with her.
2. Since Albert, the cook, was given general training three months ago, a refresher session would help reinforce the information learned. Paul should also discuss Albert's progress with the supervisor to determine if more training is necessary. He should also present the trainees with some form of recognition at the completion of their training, such as a certificate, which will let them know they are doing their jobs well and that management values the training program.

15-23 Discussion Questions

1. The benefits of teaching food safety in an establishment include:
   - ▷ Avoiding the costs associated with a foodborne-illness outbreak, including legal fees and medical bills
   - ▷ Preventing loss of revenue and reputation incurred when an establishment is forced to close due to a foodborne-illness outbreak
   - ▷ Improving employee morale and reducing turnover
2. The three key elements of training are presentation, application, and feedback. Once content is presented, the learner must have the opportunity to practice, apply, or respond to the content in order to retain it. While learners are practicing or applying the content, they must receive feedback (positive or negative reinforcement) on their performance. The feedback received must be specific and immediate. Application without feedback will not be effective.
3. An establishment can determine its food safety training needs by doing the following:
   - ▷ Testing employees' food safety knowledge
   - ▷ Observing employee job performance
   - ▷ Questioning or surveying employees to identify areas of weakness
4. An objective states what the learner will be able to do after instruction is completed. Objectives need to be stated clearly and in measurable terms. Examples of clearly stated objectives include:
   - ▷ Demonstrate the proper procedure for calibrating a bimetallic stemmed thermometer using the ice-point method.
   - ▷ Identify the minimum internal cooking temperatures for meat, seafood, and poultry.
5. Methods that can be used to deliver training include:
   - ▷ Demonstration
   - ▷ Lecture
   - ▷ Role-play
   - ▷ Job aids
   - ▷ One-on-one training
   - ▷ Technology-based training

15-24 Multiple-Choice Study Questions

1. C	3. B	5. A	7. D	9. A
2. D	4. D	6. B	8. D	10. C

Apply Your Knowledge

Notes

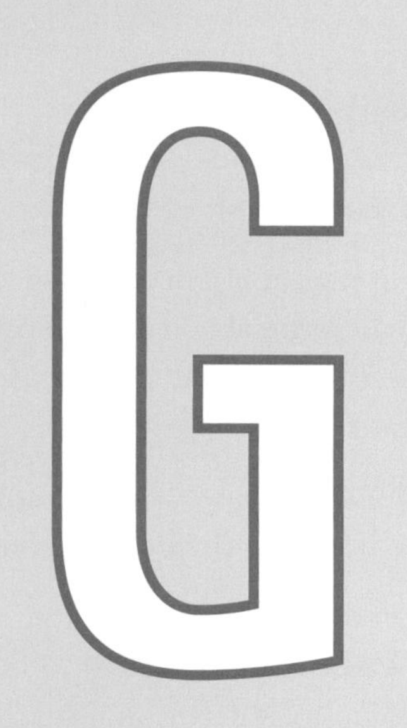

Glossary

Note: The number(s) in bold italics at the end of each entry refers to the chapter in which the term is discussed in detail.

A

Abrasive cleaners. Cleaners containing a scouring agent used to scrub off hard-to-remove soils. They may scratch some surfaces. ***12***

Acid cleaners. Acid cleaners are used on mineral deposits and other soils that alkaline cleaners cannot remove, such as scale, rust, and tarnish. ***12***

Acidity. Level of acid in a food. An acidic substance has a pH below 7.0. Foodborne microorganisms typically do not grow in highly acidic food, while they grow best in food with a neutral to slightly acidic pH. ***2***

Active managerial control. Proactive approach for addressing the five most common risk factors responsible for foodborne illness as identified by the CDC. Managers must continuously monitor and verify the procedures responsible for controlling the risks. ***10***

Air curtains. Devices installed above or alongside doors that blow a steady stream of air across an entryway, creating an air shield around open doors. Insects avoid them. Also called air doors or fly fans. ***13***

Air gap. Air space used to separate a water-supply outlet from any potentially contaminated source. A properly designed and installed sink has air gaps to prevent backflow. The air space between the floor drain and drain pipe of a sink is an example. An air gap is the only completely reliable method for preventing backflow. ***11***

Alkalinity. Level of alkali in food. An alkaline substance has a pH above 7.0. Most food is not alkaline. ***2***

Americans with Disabilities Act (ADA). Federal law requiring reasonable accommodation for access to a facility by patrons and employees with disabilities. ***11***

Application. Applying what was learned, with feedback from the instructor/trainer. ***15***

Aseptically packaged food. Food that has been sealed under sterile conditions, usually after UHT-pasteurization. ***6***

B

Backflow. Unwanted reverse flow of contaminants through a cross-connection into a potable water system. It occurs when the pressure in the potable water supply drops below the pressure of the contaminated supply. ***11***

Bacteria. Living, single-celled microorganisms that can cause food spoilage and illness. Some form spores that can survive freezing and very high temperatures. ***2***

Bacterial growth. Reproduction of bacteria by splitting in two. When conditions are favorable, bacterial growth can be rapid—doubling the population as often as every twenty minutes. Their growth can be broken down into four phases: lag phase, log phase, stationary phase, and death phase. ***2***

Bimetallic stemmed thermometer. The most common and versatile type of thermometer, measuring temperature through a metal probe with a sensor in the end. Most can measure temperatures from 0°F to 220°F (–18°C to 104°C) and are accurate to within ±2°F (±1°C). They are easily calibrated. ***5***

Biological hazard. Pathogenic microorganisms that can contaminate food, such as certain bacteria, viruses, parasites, and fungi, as well as toxins found in certain plants, mushrooms, and fish. ***1***

Biological toxins. Poisons produced by pathogens, plants, or animals. They can also occur in animals as a result of their diet. ***3***

Blast chiller. Equipment designed to cool food quickly. Many are able to cool food from 135°F to 37°F (57°C to 3°C) within ninety minutes. ***11***

Boiling-point method. Method of calibrating a thermometer based on the boiling point of water. ***5***

Booster heater. Water heater attached to hot-water lines leading to warewashing machines or sinks. Raises water to temperatures required for heat sanitizing of tableware and utensils (180°F [82°C]). ***11***

C

Calibration. Process of ensuring that a thermometer gives accurate readings by adjusting it to a known standard, such as the freezing point or boiling point of water. ***5***

Cantilever-mounted equipment. Equipment that is attached to a wall with a bracket, allowing for easier cleaning behind and underneath. ***11***

Carrier. Person who carries pathogens and infects others, yet never becomes ill himself. ***4***

Centers for Disease Control and Prevention (CDC). Agency of the U.S. Public Health Service that investigates foodborne-illness outbreaks, studies the causes and control of disease, publishes statistical data, and conducts the Vessel Sanitation Program. ***14***

Chemical hazard. Chemical substances that can contaminate food, such as pesticides, food additives, preservatives, cleaning supplies, and toxic metals that leach from cookware and equipment. ***1***

Chemical sanitizing. Using a chemical solution to reduce the number of microorganisms on a clean surface to safe levels. Items can be sanitized by immersing in a specific concentration of sanitizing solution for a required period of time, or by rinsing, swabbing, or spraying the items with a specific concentration of sanitizing solution. ***12***

Chemical toxins. Poisons found in some cleaning agents and pesticides, as well as the by-products of toxic-metal reactions. ***3***

Chlorine. A commonly used chemical sanitizer, due to its low cost and effectiveness. It kills a wide range of microorganisms. ***12***

Ciguatera poisoning. Illness that occurs when a person eats fish that has consumed the ciguatera toxin. This toxin occurs in certain predatory tropical reef fish, such as amberjack, barracuda, grouper, and snapper. ***3***

Clean. Free of visible soil. It refers only to the appearance of a surface. ***1, 12***

Cleaning. Process of removing food and other types of soil from a surface. ***12***

Cleaning agents. Chemical compounds that remove food, soil, rust stains, minerals, or other deposits from surfaces. ***12***

Cold-holding equipment. Equipment specifically designed to keep cold food at an internal temperature of 41°F (5°C) or lower. ***9***

Cold paddle. Plastic paddle that can be filled with water and frozen. When used to stir hot food, it cools the food quickly. ***8***

Contact spray. Spray used to kill insects on contact. Usually used on groups of insects, such as clusters of roaches and nests of ants. ***13***

Contamination. Presence of harmful substances in food. Some contaminates occur naturally, while others are introduced by humans or the environment. ***1***

Corrective action. Predetermined step taken when food does not meet a critical limit. ***10***

Coving. Curved, sealed edge placed between the floor and wall to eliminate sharp corners or gaps that would be impossible to clean. Coving also eliminates hiding places for pests and prevents moisture from deteriorating walls. ***11***

Critical control point (CCP). In a HACCP system, the points in the process where you can intervene to prevent, eliminate, or reduce identified hazards to safe levels. ***10***

Critical limit. In a HACCP system, the minimum or maximum limit a critical control point (CCP) must meet in order to prevent, eliminate, or reduce a hazard to an acceptable level. ***10***

Cross-connection. Physical link through which contaminants from drains, sewers, or other waste-water sources can enter a potable water supply. A hose connected to a faucet and submerged in a mop bucket is an example. ***11***

Cross-contamination. The transfer of microorganisms from one surface or food to another. ***1***

Death phase. The phase in bacterial growth in which the number of bacteria dying exceeds the number growing, resulting in a population decline. ***2***

Demonstration. Process of illustrating a skill or task in front of another person or a group. ***15***

Detergent. Cleaning agent designed to penetrate and soften soil to help remove it from a surface. ***12***

Dry storage. Storage used to hold dry and canned food at temperatures between 50°F and 70°F (10°C and 21°C) and at a relative humidity of fifty to sixty percent. ***7***

E

Electronic insect eliminator ("zapper"). Mechanical device that uses light to attract flying insects to an electrically charged grid that kills them. ***13***

Environmental Protection Agency (EPA). Federal agency that sets standards for environmental quality—including air and water quality—and regulates pesticide use and waste handling. ***14***

Evaluation. Judging the performance of training participants against learning objectives. ***15***

F

FAT TOM. Acronym for the conditions needed by most foodborne microorganisms to grow: Food, Acidity, Temperature, Time, Oxygen, Moisture. ***2***

FDA Food Code. Science-based reference for retail food establishments on how to prevent foodborne illness. These recommendations are written by the FDA to assist state health departments in developing regulations for a foodservice inspection program. ***1, 14***

Feedback. Evaluation given to employees about their performance, including constructive criticism given to correct a mistake, or praise to reinforce proper performance of a skill or procedure. ***15***

Finger cot. Protective covering used to cover a properly bandaged cut or wound on the finger. ***4***

First in, first out (FIFO). Method of stock rotation in which products are shelved based on their use-by or expiration dates, so oldest products are used first. ***7***

Flood rim. Spill-over point of a sink. ***11***

Flow of food. Path food takes through an establishment, from purchasing and receiving, through storing, preparing, cooking, holding, cooling, reheating, and serving. ***5***

Food allergy. The body's negative reaction to a particular food protein. ***3***

Food and Drug Administration (FDA). The federal agency that writes the Food Code. The FDA also inspects foodservice operations that cross state borders (interstate establishments such as food manufacturers and processors, and planes and trains). In addition, the FDA shares responsibility with the USDA for inspecting food-processing plants. ***14***

Food bar. Self-service buffet at which patrons can choose what they want to eat as they serve themselves. ***9***

Food-contact surface. Surface that comes into direct contact with food, such as a cutting board. ***1***

Food-grade sealant. Nontoxic sealant used to seal equipment to a countertop or a masonry base. ***11***

Food irradiation. Process of exposing food to an electron beam or gamma rays to reduce pathogenic and spoilage microorganisms. Also known as cold pasteurization. ***2***

Food Safety and Inspection Service (FSIS). Agency of the USDA that inspects and grades meat, meat products, poultry, dairy products, egg and egg products, and fruit and vegetables shipped across state boundaries. ***14***

Food safety management system. Group of programs and procedures designed to control hazards throughout the flow of food. ***10***

Food security. The prevention or elimination of the deliberate contamination of food. ***3***

Foodborne illness. Disease carried or transmitted to people by food. ***1***

Foodborne-illness outbreak. According to the CDC, an incident in which two or more people experience the same illness after eating the same food. ***1***

Foodborne infection. Result of a person eating food containing pathogens, which then grow in the intestines and cause illness. Typically, symptoms of a foodborne infection do not appear immediately. ***2***

Foodborne intoxication. Result of a person eating food containing toxins that cause an illness. The toxins may have been produced by pathogens found on the food or may be the result of a chemical contamination. The toxins might also be a natural part of the plant or animal consumed. Typically, symptoms of foodborne intoxication appear quickly, within a few hours. ***2***

Foodborne toxin-mediated infection. Result of a person eating food containing pathogens, which then produce illness-causing toxins in the intestines. ***2***

Foot-candle. Unit of lighting equal to the illumination one foot from a uniform light source. ***11***

Frozen storage. Storage typically designed to hold food at 0°F (–18°C) or lower. Some types of food require a different temperature. ***7***

Fungi. Ranging in size from microscopic, single-celled organisms to very large, multicellular organisms, fungi most often cause food spoilage. Molds, yeasts, and mushrooms are examples of fungi. ***2***

G

Gastrointestinal illness. Illness relating to the stomach or intestine. ***4***

Glue board. Pest-control device in which mice are trapped by glue and then die from exhaustion or lack of water or air. They are also used to identify the type of cockroaches that might be present. ***13***

H

HACCP plan. Written document based on HACCP principles describing procedures a particular establishment will follow to ensure the safety of food served. ***10***

Hair restraint. Device used to keep a foodhandler's hair away from food and to keep the individual from touching it. ***4***

Hand sanitizer. Liquid used to lower the number of microorganisms on the skin surface. Hand sanitizers should be used after proper handwashing, not in place of it. ***4***

Handwashing station. Sink designated for handwashing only. Handwashing stations must be conveniently located in restrooms, food-preparation areas, service areas, and warewashing areas. ***11***

Hard water. Water containing minerals such as calcium and iron in concentrations higher than 120 parts per million (ppm). ***12***

Hazard analysis. Process of identifying and evaluating potential hazards associated with food in order to determine what must be addressed in the HACCP plan. ***10***

Hazard Analysis Critical Control Point (HACCP). System designed to keep food safe throughout its flow through an establishment. HACCP is based on the idea that if hazards are identified at specific points in a food's flow, the hazards can be prevented, eliminated, or reduced to safe levels. ***10***

Hazard Communication Standard (HCS). OSHA standard, also known as Right-to-Know or HAZCOM, requiring employers to tell their employees about potential chemical hazards at the establishment. It also requires employers to train employees on how to use chemicals safely. ***12***

Health inspector. City, county, or state employee who conducts foodservice inspections in most states. Inspectors generally are trained in food safety, sanitation, and public health principles and methods. Also called sanitarians, health officials, or environmental health specialists. ***14***

Heat sanitizing. Using heat to reduce the number of microorganisms on a clean surface to safe levels. The most common way to heat-sanitize tableware, utensils, or equipment is to submerge them in or spray them with hot water. ***12***

Heat-treated. Food that has been cooked, partially cooked, or warmed. ***1***

Hepatitis A. Disease-causing inflammation of the liver. It is transmitted to food by poor personal hygiene or contact with contaminated water. ***4***

High-risk population. People susceptible to foodborne illness due to the effects of age or health on their immune systems, including infants and preschool-age children, pregnant women, older people, people taking certain medications, and those with certain diseases or weakened immune systems. ***1***

Histamine. Biological toxin associated with temperature-abused scombroid fish that causes scombroid poisoning. ***3***

Host. Person, animal, or plant on which another organism lives and takes nourishment. ***2***

Hot-holding equipment. Equipment such as chafing dishes, steam tables, and heated cabinets specifically designed to hold potentially hazardous food at 135°F (57°C) or higher. ***9***

Hygrometer. Instrument used to measure relative humidity in storage areas. ***7***

I

Ice-point method. Method of calibrating thermometers based on the freezing-point of water. ***5***

Ice-water bath. Method of cooling food in which a container holding hot food is placed into a larger container of ice water. The ice water surrounding the hot food container disperses the heat quickly. ***8***

Immune system. The body's defense system against illness. People with compromised immune systems are more susceptible to foodborne illness. ***1***

Infected lesion. Wound or injury contaminated with a pathogen. ***4***

Infestation. Situation that exists when pests overrun or inhabit an establishment in large numbers. ***13***

Integrated pest management (IPM). Program using prevention measures to keep pests from entering an establishment and control measures to eliminate any pests that do get inside. ***13***

Iodine. Sanitizer effective at low concentrations and not as quickly inactivated by soil as chlorine. It might stain surfaces and is less effective than chlorine. ***12***

J

Jaundice. Yellowing of the skin and eyes that could indicate a person is ill with hepatitis A. ***4***

Job aids. Materials or visual reminders used to deliver training content to employees. ***15***

L

Lag phase. Phase in bacterial growth in which bacteria are first introduced to a new environment. In this phase, bacteria go through an adjustment period in which their numbers are stable as they prepare to grow. To control the growth of bacteria, prolong the lag phase as long as possible. ***2***

Lecture. Prepared oral presentation used to deliver content to a group. ***15***

Log phase. Phase in bacterial growth in which conditions are favorable for bacteria to multiply very rapidly. Food quickly becomes unsafe during this phase. ***2***

M

Master cleaning schedule. Detailed schedule that lists all cleaning tasks in an establishment, when and how they are to be performed, and who will do the cleaning. ***12***

Material Safety Data Sheets (MSDS). Sheets supplied by the chemical manufacturer listing the chemical and its common names, its potential physical and health hazards, information about using and handling it safely, and other important information. OSHA requires employers to store these sheets so they are accessible to employees. ***12***

Microorganisms. Small, living organisms that can be seen only with the aid of a microscope. Four types of microorganisms with the potential to contaminate food and cause foodborne illness are bacteria, viruses, parasites, and fungi. ***2***

Minimum internal temperature. Required cooking temperature the internal portion of food must reach—specific to the type of food being cooked—in order to sufficiently reduce the number of illness-causing microorganisms that might be present. ***8***

Mobile unit. Portable foodservice facilities, ranging from concession vans to full field kitchens capable of preparing and cooking elaborate meals. ***9***

Modified atmosphere packaging (MAP). Packaging process by which air is removed from a food package and replaced with gases, such as carbon dioxide and nitrogen, to help extend the product's shelf life. ***6***

Mold. Type of fungus that causes food spoilage. Some produce toxins that can cause foodborne illness. ***2***

Monitoring. In a HACCP system, the process of analyzing whether critical limits are being met and things are being done right. ***10***

N

National Marine Fisheries Service (NMFS). Agency of the U.S. Department of Commerce that provides a voluntary inspection program that includes product standards and sanitary requirements for fish processing operations. ***14***

NSF International. Organization that develops and publishes standards for sanitary equipment design. They also assess and certify that equipment has met these standards. Restaurant and foodservice managers should look for an NSF International mark (or UL EPH product mark) on commercial foodservice equipment. ***11***

O

Objective. Statement of what a trainee will be able to do after training or instruction is completed. ***15***

Occupational Safety and Health Administration (OSHA). Federal agency that regulates and monitors workplace safety. ***12***

Off-site service. Service of food to someplace other than where it is prepared or cooked, including catering and vending. ***9***

P

Parasite. Organism that needs to live in a host organism to survive. Parasites can live in many animals that humans use for food, including cows, chickens, pigs, and fish. ***2***

Pathogens. Disease-causing microorganisms. ***2***

Personal hygiene. Sanitary health habits that include keeping body, hair, and teeth clean, maintaining good health, wearing clean clothes, and washing hands regularly, especially when handling food and beverages. ***4***

Pest control operator (PCO). Licensed professional who uses safe, up-to-date methods to prevent and control pests. ***13***

Pesticide. Chemical used to control pests, usually insects. ***13***

pH. Measure of a food's acidity or alkalinity. The pH scale ranges from 0 to 14.0. A pH above 7.0 is alkaline, while a pH below 7.0 is acidic. A pH of 7.0 is neutral. Pathogenic bacteria grow well in food with a pH between 4.6 and 7.5 (slightly acidic to neutral). ***2***

Physical hazard. Foreign objects that can accidentally get into food and contaminate it, such as hair, dirt, metal staples, and broken glass, as well as naturally occuring objects, such as bones in filets. ***1***

Plant toxins. Poisons found naturally in some plants. ***3***

Pooled eggs. Eggs that have been cracked open and combined in a common container. ***8***

Porosity. Extent to which water and other liquids are absorbed by a substance. Term usually used in relation to flooring material. ***11***

Potable water. Water that is safe to drink or to use as an ingredient in food. ***11***

Potentially hazardous food. Food in which microorganisms can grow rapidly. Potentially hazardous food has a history of being involved in foodborne-illness outbreaks, has potential for contamination due to production and processing methods, and has characteristics that generally allow microorganisms to grow rapidly. Potentially hazardous food is often moist, contains protein, and has a neutral or slightly acidic pH. ***1***

Presentation. Delivery of content to the learner, which can be accomplished through a variety of methods. ***15***

Pulper. Device used to grind food and other waste into small parts that are flushed with water, which is then removed. The processed, solid wastes weigh less and are more compact for easier disposal. ***11***

Quaternary ammonium compounds (quats). Group of sanitizers all having the same basic chemical structure. They work in most temperature and pH ranges, are noncorrosive, and remain active for short periods of time after they have dried. However, quats may not kill certain types of microorganisms, and they leave a film on surfaces. ***12***

R

Ready-to-eat food. Properly cooked food, as well as raw, washed whole or cut fruit and vegetables (including those that have had their rinds, peels, husks, or shells removed). ***1***

Reasonable care defense. Defense against a food-related lawsuit stating that an establishment did everything that could be reasonably expected to ensure that the food served was safe. ***1***

Record keeping. In a HACCP system, the process of collecting documents that allow you to show you are continuously preparing and serving safe food. ***10***

Refrigerated storage. Storage used for holding potentially hazardous food at an internal temperature of 41°F (5°C) or lower. Some jurisdictions allow food in refrigerators to be held at an internal temperature of 45°F (7°C) or lower. Check with the local regulatory agency for specific regulations. ***7***

Regulations. Laws determining standards of behavior. Restaurant and foodservice regulations are typically written at the state level and based on the FDA Food Code. ***14***

Residual spray. Type of pesticide spray that leaves behind a film that insects absorb as they crawl across it. Used in cracks and crevices like those along baseboards, these sprays can can be liquid or a dust, such as boric acid. ***13***

Resiliency. Ability of a surface to react to a shock without breaking or cracking, usually used in relation to a flooring material. ***11***

Role-play. Training method in which trainees enact a situation in order to try out new skills or apply new knowledge. ***15***

S

Sanitary. State that exists when the number of pathogens on a clean surface has been reduced to safe levels. ***1, 12***

Sanitizer. Compound used to reduce the number of pathogens on a clean surface to safe levels. ***12***

Sanitizing. Process of reducing the number of microorganisms on a clean surface to safe levels. ***12***

Scombroid poisoning. Illness that occurs when a person eats a scombroid fish that has been time-temperature abused. Scombroid fish include tuna, mackerel, bluefish, skipjack, and bonito. ***3***

Service sink. Sink used exclusively for cleaning mops and disposing of waste water. At least one service sink or one curbed drain area is required in an establishment. ***11***

Shelf life. Recommended period of time food may be stored and remain suitable for use. ***7***

Shellstock identification tag. Tag that accompanies each container of live, molluscan shellfish, on which the delivery date must be written. Tags are to be kept on file for ninety days after the last shellfish was used. ***6***

Single-use gloves. Disposable gloves designed for one-time use that provide a barrier between hands and the food they come in contact with. ***4***

Single-use item. Disposable tableware or packaged food designed to be used only once, including plastic flatware, paper or plastic cups, plates, and bowls, as well as single-serve food and beverages. ***9***

Single-use paper towel. Paper towel designed to be used once, then discarded. ***4***

Slacking. Process of gradually thawing frozen food in preparation for deep frying. ***8***

Sneeze guard. Food shield used on food bars. They are usually placed fourteen inches above the food and extended seven inches beyond the food. ***9***

Solvent cleaners. Alkaline detergents, often called degreasers, that contain a grease-dissolving agent. ***12***

Sous vide* food.** Food vacuum-packed in individual pouches, partially or fully cooked, and then chilled. This food is often heated for service in the establishment. ***6

Spoilage microorganism. Foodborne microorganism that causes food to spoil, but typically does not cause foodborne illness. ***2***

Spore. Alternative form for some bacteria, with a thick wall to protect it from adverse conditions, such as high and low temperatures, low moisture, and high acidity. Capable of turning back into a vegetative microorganism when conditions again become favorable. ***2***

Stationary phase. Phase of bacterial growth in which just as many bacteria are growing as are dying. Follows the log phase of bacterial growth. ***2***

T

Technology-based training. Training programs delivered via a computer or other technology. ***15***

Temperature danger zone. Temperature range between 41°F and 135°F (5°C to 57°C) within which most foodborne microorganisms rapidly grow and reproduce. ***2***

Temporary unit. Establishment operating in one location for no more than fourteen consecutive days in conjunction with a special event or celebration. They usually serve prepackaged food or food requiring only limited preparation. ***9***

Thermometer. Device for accurately measuring the internal temperature of food, the air temperature inside a freezer or cooler, or the temperature of equipment. ***5***

Time-temperature abuse. Allowing food to remain too long at temperatures favorable to the growth of foodborne microorganisms. ***1***

Time-temperature indicator (TTI). Time and temperature monitoring device attached to a food shipment to determine if the product's temperature has exceeded safe limits during shipment or subsequent storage. ***5***

Toxic-metal poisoning. Illness caused when toxic metals are leached from utensils or equipment containing them. ***3***

Toxins. Poisons produced by pathogens, plants, or animals. Most occur naturally and are not caused by the presence of microorganisms. Some occur in animals as a result of their diet. Many chemicals are also toxic. ***2, 3***

Training need. Gap between what employees are required to know to perform their jobs and what they actually know. ***15***

Training program. Structured sequence of events that leads to learning. ***15***

Tumble chiller. Equipment designed to cool food quickly. Prepackaged hot food is placed into a drum rotating inside a reservoir of chilled water. The tumbling action increases the effectiveness of the chilled water in cooling the food. ***11***

Two-stage cooling. Criteria by which cooked food is cooled from 135°F (57°C) to 70°F (21°C) within two hours and from 70°F (21°C) to 41°F (5°C) or lower in an additional four hours, for a total cooling time of six hours. ***8***

U

Ultra-high temperature (UHT) pasteurization. Process of heat-treating food at a very high temperature for a short time to kill microorganisms. The food is often then packaged under sterile conditions. ***6***

Underwriters Laboratories (UL). Provides sanitation classification listings for equipment found in compliance with NSF International standards. Also lists products complying with their own published environmental and public health standards. ***11***

U.S. Department of Agriculture (USDA). Federal agency responsible for the inspection and quality grading of meat, meat products, poultry, dairy products, eggs and egg products, and fruit and vegetables shipped across state lines. ***14***

V

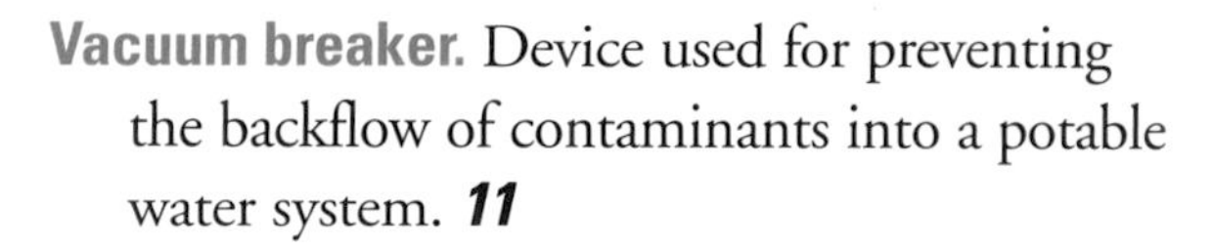

Vacuum breaker. Device used for preventing the backflow of contaminants into a potable water system. ***11***

Vacuum-packed food. Food processed by removing air from around it while sealed in a package. This process increases the product's shelf life. ***6***

Vegetative microorganisms. Bacteria in the process of reproducing (growing) by splitting in two. ***2***

Vending machine. Machines dispensing hot and cold food, beverages, and snacks. ***9***

Verification. In a HACCP system, the process of confirming that critical control points and critical limits are appropriate, that monitoring is alerting you to hazards, that corrective actions are adequate to prevent foodborne illness from occurring, and that employees are following established procedures. ***10***

Virus. Smallest of the microbial food contaminants, viruses rely on a living host to reproduce. Some survive freezing and cooking temperatures. They usually contaminate food through a foodhandler's improper personal hygiene. ***2***

W

Warranty of sale. Rules stating how food must be handled in an establishment. ***1***

Water activity (a_w). Amount of moisture available in food for microorganisms to grow. Potentially hazardous food items typically have water-activity values of 0.85 or above. ***2***

Yeast. Type of fungus that causes food spoilage. ***2***

Index

A

B

C

D

G

H

M

R

S

T

U

W